MODELLING and QUANTITATIVE METHODS in FISHERIES

Malcolm Haddon

CHAPMAN & HALL/CRC

A CRC Press Company
Boca Raton   London   New York   Washington, D.C.

Library of Congress Cataloging-in-Publication Data

Haddon, Malcolm.
    Modelling and quantitative methods in fisheries / Malcom Haddon.
        p.   cm.
    Includes bibliographical references (p.).
    ISBN 1-58488-177-1
    1. Fisheries—Mathematical models. I. Title.
SH331 .5 .M48 H34 2001
333.95′611′015118—dc21

20010286
CIP

Visit the CRC Press Web site at www.crcpress.com
Revised Printing

© 2001 by Chapman & Hall/CRC

No claim to original U.S. Government works
International Standard Book Number 1-58488-177-1
Library of Congress Card Number 20010286
Printed in the United States of America   3 4 5 6 7 8 9 0
Printed on acid-free paper

Contents

List of Example Boxes

To

K.J. MELDA and Wolf,

for demonstrating that purpose is not everything.

Preface

This book aims to introduce some of the methods used by quantitative ecologists and modellers. The emphasis throughout the book is focused upon fisheries and models used in fisheries, but the ideas relate to many other fields of endeavour. The first few chapters, on simple population models, on parameter estimation (principally using maximum likelihood methods), and on computer intensive methods, should be of interest to all population ecologists. Those chapters on growth, recruitment, and explicit fisheries models are obviously focussed on the needs of fisheries scientists.

From 1995 to 1997, I was fortunate enough to be in a position to develop and present short and intensive courses on modelling and quantitative methods to a large number of fisheries scientists and others at fisheries laboratories and universities around Australia. I am grateful to the Australian Fisheries Research and Development Corporation (FRDC) for funding this project and giving me the opportunity to meet so many of Australia's fisheries community. Unfortunately, there was no single text that covered the details of what I felt was necessary for an adequate grounding in the skills required for the quantitative study of marine populations. The course notes I prepared were a first attempt to fill that gap but were designed to complement the presentations given in the short courses rather than as a stand-alone product. For this book, the material has been completely rewritten and expanded with the inclusion of many active examples. While this rewrite greatly slowed the production of the final book, a reader should now be able to pursue a course of independent study of the material and others could use this book as the foundation for a formal course in quantitative fisheries science.

The main objective of this book is to provide a working resource that guides the reader towards an understanding of some of the analytical methods currently being used in quantitative biology and fisheries science. While a theoretical treatment has been provided, a major aim was to focus on understanding the details of how to perform the analyses described. An integral part of this description was to include Microsoft Excel workbooks relating to each example and problem discussed. Excel was chosen because of its flexibility, general availability, and relative ease of use. The appendix on the use of Excel in fisheries should provide sufficient details for those who are not versed in using this program. For maximum benefit the example boxes scattered through the text should be constructed and perhaps modified as the reader becomes more confident. Doing something leads to a much better understanding than just reading about it. These workbooks, and other examples, can be found on the download pages of the following web addresses: www.utas.edu.au/docs/tafi/TAFI_Homepage.html and www.crcpress.com/us/ElectronicProducts/downandup.asp. The files should be downloaded but try to construct them before considering the finished versions. The reader should try to use them only as a check on what was produced or if he or she becomes deeply stuck. I have tried to use real examples wherever possible in the belief that these are more acceptable.

When I was at school one of my mathematics teachers used to write an equation on the board and then exclaim, "... as we all know..." while writing a derived equation underneath. He clearly believed that the algebraic transition from one equation to the other should have been very clear to his pupils. Unfortunately, it was often the case that only a few people understood. While it is true that omitting the details of the steps between the algebraic changes leads to brevity it certainly does not improve clarity or ease of understanding. In this book, whenever the details of a set of equations or model are given, an attempt has usually been made to avoid omitting any of the steps needed to understand their derivation equations.

For most of the time it took, I enjoyed writing this book. I hope it helps people move forward in their work.

Malcolm Haddon

Hobart, January 2001

Preface to the Revised Printing

This is a revised printing of the book originally published in 2001. Shortly after publication, Dr Michael Fogarty pointed out that the first three figures in my chapter 1, the first six figures in chapter 2, and Appendix 9.1, were closely similar to diagrams and a proof contained in some of his unpublished course notes. Dr. Fogarty presented this course in New Zealand in December 1991. Instead of discussing this claim I should have moved immediately to remove any appearance of impropriety, and I apologise to Dr Fogarty for giving him this perception. To remove the impression that, even accidentally, I would use anyone else's material without acknowledgement I have now either removed these items or the disputed details and have appropriately recast the preliminary material in chapters one and two and Appendix 9.1. I have also taken the opportunity of correcting some confusions and errors in other chapters.

Malcolm Haddon

Hobart, May 2002

1

Fisheries, Population Dynamics, and Modelling

1.1 THE FORMULATION OF FISH POPULATION DYNAMICS

Fish populations can undergo many changes in response to being fished. These include changes to total numbers, total biomass, size-frequency distributions, age-structure, and spatial distributions. Fisheries science has naturally developed into using mathematical and statistical descriptions of these processes in attempts to understand the dynamics of exploited populations. The underlying assumption is that if we can understand how populations respond to different perturbations then we should be able to manage those fisheries according to our chosen objectives.

Unhappily, the astonishing local abundance of some fish species in the wild can lead fishers to believe they can have no impact on stocks. This unfortunate intuition is nicely illustrated by one of the more famous scientists from the 19th century. Thomas Huxley was impressed by the sheer abundance of some fish populations. In a paper published in 1881, he calculated the number of fish in a single school of North Sea herring.

In these shoals the fish are closely packed, like a flock of sheep straying slowly along a pasture, and it is probably quite safe to assume that there is at least one fish for every cubic foot of water occupied by the shoal. If this be so, every square mile of such a shoal, supposing it to be three fathoms deep, must contain more than 500,000,000 herrings (Huxley, 1881) [1 fathom = 1.83 m].

Huxley was explicit about his belief that human fishing could not have a negative impact upon marine fish stocks. In a speech made in 1883, he claimed that most fish populations were so numerous and fecund that they could not be affected by the limited activities of human fishing.

I believe then that the cod fishery, the herring fishery, pilchard fishery, the mackerel fishery, and probably all the great sea fisheries are inexhaustible: that is to say that nothing we do seriously affects the numbers of fish. And any attempt to regulate these fisheries seems consequently, from the nature of the case to be useless (Huxley, 1884).

1

Such arguments from astonishment are still met with today and have been referred to as the inexhaustibility paradigm (Mace, 1997). That some people fail to grasp that unrestrained fishing can impact on fished populations is remarkable given the weight of evidence to the contrary. The sad pseudo-experiment of stopping all commercial fishing in the North Sea during the years of the First World War demonstrated conclusively that catch levels were already too high around Europe. The respite from fishing during the war years allowed stocks to recover so that catch rates of large fish after the war were much higher than before. Unfortunately, this improvement did not last long once unrestrained fishing resumed. Sadly, this ghastly experiment was repeated during the Second World War with the same results (Smith, 1988). Despite a great deal of evidence, the debate on why assessment and management of commercial fish stocks were required continued for many decades (Hardy, 1959; Smith, 1994).

Many developments in fisheries science assisted the change in perception but it was at least three decades into the 20th century before mathematical treatments of aspects other than simple summaries of catch-per-unit-effort (CPUE) were considered. Russell (1931) clarified the "overfishing problem" with a simple, almost qualitative, algebraic expression.

> It is my aim here to formulate in a simplified and general way, and without mathematical treatment, the broad facts of the case, to state in simple language those elementary principles that are at the back of everyone's mind who deals with the problem of the rational exploitation of the fisheries (Russell, 1931, p.3).

Russell started by recognizing that a stock could be divided into animals of a size liable to capture (already recruited to the fishery) and those smaller than this limit. He also considered only complete stocks so that emigration and immigration were irrelevant. Russell focused on what would induce an increase in the population and what would lead to a decrease. He summarized stock biomass dynamics as

$$S_{i+1} = S_i + (A + G) - (C + M) \qquad (1.1)$$

where S_i is the stock biomass in year i, A is the sum of the initial weights of all individuals recruiting to the stock each year, G is the sum of the growth in biomass of individuals already recruited to the stock, C is the sum of weights of all fish caught, and M is the sum of the weights of all fish which die of natural causes during the year. Nowadays we might use different letters [perhaps: $B_{i+1} = B_i + (R + G) - (F + M)$] to those used by Russell, but that is a trivial difference (Krebs, 1985). We must be careful not to confuse the M used here, with the symbol generally used for the instantaneous natural mortality rate (see Chapter 2). The essential aspect of fish stock dynamics, described by Russell, was that the stock biomass had gains (recruitment and individual growth) and losses (natural and fishing mortality). Russell said of his simple formulation:

> This is self-evident, and the sole value of the exact formulation given above is that it distinguishes the separate factors making up gain and loss respectively, and is therefore an aid to clear thinking (Russell, 1931, p.4).

Russell's work had a great deal of influence (Beverton and Holt, 1957; Hardy, 1959; Krebs, 1985). Beverton and Holt (1957) pointed to other workers who had identified the basic principles of the dynamics of exploited fish populations before Russell (Petersen, 1903, and Baranov, 1918). However, Russell appears to have had a more immediate influence, with the others being of more historical interest. Baranov's work, especially, was very advanced for his day but it was published in Russian and only began to be recognized for its value much later (Ricker, 1944; 1975). Russell was almost dismissive of his own statements, but categorizing or estimating the factors identified by Russell (lately within age- or size- or spatially structured models) has been the main focus of fisheries scientists ever since. Methods of modelling the details of these processes have varied greatly but the underlying factors conveyed in Eq. 1.1 are standard.

The obvious factors missing from Russell's formulation are the effects of other species (competitors, predators, etc.) and of the physical environment in which the species live, which can include everything from *el niño* effects to pollution stress (Pitcher and Hart, 1982). It is still the case in most fish stock assessments that the effects of other species and the physical environment are largely ignored. However, there are currently movements towards encouraging management of ecosystems and multi-species management that are challenging that view (Pitcher, 2001; Pauly *et al*, 2001). Ecosystem management is becoming a trendy phrase in resource management and it may become a political reality before the technical ability is developed to understand ecosystem dynamics in any way useful to management. If resource management is to be guided by science rather than public opinion then ecosystem management may prove the greatest challenge yet for fisheries scientists.

The intuitions behind much of quantitative fisheries science are mostly the same now as earlier in the 20th century. The rising interest in multi-species and ecosystem management, with the need for a precautionary principle and perhaps for marine protected areas, can be seen as a move to adopt a new set of intuitions about fished stocks and our interactions with them. These ideas are still under active development and are currently based upon extensions of methods used with single species (Garcia and Grainger, 1997; Walters and Bonfil, 1999). A completely different direction is being followed by users of simulation models such as EcoPath or EcoSim (introductions to EcoPath can be found on the WWW, for EcoSim see Walters *et al*, 1997). Despite all of this, in this present work we will be concentrating on single species systems, but some of the possible effects of environmental variability on such things as recruitment and growth (seasonal changes) will be included.

1.2 THE OBJECTIVES OF STOCK ASSESSMENT

Understanding the variations exhibited in the catches of different fisheries (Fig. 1.1) is a major objective for fisheries scientists. By referring to the yield from a fishery as its production there is a potential for confusion. Care must be taken not to mix up a stock's production or yield with its productivity. The two would only tend to be the same if a fishery were being harvested in a sustainable manner.

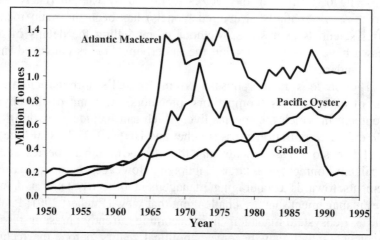

Figure 1.1. The yield of three different fisheries around the World from 1950 to 1993 (data from FAO, 1995). Pacific oyster refers to all oysters reported from the Pacific region and illustrates an increasing trend; Atlantic mackerel might include snoeks and cutlassfishes and exhibits a rise to a relatively stable fishery. Finally, Gadoids includes cods, hakes, and haddocks from the Arctic Atlantic and illustrates a fishery rising to a peak and then declining to a much lower level.

Fishing industries have the potential for fishing a stock too hard and bringing about a reduction in the potential sustainable harvest or even a stock collapse (fishing becomes inefficient and uneconomic). Variations in the yield from a fishery arise through the combined effects of variations in effort, in recruitment, and in natural mortality and growth. Understanding which aspects of production are driving a fishery is the primary task of a stock assessment. The fisheries illustrated in Fig. 1.1. are, in some cases, combinations of species, which could confuse the situation, but similar patterns of increasing harvest levels followed by declines or relative stability can be seen in particular species (Fig. 1.2).

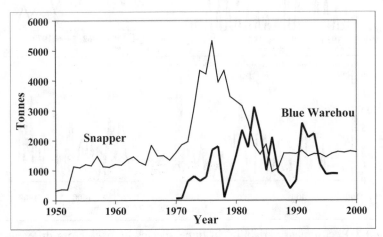

Figure 1.2. The yield of two New Zealand fisheries. The snapper (*Pagrus auratus*) is from the west of the North Island. It shows a fishery developing, the impact of pair-trawlers in the 1970s, and a subsequent serious decline to stability forced through a Total Allowable Catch. The blue warehou (*Seriolella brama*) is from the east and south of the South Island and shows a naturally variable fishery in which availability can vary between years (data from Annala *et al*, 2001).

The difficulty for the fisheries scientist is to decide whether a particular effort or catch level is sustainable for a given future time. This can be made especially difficult by the major sources of productivity varying naturally through time. It is certainly the case that in many exploited species, recruitment is the most variable element of production. Cushing stated:

> From year to year, recruitment varies between a factor of about three to more than two orders of magnitude. The response of recruitment to changes in spawning stock is obscure, probably as a consequence of the variability. But this natural variation provides the mechanism

by which the stock remains adapted to its environment. (Cushing, 1988, p.105).

Time-series of data suggest that different species can have very different patterns of recruitment (Fig. 1.3) but that all are highly variable over varying time scales.

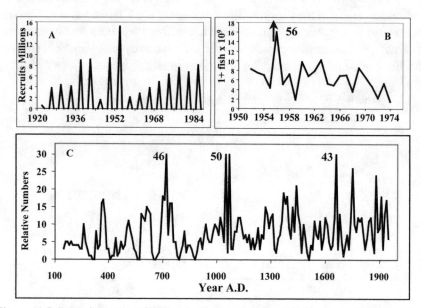

Figure 1.3 Recruitment variability in three species of fish across different time scales. Panel *A* relates to the dominant cohort of Fraser river sockeye salmon (data from Ricker, 1997). Panel *B* relates to North Sea herring the arrow refers to an exceptional recruitment in 1956 (data from Cushing, 1988). Finally, panel *C* relates to 1+ and older Pacific hake, note the exceptionally long time line extending back nearly 2000 years (data from Soutar and Isaacs, 1969). 1+ refers to fish that are between one and two years old. As with Panel *B*, the numbers next to the spikes in the diagram relate to exceptional time periods.

The biotic and abiotic factors affecting recruitment variation will thus strongly influence the resilience of those populations to perturbations (especially disturbances due to harvesting). Generally, any influence the biotic environment has will be upon the natural mortality rates (including those of pre-recruits, thereby affecting recruitment success). These effects could include predation, disease, parasitism, and availability of food and intra- and inter-specific competition. High recruitment success can appear to lead to reduced growth rates for the cohorts concerned, presumably due to competition (e.g. Punt *et al*, 2001).

If the natural mortality term (M) is taken to place a species in its broader ecological context (not forgetting growth effects), we should remember to question the commonly used assumption of a constant level of natural mortality through time.

Unfortunately, even into the 1980s, many mathematical treatments of fish population dynamics were limited to determining the expected behaviour of fish stocks that had attained equilibrium in relation to the fishing pressure being experienced. "Equilibrium" just meant that the population is assumed to have reached a stable balance between numbers surviving and those dying. It is unfortunate that emphasis was focused upon equilibrium conditions because even if a stock appears to have reached equilibrium it will undoubtedly be, at most, a dynamic equilibrium. If CPUE is declining through time then, because equilibrium analyses assume that all catch levels are sustainable, such analyses are invariably less conservative than dynamic non-equilibrium analyses.

Different exploited fish populations may express a wide range of dynamic behaviours (Figs. 1.1 and 1.2). One of the key goals of stock assessment is to understand both the natural variation found in exploited populations and how harvesting affects their dynamics. This clearly requires an understanding of the productive stock (stock structure) as well as the individual components of productivity (recruitment processes, individual growth, and mortality processes). We will thus consider each of the components of productivity.

We will not explicitly consider stock structure but it is consistent with Russell's formulation that we can regard two populations as being from different stocks if their growth and natural mortality characteristics are significantly different. These two aspects of stock dynamics have a large influence on productivity and so the dynamics of the two populations will be different and ideally they should really be managed separately for maximum efficiency and stock sustainability (Haddon and Willis, 1995). Stock determination is one reason that studies of the biology of exploited species, instead of just their population dynamics, can have great value to fisheries management decisions.

Classical fisheries science and management has a very poor track record peppered with numerous fishery collapses (Pitcher, 2001; and references therein). While this cannot be denied we must remember the politico-economic system under which fisheries research is undertaken and used. "If fisheries scientists have failed, it is in not educating those who make decisions in fisheries management to work within the limits of what is biologically possible instead of within the bounds set by what is economically required." (Haddon, 1980, p. 48). Awareness of the uncertainty inherent in all fisheries assessments is growing but this is still not always reflected in management decisions. Whichever path is taken by natural resource managers in the future, a knowledge of the strengths and weaknesses of the kinds of mathematical models used will assist in using them more effectively.

1.3 CHARACTERISTICS OF MATHEMATICAL MODELS

1.3.1 General Properties

We have considered just a few of the properties of wild aquatic populations that affect what we can know and what we should attempt to discover when we wish to manage such stocks adequately. Fisheries assessments are generally based upon mathematical models of the production processes and the populations being fished. Before considering any particular model in detail, it would be helpful to consider models in general.

Mathematical models are a subset of the class of all models. Models may take many forms ranging from a physical representation of whatever is being modelled (e.g., a ball and stick model of DNA as produced by Watson and Crick, 1953), diagrammatic models (such as a geographical map), and the more abstract mathematical representations being discussed here. Despite this diversity, all models share certain attributes. All models constitute an abstraction or simulation by the modeller of the process/phenomenon being modelled.

1.3.2 Limitations Due to the Modeller

As models are never perfect copies of the thing being modelled, there must be some degree of abstraction or selection of what the modeller considers to be essential properties of whatever is being modelled. A fundamental idea behind modelling is therefore to identify the properties that must be included in order that the behaviour of the model may be expected to exhibit a close approximation to the observable behaviour of the system being studied. This selection of what are considered to be the important properties of a system also permits the modeller to emphasize particular aspects of the system being modelled. A road map shows roads greatly magnified in true geographical scale because those are what are being emphasized. The selection of what aspects of a system to include in a model is what determines whether a model will be generally applicable to a class of systems, or is so specialized that it is attempting to simulate the detailed behaviour of a particular system (for system one might read stock or population). By selecting particular parts of a natural system the model is being limited in what it can describe. The assumption is that it will provide an adequate description of the process of interest and that those aspects not included will not unexpectedly distort the representation of the whole (Haddon, 1980).

1.3.3 Limitations Due to Model Type

A model can be physical, verbal, graphical, or mathematical; however, the particular form chosen for a model imposes limits on what can be described. For example, if one tries to produce a verbal description of a dynamic population process, one is invariably limited in how well one can capture or express the properties of the populations being described. This limitation is not necessarily due to any lack of expository skills of the narrator. Instead, it is because spoken languages do not appear to be well designed for describing dynamic processes, especially where more than one variable is changing through time or relative to other variables. Fortunately, mathematics provides an excellent alternative for describing dynamic systems.

1.3.4 The Structure of Mathematical Models

There are many types of mathematical model. They can be characterized as descriptive, explanatory, realistic, idealistic, general, or particular; they can also be deterministic, stochastic, continuous, and discrete. Sometimes they can be combinations of some or all of these things. With all these possibilities, there is a great potential for confusion over exactly what role mathematical models can play in scientific investigations. To gain a better understanding of the potential limitations of particular models, we will attempt to explain the meaning of some of these terms and introduce the arguments used for and against application of the various approaches.

Mathematical population models are termed **dynamic** because they represent the present state of a population/fishery in terms of its past state or states with the potential to describe future states. For example the Schaefer model (Schaefer, 1957) of stock biomass dynamics (of which we will be hearing more) can be partly represented as

$$B_{t+1} = B_t + rB_t\left(1 - \frac{B_t}{K}\right) - C_t \qquad (1.2)$$

Here the variables are C_t, the catch during time t, and B_t, the stock biomass at the end of time t (B_t is also an output of the model). The model parameters are r, representing the population growth rate of biomass (production), and K, the maximum biomass that the system can attain (these parameters come from the logistic model from early mathematical ecology;

see Chapter 2). By examining this relatively simple model one can see that expected biomass levels at one time ($t + 1$) are directly related to catches and the earlier biomass (time = t; they are serially correlated). The degree of impact of the earlier biomass on population growth is controlled by the combination of the two parameters r and K. By accounting for the serial correlations between variables from time period to time period such dynamic state models differ markedly from traditional statistical analyses of population dynamics. Serial correlation removes the assumption of sample independence required by more classical statistical analyses.

1.3.5 Parameters and Variables

At the most primitive level, models are made up of variables and parameters. Parameters are the things that dictate quantitatively how the variables interact. They differ from a model's variables because the parameters have to be estimated if a model is to be fitted to observed data. A model's variables must represent something definable or measurable in nature (at least in principle); ideally they must represent something that is real. Parameters modify the impact or contribution of a variable to the model's outputs, or are concerned with the relationships between the variables within the model (Fig. 1.4).

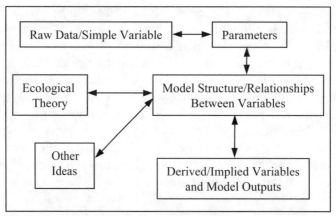

Figure 1.4 Schematic diagram of the relationships between raw data, fitted parameters, variables, ecological theory, and the model outputs, which make up a mathematical model. A model need not be fitted to data but it still requires parameter values.

In any model, such as Eq. 1.2, we must either estimate or provide values for the parameters. With the variables, one either provides observed values for them (e.g., a time series of catches) or they are an output from the model. Thus, in Eq. 1.2, given a time series of observed catches plus estimates of parameter values for r and K, a time series of biomass values, B_t, is implied by the model as an output. As long as one is aware of the possibilities for confusion that can arise over the terms "observe", "estimate", "variable", "parameter", and "model output", then one can be more clear about exactly what one is doing while modelling a particular phenomenon. The relation between theory and model structure is not as simple as represented (Fig. 1.4). Background knowledge and theory ("Other Ideas" in Fig. 1.4) may be the drive behind the selection of a model's structure. The relationships proposed between a set of variables may constitute a novel hypothesis or theory about the organization of nature.

1.4 TYPES OF MODEL STRUCTURE

1.4.1 Deterministic/Stochastic

We can define a model parameter as *a quantitative property (of the system being modelled) that is assumed either to remain constant over the period for which data is available, or to be modulated by environmental variation.* Roughly speaking, models in which the parameters remain constant on the time scale of the model's application are referred to as **deterministic**. With a given set of inputs, because of its set of constant parameters, a deterministic model will always give the same outputs. Because the relationships between the model variables are fixed (constant parameters), the output from a given input is "determined" by the structure of the model. One should not be confused by situations where parameters in a deterministic model are altered sequentially by taking one of an indexed set of predetermined values (e.g., a recruitment index or catchability index may alter and be changed on a yearly basis). In such a case, although the parameters are varying they are doing so in a repeatable, deterministic fashion (constant over a longer time scale), and the major property that a given input will always give the same output still holds.

Deterministic models contrast with **stochastic** models in which at least one of the parameters varies in a random or unpredictable fashion over the time period covered by the model. Thus, given a set of input values the particular output values will be uncertain. The parameters that vary will take on a random value from a predetermined probability distribution (either from one of the classical probability density functions – pdf, or a custom distribution). Thus, for example, in a model of a stock, each year the recruitment level may attain a mean value plus or minus a random amount determined by the nature of a random variate

$$R_y = R_{\bar{y}} + N\left(0, \sigma^2\right) \tag{1.3}$$

where R_y is the recruitment in year y, $R_{\bar{y}}$ is the average recruitment, and $N(0,\sigma^2)$ is the notation used for a random variable whose values are described in this example by a normal distribution with mean zero (i.e., has both positive and negative values), and variance σ^2.

Given a set of input data, a deterministic model expresses all of its possible responses. However, there would be little point in running a stochastic model only once. Instead, stochastic models form the basis of so-called Monte Carlo simulations where the model is run repeatedly with the same input data but for each run new random values are produced for the stochastic parameters (e.g., as per Eq. 1.3). For each run a different output is produced and these are tabulated or graphed to see what range of outcomes could be expected from such a system. Even if the variation intrinsic to a model is normally distributed it does not imply that a particular output can be expected to be normally distributed about some mean value. If there are non-linear aspects in the model, skew and other distortions may arise. We will be looking more closely at this phenomenon when discussing stock-recruitment relationships and Monte Carlo models.

Future projections, risk assessment, and the determination of the possible impact of uncertainty in one's data, all require the use of Monte Carlo modelling. One of the objectives of this book is to enable readers to attain a level of experience such that they may create and run Monte Carlo simulations that will be of use when analyzing the stock dynamics of the fisheries for which they are responsible.

1.4.2 Continuous vs. Discrete Models

Early fishery modellers all used continuous differential equations to design their models, so the time steps in the models were all infinitesimal. This was because computers were still very much in their infancy and analytical solutions were the culture of the day. Early fishery models were thus formed using differential calculus (Jeffrey, 1969), and parts of their structures were determined more by what could be solved analytically than because they reflected nature in a particular accurate manner. At the same time the application of these models reflected or assumed equilibrium conditions. Fortunately, we can now simulate a population using easily available computers and software, and we can use more realistic, or more detailed, formulations. While it may not be possible to solve such models analytically (i.e., if the model formulation has that structure its solution must be this), they can usually be solved numerically (informed and improving trial and error). Although both approaches are still used, one big change in fisheries science has been a move away from continuous differential equations towards difference equations, which attempt to model a system as it changes through discrete intervals (ranging from infinitesimal up to yearly time steps). Despite the increases in complexity, all of these models retain, in essence, the structure of Russell's (1931) formulation.

There are other aspects of model building that can limit what behaviours can be captured or described by a model. The actual structure or form of a model imposes limits. For example, if a mathematical modeller uses difference equations to describe a system, the resolution of events cannot be finer than the time intervals with which the model is structured. This obvious effect occurs in many places. For example, in models that include a seasonal component the resolution is quite clearly limited depending on whether the data are for weeks, months, or some other interval.

1.4.3 Descriptive/Explanatory

Whether a model is discrete or continuous, and deterministic or stochastic, is a matter of model structure and clearly influences what can be modelled. But when we consider the purpose for which a model is to be used, we should consider other properties relating to its design that also influence the potential scope and use of the model. For a model to be descriptive all it needs to do is mimic the empirical behaviour of the variables making up the observed data. A fine fit to individual growth data, for example, may usually be obtained by using polynomial equations

$$y = a + bx + cx^2 + dx^3 .... + mx^n \qquad (1.4)$$

in which no attempt is made to interpret the various parameters used (usually one would never use a polynomial greater than order six with order two or three being more common). Such descriptive models can be regarded as "black boxes", which provide a deterministic output for a given input. It is not necessary to know the workings of such models; one could even use a simple look-up table which produced a particular output value from a given input value by literally looking up the output from a cross-tabulation of values.

Such "black box" models would be descriptive and nothing else. We need not take notice of any assumptions or parameters used in their construction. Such purely descriptive models need not have elements of realism about them except for the variables being described (Fig. 1.5).

Explanatory models also provide a description of the empirical observations of interest but in addition they attempt to provide some justification or explanation for why the particular observations noted occurred instead of a different set. With explanatory models it is necessary to take into account the assumptions, parameters, as well as the variables that make up the model (Fig. 1.5). By attempting to make the parameters, variables, and how the variables interact, reflect nature, explanatory models attempt to simulate events in nature. A model is explanatory if it contains theoretical constructs (assumptions, variables, and/or parameters), which purport to relate to the structure of nature and not just to how nature behaves.

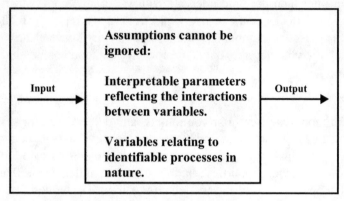

Figure 1.5 An explanatory model provides a description of the observable data but also reflects, mimics, or simulates the manner in which the modeller considers nature to be structured. It provides a reason for expecting the relationship between inputs and outputs to remain as the model predicts. In a purely descriptive model, the contents of the box are not particularly important.

1.4.4 Testing Explanatory Models

Explanatory models are, at least partly, hypotheses or theories about the structure of nature and how it operates. They should thus be testable against observations from nature. The question arises: How do we test explanatory models? We can ask: Does fitting a model to data constitute a test of the model? Clearly, if the expected values for the observed data, predicted by a model, account for a large proportion of the variability within the observed data, then our confidence that the model adequately describes the observations can also be great. But the initial model fitting does not constitute a direct test of the structure of the model. A good fit to a model does not test whether the model *explains* observed data; it only tests how well the model *describes* the data (Haddon, 1980). The distinction between explanation and description is very important and requires emphasis (which is why this sentence is here). A purely descriptive or empirical model could provide just as good a fit to the data which hopefully makes it clear that we need further, independent observations against which to really test the model's structure. What requires testing is not only whether a model can fit a set of observed data (i.e., not only the quality of fit) but also whether the model assumptions are valid and whether the interactions between model variables, as encoded in one's model, closely reflect nature.

Comparing the now fitted model with new observations does constitute a test of sorts. Ideally, given particular inputs, the model would provide a predicted observation along with confidence intervals around the expected result. An observation would be said to refute the model if the model predicted that its value was highly unlikely given the inputs. But with this test, if there is a refutation, there is no indication of what aspect of the model was at fault. This is because it is not a test of the model's structure but merely a test of whether the particular parameter values are adequate (given the model structure) to predict future outcomes! Was the fitting procedure limited because the data available for the fitting did not express the full potential for variation inherent in the population under study? Was it the assumptions or the particular manner in which the modeller has made the variables interact that was at fault? Was the model too simple, meaning were important interactions or variables left out of the structure? We cannot tell without independent tests of the assumptions or of the importance of particular variables.

If novel observations are in accord with the model then one has gained little. In practice, it is likely that the new data would then be included with the original and the parameters re-estimated. But the same could be said about a purely empirical model. What are needed are independent tests that the structure chosen does not leave out important sources of variation; to test this requires more than a simple comparison of expected outputs with real observations.

While we can be content with the quality of fit between our observed data and that predicted from a model, we can never be sure that the model we settle on is the best possible. It is certainly the case that some models can appear less acceptable because alternative models may fit the data more effectively. The process of fitting a model can have the appearance of using the quality of fit as a test between different models. We can illustrate that this is not necessarily the case by considering that we could always produce a purely descriptive model with very many parameters which provides an extremely good fit but this would not be considered a "better" model relative to a more realistic one (Fig. 1.6).

Discussion over which curve or model best represents a set of data depends not only upon the quality of fit but also upon other information concerning the form of the relationship between the variables (Fig. 1.6). Clearly, in such cases, criteria other than just quality of numerical fit must be used to determine which model should be preferred.

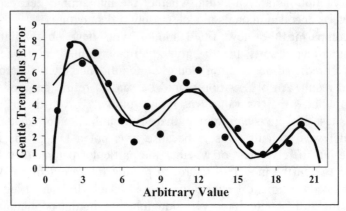

Figure 1.6 Artificial data generated from a straight-line decline (Y = 6 - 0.25 X) plus normal random error [N(0,2)]. A straight line fitted to this data gives: Y = 6.1583 - 0.2429x, describing 44.1% of the variation in the data. The thick curved line that extrapolates severely downward beyond the data, is a sixth order polynomial equation with four more (Y = -0.000045x^6 + 0.0031x^5 - 0.08357x^4 + 1.0714x^3 - 6.7133x^2 + 18.0600x - 8.9261) parameters than the straight line. It describes 88.25% of the variation in the data and is clearly a better fit, at least when not extrapolated. The fine curved line is a straight line with an intrinsic cycle [Y = a + bx + C sin((2π(x-s))/D)], which has three more parameters than the straight line. In this case, however, we know the straight line to be the model that better represents the underlying process, the cyclicity is merely apparent. In Chapter 3 we consider methods for assessing whether increasing the number of parameters in a model is statistically justifiable, though with data such as this only further work could enlighten one as to the optimum model.

Any explanatory model must be biologically plausible. It might be possible to ascribe meaning even to the parameters of a completely arbitrary model structure. However, such interpretations would be *ad hoc* and only superficially plausible. There would be no expectation that the model would do more than describe a particular set of data. An explanatory model should be applicable to a new data set, although perhaps with a new set of particular parameters to suit the new circumstances.

Precision may not be possible even in a realistic model because of intrinsic uncertainty either in our estimates of various parameters (observation error) or in the system's responses, perhaps to environmental variation (process error). In other words, it may not be possible to go beyond certain limits with the precision of our predicted system outcomes (the quality of fit may have intrinsic limits).

1.4.5 Realism/Generality

Related to the problem of whether or not we should work with explanatory models is the problem of realism within models. Purely descriptive models need have nothing realistic about them at all. But it is an assumption that if one is developing an explanatory model, then at least parts of it have to be realistic. For example, in populations where ages or sizes can be distinguished, age- or size- structured models would be considered more realistic than a model which lumped all age or size categories into one. But a model could be a combination of real and empirical.

For a model to be general would mean that it would have a very broad domain of applicability; it could be applied validly in very many circumstances. There have been many instances in the development of fisheries where a number of particular models describing a particular process (e.g., growth) have been subsumed into a more general mathematical model of which they are special cases (see Chapter 8). Usually this involves increasing the number of parameters involved, but nevertheless, these new models are clearly more mathematically general. It is difficult to draw conclusions over whether such more general equations/models are less realistic. That would be a matter of whether the extra parameters can be realistically interpreted or whether they are simply *ad hoc* solutions to combining disparate equations into one that is more mathematically general. With more complex phenomena, such as age-structured models, general models do not normally give as accurate predictions as more specialized models tuned to the particular situation. It is because of this that modellers often consider mathematically general models to be less realistic when dealing with particular circumstances (Maynard-Smith, 1974).

1.4.6 When is a Model a Theory?

All models may be considered to have theoretical components, even supposedly empirical models. It becomes a matter of perception more than model structure. With simple models, for example, the underlying assumptions can begin to take on the weight of hypothetical assertions. Thus, if one were using the logistic equation to describe the growth of a population, it imports the assumption that density-dependent compensation of the population growth rate is linearly related to population density. In other words, the negative impact on population growth of increases in population size is linearly related to population size. This can be regarded either as a domain assumption (that is, the model can only apply validly to situations where density-dependent effects are linearly related to population density), or as a theory (non-linear density-dependent effects are unimportant in the system being modelled). It is clearly a matter of perception as to which of these two possibilities obtains. This is a good reason one should be explicit concerning the interpretation of the assumptions in one's model.

If one were to restrict oneself purely to empirical relationships, the only way in which one's models could improve would be to increase the amount of variance in the observations accounted for by the model. There would be no valid expectation that an empirical model would provide insights into the future behaviour of a system. An advantage of explanatory/theoretical models is that it should be possible to test the assumptions, relationships between variables, and error structures, independently from the quality of fit to observed outcomes. Of course, one way of determining whether there is something wrong with one's model is to compare expected outcomes with observed outcomes.

It should, therefore, be possible to present evidence in support of a model, which goes beyond the quality of fit. Those models where the proposed structure is not supported in this way may as well be empirical.

2

Simple Population Models

2.1 INTRODUCTION

2.1.1 Biological Population Dynamics

A biological population is a collection of individuals and, even if a population is not recognized as an entity in its own right, it has identifiable emergent properties not possessed by individual organisms. These properties include the population's size, growth, immigration, and emigration rates, its age-structure, and how it arranges its members within its geographical distribution. The dynamic behaviour of a population relates to changes in its size and other properties through time. One of the aims of any population modelling exercise is to attempt to describe and possibly explain aspects of this dynamic behaviour.

Mathematical equations used to model biological populations are only an abstract representation of the population of interest. This requires emphasis because the equations in many population models can exhibit dynamic behaviours that biological populations either do not or could not exhibit. They are, after all, only models. For example, a mathematical model might predict that under some circumstances the modelled population was made up of a negative number of organisms or produced a negative number of recruits. Such obvious discrepancies between the behaviour of the mathematical equations and possible biological behaviours are of little consequence because they are easily discovered and avoided. Unfortunately, purely mathematical behaviours can also arise which are less obvious in their effects. It is thus sensible to understand the dynamic behaviour of any equations used in a modelling exercise to avoid ascribing non-sensible behaviours to innocent populations.

2.1.2 The Dynamics of Mathematical Models

The purpose of this chapter is to give an explicit, though brief, introduction to the properties of models and how their dynamic behaviours are dependent upon both their particular mathematical form and the particular values given to their parameters. To do this we will consider some of the mathematical models that are commonly used in both ecology and fisheries to describe population dynamics. We will also be introducing the distinction between models that have no age structure and those that do. It will be seen that this is a natural progression, as certain simple or whole-population models can be combined in particular ways to produce an age-structured model.

First, we will consider exponential population growth and its relationship to both the logistic model of population growth, commonly used in basic fisheries models, and to age-structured models, such as those used in the analysis of yield-per-recruit. This will lead us to a more detailed study of the logistic population growth model. As we shall see, the discrete logistic model can exhibit a wide range of dynamic behaviours ranging from monotonically damped equilibria (a smooth rise to a stable equilibrium population size) to chaotic behaviour (completely unpredictable behaviour). The simple age-structured models we will consider will demonstrate that if information about the internal structure of a population is available we can investigate how the productivity of any population is distributed between new recruits to the population and the growth of individuals already in the population.

2.2 ASSUMPTIONS - EXPLICIT AND IMPLICIT

2.2.1 All Assumptions Should be Explicit

A model can be viewed as a purely abstract mathematical system or we can make the step of relating its variables and parameters to the real World. An explicit listing of a model's assumptions about its relation to reality, along with their implications, should be standard practice but is not as common as it ought to be. The most important assumptions of the first model we will consider here are (Slobodkin, 1961; Pianka, 1974)

1. **All animals in the population have identical ecological properties.** *This implies that morphology, genetics, and behaviour have no effect on population dynamics. Age-structured models are one way to alter this assumption. In addition, we are dealing with a single population or stock, that is, either there is no immigration ($I = 0$) or emigration ($E = 0$), or alternatively, ($E = I$) that immigration equals emigration; we are ignoring biogeography.*

2. **There are no significant time delays in population processes.** *The whole population responds immediately to any changes in population*

size irrespective of season, geographical scale or distance. Age-structured models also affect this assumption, at least by introducing time lags between reproduction and growth.

3. **The parameters of the model are constants.** *This ignores seasonal and environmentally induced natural variations in such things as maximum population size and maximum population growth rate.*

All of these assumptions about the relation between the model and reality are unlikely or unrealistic but, for the moment, they are necessary if we are to keep the model simple (Slobodkin, 1961). Wherever we have used the term "population" in the three assumptions above, we could have used the term "fish stock". The idea is that the model is concerned with a self-sustaining population that does not depend upon immigration to maintain its size. By making our assumptions explicit, we are helping to determine how it might be possible to interpret the terms of our mathematical model. The particular assumptions here could have been described differently, providing a different emphasis, e.g. in fisheries the assumption of a single stock would commonly be given explicitly, but the overall intent is clear. The assumptions also determine the "domain of applicability" (meaning the range of situations to which the model is expected to relate; Lakatos, 1970). It is excellent practice to inspect statements about the assumptions of models in the literature to see if one can add to the list given.

2.3 DENSITY - INDEPENDENT GROWTH

2.3.1 Exponential Growth

Population growth at its abstract simplest can be imagined as a population growing in an unlimited environment. An example, which might approximate this theoretical possibility of unrestricted growth, could arise where resources may not yet be limiting for a colonizing species when it first arrives in a new and empty location. Large parts of the world's human population are still growing as if this were the case but sadly they will soon discover that resources are, in fact, limited. However, a real world example might be the populations of organisms that first colonized the new volcanic Island of Surtsey, which started to form in November 1963 off Iceland, or the remains of the volcanic island of Krakatoa, which blew up in August 1883 (Krebs, 1985). The initial, accelerating growth exhibited by such populations is described as exponential growth.

If birth and death rates are constant at all population sizes it implies a constant proportional increase in population size each time period. The rate at which a population's size changes can be described by

$$\frac{dN}{dt} = (b-d)\,N \quad \equiv \quad rN \tag{2.1}$$

In this differential equation, dN/dt translates as the rate of change of the population size N relative to time t, b is the birth rate, and d is the death rate. In ecology textbooks this is commonly rendered as the equivalent shorter equation where $r = (b - d)$ and r is often termed the intrinsic rate of increase, the "instantaneous rate of population growth", or even the "per-capita rate of population growth" (Krebs, 1985, p. 212). The point to note is that the per capita growth rate (b - d), or r, is a constant and is assumed independent of population size N. Because of this independence, this type of growth is termed "density-independent". Good summaries of population growth from a more purely ecological viewpoint can be found in many ecological texts (e.g., Begon and Mortimer, 1986; Caughley, 1977; Christiansen and Fenchel, 1977; Krebs, 1985; Pianka, 1974).

There is a possible source of confusion over the term "rate". A constant birth rate at all population sizes does not mean that the same absolute number of offspring will be produced at all population sizes, but rather that the population increases by the same proportional rate at all sizes. Thus, a growth rate of 0.1 implies a population of 100 will increase by 10, while a population of 1000 will increase by 100. As population size increases, the absolute number by which the population grows in each time interval will also increase, but the proportional increase will stay constant.

Equation (2.1) is a differential equation relating the rate of change of population size to time. This can be integrated to produce an equation describing the expected numbers in the population at any time after some given starting population size and time. The solution to Eq. 2.1 is

$$N_t = N_0 e^{(b-d)t} \quad \equiv \quad N_0 e^{rt} \quad \equiv \quad N_{t-1} e^{r} \tag{2.2}$$

where N_t is the expected population size at time t and N_0 is the population size at time zero (the starting population size). Because the irrational constant e is used (Jeffrey, 1969), this model is termed exponential growth and that phrase often brings to mind a continuously increasing population. However, depending upon the balance between births and deaths, exponential growth can describe a population going to extinction, staying stable, or growing rapidly (Fig. 2.1; Berryman, 1981; Krebs, 1985). In simple whole-population models, Eq. 2.2 is of limited use. Knowledge of its properties is worthwhile, however, as it is used within age-structured models that follow the fate of individual cohorts whose numbers can only decline after birth.

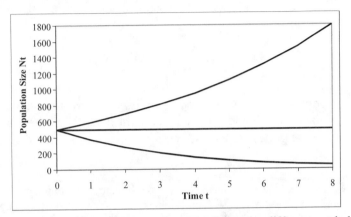

Figure 2.1 Population growth curves from Eq. 2.2 under different net balances of birth and death rates (Berryman, 1981). When the nett reproductive rate is positive the exponential increase produces an accelerating curve leading to ever increasing population sizes. When there are more deaths than births an exponential decrease produces a decelerating curve which approaches zero ever more slowly (Example Box 2.1). Obviously, when births balance deaths, an equilibrium ensues.

Note that exponential increase produces an accelerating curve in which the proportional increase in population size is constant each time period (Eq. 2.2, Fig. 2.1) but where the absolute numbers of individuals entering the population each time period increases more quickly with time. In the case of exponential decrease this is a decelerating curve where the proportional decrease is constant throughout but the actual number dying each time period reduces so the population approaches zero ever more slowly with time. This distinction between constant proportional rates of change and increasing or decreasing absolute numbers of individuals is an important one to grasp when considering the exponential model of population changes. The intuitions concerning this phenomenon may come more easily if one recognizes that this process is analogous to compound interest in financial terms. Positive interest rates lead to exponential growth and this is a common property for anything that is increased or decreased by a constant proportion through a series of time periods.

2.3.2 Standard Transformations

The mathematical form of exponential population growth leads to a number of properties that can simplify our representation of population growth processes. The most important property for the practice of population dynamics is that a natural logarithm transformation linearizes the pattern of growth (natural logs are logs to base e). Thus,

$$Ln\left(N_0 e^{rt}\right) = Ln\left(N_0\right) + Ln\left(e^{rt}\right) = Ln\left(N_0\right) + rt \qquad (2.3)$$

In a population growing exponentially, a plot of the natural log of numbers against time should produce a straight line, the gradient would be an estimate of the growth rate r and the intercept an estimate of $Ln(N_0)$ (see Eq. 2.3).

Example Box 2.1 An Excel sheet set up to calculate three exponential curves. Each population starts with 500 individuals, and how it grows depends on whether the birth rate (Row 1) is greater than the death rate (Row 2). The equation in column B must be copied into columns C and D, and down to whatever time is required. By calculating down to a time of eight (Row 21) and plotting the numbers for the three populations (columns B, C, and D) against time (column A), one should be able to generate the equivalent to Fig. 2.1. The growth model is Eq. 2.2. Investigate the properties of this growth model by varying any of the parameters in B1:D3. Try setting the vertical axis of the plot of exponential growth to a logarithmic scale or plot the natural log of numbers against time [put =Ln(B5) into column E, etc.].

	A	B	C	D
1	b – births	0.1	0.1	0.1
2	d – deaths	0.04	0.1	0.125
3	N₀	500	500	500
4	Time	Pop 1	Pop 2	Pop 3
5	0.0	=B$3*EXP((B$1-B$2)*$A5)	500	500
6	0.5	=B$3*EXP((B$1-B$2)*$A6)	500	493.8
7	1.0	=B$3*EXP((B$1-B$2)*$A7)	500	487.7
8	1.5	'	500	481.6
9	2.0	'	500	475.6
10	Cont. down	Copy down	Copy down	Copy down

2.3.3 Why Consider Equilibrium Conditions?

Natural populations being harvested rarely appear to be in equilibrium.

Nevertheless, to understand a mathematical model's properties it is usual to consider under what conditions equilibrium could be attained.

At equilibrium, by definition, the rate of change of population size with respect to time will be zero ($dN/dt = 0$) and with Eq. 2.2 that could occur under two conditions. The trivial case is when $N = 0$, i.e., the population is extinct (because with no immigration, at all subsequent times N will equal 0). The more interesting case biologically is when the birth rate exactly equals the death rate and both are positive. With Eq. 2.2 this is what is termed an astable equilibrium in which any perturbation to the birth or death rates will disrupt the equilibrium and lead to either an exponential increase or a decrease towards extinction.

If the birth and death rates stay constant but there is a perturbation to the population size (for whatever reason), a new equilibrium population size will result. The key point being that the population does not return to its pervious "equilibrium." At equilibrium, $N_{t+1} = N_t$, so there can be no changes in numbers from time t to time $t+1$. Hopefully, it is clear that there can be an astable equilibrium at any population size as long as births equal deaths. When not in equilibrium the populations will either increase to infinity or contract to extinction at a rate dependent upon how dissimilar births are from deaths.

2.4 DENSITY - DEPENDENT MODELS

2.4.1 An Upper Limit and Persistence

Few general characteristics can be ascribed to all biological populations. However, as Linnaeus implied, Malthus pointed out, and the world's human population will soon discover, no population can grow indefinitely because all populations live in limited environments. Thus, positive exponential growth can only be a short-term phenomenon. Another general property of populations is that most are believed to persist despite random environmental perturbations (although long term persistence is not guaranteed as the current spate of extinctions around the world testifies). Unfortunately, our first model, Eq 2.2, allows for very little other than indefinite growth or extinction. While it is true that some "weedy" species exhibit a boom and bust lifestyle (Andrewartha and Birch, 1954), if there is to be population self-regulation then an obvious option is to alter the simple exponential model to account for the general properties of an upper limit and persistence.

2.4.2 The Logistic Model of Growth

One of the simplest models to be derivable from the exponential growth model is that known as the logistic model. It is worth noting that in the past a number of fisheries were managed using equilibrium analyses of continuous models based upon the logistic (details will be given when we consider surplus-production models in Chapter 10). It has also been suggested that at least part of the sad history of failures in fishery advice and management stem from this combination (Larkin, 1977); this suggestion is often followed by numerous insults at the logistic and its implications, as well as the people who use it. Nevertheless, the invective against the logistic is misplaced. The logistic is simply a convenient model of linear density-dependent effects (similar fisheries advice would have come from any equilibrium analysis using a linear model of population regulation). Beverton and Holt (1957) pointed out the weaknesses inherent in using the logistic population growth curve but they recognized that when detailed information was lacking then this approach might have value as long as the insights were treated with reservations. They stated:

> It is when such detailed information does not exist that the sigmoid curve theory, by making the simplest reasonable assumption about the dynamics of a population, is valuable as a means of obtaining a rough appreciation from the minimum of data. (Beverton and Holt, 1957, p.330).

The exponential growth model was density-independent in its dynamics, that is, the birth and death rates were unaffected by the population size. If population growth was regulated such that the difference between births and deaths was not a constant but became smaller as population size increased and greater as the population decreased, then the likelihood of runaway population growth or rapid extinction could be greatly reduced. Such a modification of the dynamic response with respect to density is what makes such models "density-dependent". In a density-dependent model, the population rate of increase (as a balance between births and deaths) alters in response to population size changes (Fig. 2.2).

A decline in the rate of increase in the density-dependent model can be brought about by a decline in the birth rate with population size, or an increase in the death rate, or a combination of the two. We will consider the general case where both rates are affected by population size but the same outcomes would derive from just one being affected.

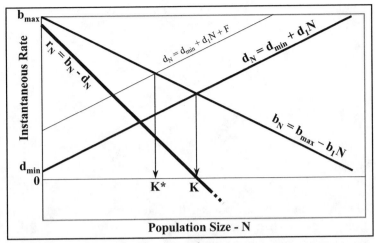

Figure 2.2 Comparison of the influence of population size on the instantaneous rate of increase in a density-dependent model (after Pianka, 1974, p. 86; this is very fundamental population biology and Pianka's discussion of these relationships is highly recommended). The death and birth rates are both described by linear equations as illustrated (see Eq. 2.4), hence the term linear density-dependence. The junction of the two is where births balance deaths leading to the equilibrium population size K. If fishing were able to take a constant amount from the stock this would be equivalent to an added density-independent increment to the death rate (the fine line $d_N = d_{min} + d_1N + F$) obviously leading to a new equilibrium at K^* (see Example Boxes 2.2 and 2.3, and Fig. 2.5). The modification this would make to the reproductive rate r_N is not shown (the r_N line would pass through K^*). If a constant proportion were taken this would be represented by a change in the gradient terms.

The simplest model to include density dependence would be where the birth and death rates are linearly related to population size (Fig. 2.2; Pianka, 1974). For this to be the case instead of the rates being a simple constant we would have to include population size and some modifying parameter in a linear equation

$$b_N = b_{max} - b_1N$$
$$d_N = d_{min} + d_1N$$

(2.4)

where b_{max} and d_{min} are the birth and death rates when population size, N, is very small and b_1 and d_1 are parameters that scale the rates at which the birth and death rates change with population size. Equation 2.4 implies that births decrease linearly as N increases and the death rate increases linearly as population increases (Fig. 2.2). These changes introduce the possibility of population regulation. If the terms in Eq. 2.4 are substituted for b and d in Eq. 2.1, our original model becomes

$$\frac{dN}{dt} = \left[\left(b_{max} - b_1 N \right) - \left(d_{min} + d_1 N \right) \right] N \qquad (2.5)$$

which can be rearranged

$$\frac{dN}{dt} = \left[\left(b_{max} - d_{min} \right) - \left(b_1 + d_1 \right) N \right] N = \left(b_{max} - d_{min} \right) N - \left(b_1 + d_1 \right) N^2 \quad (2.6)$$

This has the effect of putting the intercepts $(b_{max} - d_{min})$ = rate of population increase when population density is very low, together in one term. We can explore the properties of this model by determining the conditions under which equilibria exist (if any). Equilibria exist where dN/dt = 0, thus

$$\left(b_{max} - d_{min} \right) N = \left(b_1 + d_1 \right) N^2 = \left(b_{max} - d_{min} \right) = \left(b_1 + d_1 \right) N^* \quad (2.7)$$

and therefore (Pianka, 1974)

$$N^* = K = \frac{\left(b_{max} - d_{min} \right)}{\left(b_1 + d_1 \right)} = \frac{r}{\left(b_1 + d_1 \right)} \qquad (2.8)$$

where N^* or K is the equilibrium population size often called the "carrying capacity" in the logistic equation, referring to the hypothetical maximum number or biomass that the environment can maintain if no changes occur The use of K to represent this upper limit appears to stem from Pearl and Reed (1922), who were the first to use the term K for this purpose. By including these density-dependent effects, this population model is using a form of negative feedback. Thus, as the population size increases, the birth rate decreases (possible mechanisms could include increased competition or reduced growth, which would affect fecundity or reduce energy allocated to reproduction) and/or increases in the death rate (possibly due to starvation, cannibalism, or predator aggregation).

If we simplify the model (re-parameratize by amalgamating constants) by letting $r = (b_{max} - d_{min})$, then from Eq. 2.6 the rate of population change becomes

$$\frac{dN}{dt} = rN - (b_1 + d_1)N^2 \quad = \quad rN\left(1 - \frac{(b_1 + d_1)N}{r}\right) \qquad (2.9)$$

making the further substitution of $K = r/(b_1 + d_1)$ i.e., $1/K = (b_1 + d_1)/r$ leads to

$$\frac{dN}{dt} = rN\left(1 - \frac{N}{K}\right) \qquad (2.10)$$

which is the more usual form of the well-known logistic equation (Krebs, 1985). The derivation of K shows why the two parameters, r and K, are always strongly correlated. Equation 2.10 tells us about the rate of change of population size (related to stock production). By searching for where this is maximum we find that the maximum rate of population change occurs when $N = K/2$ (Fig. 2.3).

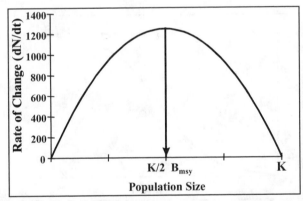

Figure 2.3 Plot of the equilibrium rate of change of population size versus population size (production vs. stock size curve; Schaefer, 1954). Maximum productivity occurs at half K (in terms of biomass, this is B_{msy}), equilibrium occurs at zero and K. The symmetry of the production curve about K/2 is unrealistic. Permitting asymmetry of the production curve will be considered in later chapters.

We conclude that if the population is growing according to the logistic equation, it will grow at its fastest rate when it is at half the theoretical equilibrium population size. In addition, the population growth rate (dN/dt) will be zero when the population is extinct and when it is at its maximum equilibrium size, the carrying capacity. Integrating Eq. 2.10 leads to the continuous solution to the logistic equation, giving the expected population size N_t at time t after some starting time and population size N_0. Setting $\gamma =$

$(K - N_0)/N_0$, K as the symbol used in ecological literature to refer to the equilibrium population size, t is time, and where r is the maximum rate of population change, then

$$N_t = \frac{K}{1 + \dfrac{(K - N_0)}{N_0} \dfrac{1}{e^{rt}}} = \frac{K}{1 + \gamma.e^{-rt}} \tag{2.11}$$

Following the population trajectory through time for this logistic model generates the familiar S-shaped curve (Fig. 2.4; Pearl and Reed, 1922). First, there is accelerating growth with a rapid increase in population size before the compensation of density-dependent regulation has much effect (when $(1-N/K) \sim 1.0$). The acceleration slows until a maximum rate of increase is reached at half the maximum population size. After that, there is a deceleration of growth rates in a symmetric way to the manner in which it accelerated [due to the symmetry of the production curve (Fig. 2.3)], and the asymptotic population size is eventually reached when $(1-N/K) \sim 0.0$.

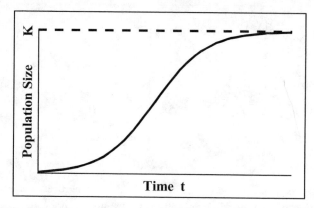

Figure 2.4 Population trajectory when growing according to the logistic curve (Eq. 2.11). The top dashed line represents the asymptotic carrying capacity K. Maximum growth rate is at $K/2$, where the inflection exists in the curve. Note the classic *sigmoid* shape described by Beverton and Holt (1957).

One aspect of the continuous logistic curve that should be noted is the smoothness of the population growth trajectory. There are no oscillations or population overshoots beyond the asymptotic value. This is due to the differential equations dealing with infinitesimals and the instantaneous response to any change in population size that this implies.

2.4.3 Discrete Logistic Model

Differential equations implicitly introduce the assumption of no significant time delays. The assumption that the population can respond immediately to changed population size appears unrealistic. It might be approximated in a large homogeneous population where births and deaths occurred continuously through time and any generations overlapped completely. It is also possible to add explicit time delays (Nicholson, 1958) to continuous differential equations. However, for populations in a seasonal environment, especially with non-overlapping generations or discrete cohorts, a discrete time model may be more appropriate. Such models are sometimes referred to as difference models or equations because they are literally formed to illustrate the difference between time intervals. A logistic model set up as a difference equation could be

$$N_{t+1} = N_t + rN_t\left(1 - \frac{N_t}{K}\right) \tag{2.12}$$

which is Eq. 2.10 converted to discrete time intervals instead of infinitesimals. If there were an extra source of mortality, such as fishing mortality, this could be included by adding an extra term

$$N_{t+1} = N_t + rN_t\left(1 - \frac{N_t}{K}\right) - C_t \tag{2.13}$$

where C_t is either a constant catch level or a time series of catches described by the subscript t.

2.4.4 Stability Properties

The exponential model of population growth was astable (any perturbation led either to extinction or runaway growth). The discrete logistic model, however, has more interesting properties. Density-dependent growth means it is more capable of compensating for increases in mortality brought on by such things as fishing. Again an equilibrium exists where $N_{t+1} = N_t$, which can be represented by a diagonal line on a phase diagram (Fig. 2.5) in which time is implicit. The logistic model has a non-linear relationship between successive generations (the curved line on the phase diagram, Fig. 2.5). The population must take a value somewhere on the curved trajectory, the actual shape of which will be determined by the particular parameter values adopted in the model.

Example Box 2.2 The discrete logistic population growth model (Eq. 2.13). Cell B1 is named "re", B2 is K, B3 is N0, and D1 as Ct. For naming cells see The Use of Excel Appendix. By plotting population size (column B) against time (column A), and varying the r, K and N_0 parameters one should be able to generate the equivalent to Fig. 2.4. Time is implicit in column B because each cell after B5 refers to the cell immediately above. That is, the cell for Nt+1 refers to the cell for Nt, immediately above and hence the time steps are implicit. Investigate the influence of a constant catch level by varying the value in D1, with D1 named Ct; be sure to increase the starting population size in B3 to be greater than the constant catch level or else the population would instantly go negative. Negative numbers are avoided by using =Max(Eq 2.13,0). For convenience, the final population size is copied to D3 to aid with the interpretation of the graph. Column C, which duplicates Nt+1, can only be copied to one row less than Nt.

	A	B	C	D
1	r_0 – growth	0.5	Ct	0
2	K – Max Pop.	1000		
3	N_0	50	N_{20}	=B25
4	Time	Nt	Nt+1	
5	0	=N0	=B6	
6	1	=Max(B5+re*B5*(1-B5/K)-Ct,0)	=B7	
7	2	=Max(B6+re*B6*(1-B6/K)-Ct,0)	=B8	
8	3	=Max(B7+re*B7*(1-B7/K)-Ct,0)	=B9	
9	4	Copy Down to at least Row 25	Copy down	
10	Continue down to row 25		to Row 24	

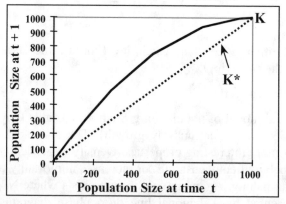

Figure 2.5 Phase diagram of the discrete logistic growth. The diagonal line of equilibrium is $N_{t+1} = N_t$. The curved line is the productivity beyond replacement (i.e., beyond the equilibrium line) at each possible population size. The equilibrium carrying capacity is at K. With an extra constant source of mortality (e.g. $C_t > 0$; see Fig. 2.2), a lower equilibrium carrying capacity K^* is obtained. Strictly, this applies only to non-oscillatory situations (see Example Boxes 2.2 and 2.3). A different productivity beyond replacement line would thus pass through K^*.

The productivity, above that required to replace the breeding population at any time, is that which, if the predictions of this simplistic model held in practice, could be cropped without damaging the population. That is, the difference between the curve of the production line and the dashed line of the equilibrium line is the hypothetical quantity that could be harvested from the population in a sustainable fashion. The maximum difference, which occurs at K/2, is known as the Maximum Sustainable Yield or MSY, and this is the origin of the idea (Schaefer, 1954, 1957). The stock biomass needed to generate this level of productivity is the B_{msy} (pronounced, B.M.S.Y). This simplistic/historical view underlies the intuitions that many people have about population productivity in fisheries. When we discuss stock-production models, we will see why this notion is too simplistic and too dangerous a view of what is possible.

Example Box 2.3 The phase diagram for the discrete logistic equation. From the spreadsheet in Example Box 2.2, plot N_{t+1} (column C) against N_t (column B) as a solid line with no dots and vary the parameters re, K and N0 until the graph approximates Fig 2.5. To include the equilibrium line, select and copy H23:I24, as below, then 'Paste Special' these data onto the graph ticking the 'New Series' and 'Categories (X values) in First Column' boxes, formatting the series to represent a solid line. At any population size, the distance between the curved line and the equilibrium line is an estimate of the surplus production. You should adjust the value of the constant catch, C_t, in D1, and observe how this affects the final equilibrium i.e., where the curved line crosses the straight line of equilibrium. To retain multiple curved lines one can copy columns B and C, as values, into columns further to the right and, as before, copy and paste special them into the graph. In this way, one can duplicate Fig 2.6.

	G	H	I	J
23		0	0	
24		1000	1000	

2.4.5 Dynamic Behaviour

Equation 2.12 (Eq. 2.13 with C_t set to zero) has a wide variety of different behaviours depending upon the value given to its *r* parameter. The fact that very complex dynamic behaviour can be obtained from a simple deterministic difference equation was highlighted by May (1973). This was surprising at the time because the model is a completely deterministic equation and the belief had been that the behaviour of deterministic equations should be capable of being understood in an analytical fashion. There are four characteristic forms of dynamic behaviour expressed by the model (Table 2.1; Fig. 2.6).

Clearly there are a number of complex behaviours that Eq 2.13 can exhibit and most of these relate to the non-linearity in density-dependent compensation for changes in population size. This is a remarkable field of research and the visual patterns that can be produced once one starts investigating chaos are undoubtedly fascinating. The detailed dynamics of a model in chaos are unpredictable. Given the state of the model at a given time, it is impossible to predict with certainty what will happen at a later point in time. However, the phase plot for the discrete logistic model illustrates the notion of a 'strange attractor' in action, and indicates that beneath the chaotic behaviour there are constraints operating on the behaviour of the model. In this case the chaotic behaviour of the various population sizes possible are constrained to lie on a parabola (Fig. 2.7; *cf.* Fig. 2.3).

Monotonically damped equilibria occur when changes in the population size are sufficiently slow that the density dependent mechanisms are able to compensate perfectly leading to smooth and orderly population growth. This can only occur if the growth rate does not rise through the limits indicated in Table 2.1. If the rate of population increase is too high then the linear density dependent compensation built into the model is inadequate to counteract for the rapid changes in population size that can occur. At this stage, we would say that non-linear density-dependent effects are being expressed.

With damped oscillatory equilibria the under and over-compensation are limited in their lack of balance so that a relatively stable equilibrium eventually arises. With stable limit cycles, the under and over-compensation for population size changes interact in such a way as to oscillate in a stable manner between multiple quasi-stable states (Fig. 2.6). Finally, when the growth rate passes the chaos threshold the degree of under and over-compensation is so great that unpredictable behaviour arises (though constrained within the bounds of the strange attractor).

Fascinating though a pursuit of chaos theory can be (Gleick, 1988; Hoffenflier, 1995), a question remains about exactly how useful knowing about such matters is going to be in the modelling of natural populations. There are few publications concerning the population dynamics of marine species that use chaos theory in their explanations. Some publications are even explicit in denying that observations of apparently random behaviour are brought about by the non-linearity introduced by over-compensation in a density-dependent model. Instead, it is claimed that the randomness in observations is brought about by population responses to stochastic environmental effects (Higgins *et al.* 1997).

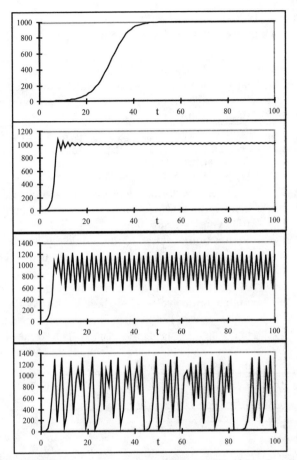

Figure 2.6 Examples of the dynamic behaviour exhibited by the discrete logistic when K was set at 1000, N_0 set to one, $C_t = 0$, and each panel is the product of a different r-value. Four different r-values are illustrated an r of 0.25 (top; monotonically damped equilibrium), r of 1.85 (second; damped oscillatory equilibrium), r of 2.5 (third; 4-way stable limit cycle), and an r of 3 (bottom; chaos). In chaos, if the starting population size N_0 is altered, even only slightly, a completely different and unpredictable outcome is produced (Example Box 2.4).

An inspection of the time-phase diagram (Fig. 2.7; in which the dynamic behaviour of a model system is illustrated) provides one with an indication of the bounds within which biological behaviour can occur as compared to mathematical behaviour. Such inspections permit one to determine which combinations of parameter values will lead to what would be pathological or impossible biological behaviour. In this way, such behaviours could be excluded by constraining the parameter values possible. It should be noted that complex dynamic behaviour in a model should therefore be considered with suspicion unless there are good biological reasons to think that such behaviour could be expected.

Table 2.1 Dynamic behaviour of Eq 2.12 given a K parameter set at 1000 and the following different r parameter values. See Fig. 2.6 for representations of the types of behaviour described. The values of r listed are only approximate but the behaviour itself may be investigated in Example Box 2.4.

r - values	Description of Behaviour
$0 < r <= 1$	**Monotonically damped equilibrium** - no oscillations, leading straight to a stable equilibrium.
$1 < r < 2.03$	**Damped oscillatory equilibrium** - oscillates but, given enough time, will return to a single equilibrium point following a perturbation.
$2.03 < r < 2.43$	**Stable limit cycles - 2** the model system oscillates in a cyclic fashion with two alternative population levels.
$2.43 < r < 2.54$	**Stable limit cycles - 4** the model system oscillates in a cyclic fashion with four alternative population levels.
$2.54 < r < 2.57$	**Stable limit cycles - >4** these cycles continue but reach higher orders, first 8, then 16, but then it becomes difficult to distinguish events from chaotic behaviour
$\sim 2.575 < r$	**Chaos** - Unpredictable behaviour that changes depending on starting conditions.

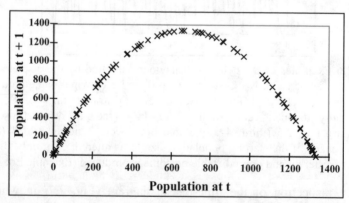

Figure 2.7 The time-phase diagram for the discrete logistic equation when the population is in a chaotic state (see Fig 2.6). The points plotted are the first 100 population sizes when the population started at a size of one, has a K of 1000, and an r of three. The order of appearance on the parabola appears to be random but the population sizes are obviously constrained to lie on a definite line. The pattern arising from this constraint is known as a strange attractor. Note the upper limit to stock size is no longer K.

Example Box 2.4 Investigation of a model's dynamic behaviour. Alter the model described in Example Boxes 2.2 and 2.3 by extending time down to 100, N_t (column B) by copying down to Row 105 and N_{t+1} (column C) down to Row 104. Plot, as points, columns B and C in a new graph in two series (B5:C55 and B56:C104). Colour the first series blue and the second series red. In this way, if the population eventually reaches equilibrium its form should be more discernible in the phase diagram as a red point. By comparing the phase diagram with a plot of N_t against time (column B against A; *cf.* Fig. 2.6), the manner in which the dynamics alters with changes to r, K, and N_0 can be followed. Using Table 2.1 as a guide, investigate the boundaries of the different dynamic behaviours.

Investigate the impact of a constant fishing mortality rate by setting C_t to >0. Is the impact the same on all the dynamical behaviours? See Section 2.5 on the Responses to Fishing Pressure for a discussion.

For an alternative model equation in column B, replace B5 + re*B5*(1-B5/K) with re*B5*exp(-K*B5) and copy down. With the alternative equation the meaning of the parameters has changed so start with *re* set to 2.0, K set to 0.0005, and N_0 set to 50. Find the boundaries between the different dynamical behaviours in this model. Is it more or less robust than the logistic?

2.5 RESPONSES TO FISHING PRESSURE

Density-dependent effects are said to compensate for population changes by altering the population growth rate. However, they may under or over compensate and so lead to oscillations (see the logistic model in Example Box 2.4). Over-compensation would be where a population's growth rate was reduced too much at high population levels and under-compensation would be where the growth rate was not reduced enough at lower population sizes. Under-compensation can lead to a population rising to levels beyond its theoretical asymptotic equilibrium. The imposition of extra density-independent mortality onto a model population can often stabilize a relatively unstable situation.

If a constant catch mortality is imposed, this can reduce the variation in a model population's behaviour through effectively offsetting any under-compensation. Of course, if there is no under-compensation then additional mortality through fishing will simply increase the mortality rate at a given population size (over-compensation). In addition, if the increased mortality is too great it will lead to a population collapse (see Example Box 2.4).

From Fig. 2.4 we can see that the maximum productivity occurs at half the maximum population size with lower productivity at either side. Thus, we would expect to be able to crop or harvest a population safely at greater levels when the biomass was only at half its equilibrium level than we could at other levels (Fig. 2.8). This is rather simplistic theory but will be pursued until the surplus-production model section where the limitations are made clearer.

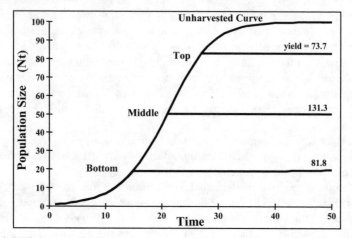

Figure 2.8 Three populations growing according to a discrete logistic population model with a regular harvest being taken from each population at different levels of standing crop, and $K = 100$, $r = 1$, and $N_0 = 1$. On a sustainable basis as in the diagram, the upper and lower populations' total yield is slightly less than half that which is possible from the middle population sizes (Pitcher and Hart, 1982). In this model, 3.51 units are harvested each time period from the top population while 6.25 and 3.8955 units can be taken from the middle and bottom populations respectively. The fourth decimal place is necessary with the bottom curve because it is very sensitive to changes in harvest rate (Fig. 2.9). The dynamics of these relationships may be examined in Example Box 2.4.

One can determine theoretical sustainable harvest rates for each of the three population levels indicated (Fig. 2.8) which allow the same amount to be harvested each time period without altering the population size. This harvest rate would equal the surplus production at the given population size. Populations that are harvested when they are near their maximum population size have more resilience to perturbations (such as fishing pressure) than populations harvested when at lower population sizes. The sensitivity to perturbations becomes extreme towards the lower population sizes (Fig. 2.9).

If harvesting occurs when the population is only about 20% of its maximum, then when altering the harvest rate within the model it is necessary to use three or four significant digits else the population starts to grow rapidly or shrinks to extinction (Fig. 2.9).

The rapid decline occurs when the harvest rate is greater than the sustainable yield at that population size. If that is the case then the situation can only become worse as each harvest event lowers the standing crop so that the productivity becomes even less and the decline accelerates. The opposite occurs if the harvest rate is less than the surplus production. In this case, each harvest event leaves the standing crop slightly larger than the previous period and hence more productive and so likely to increase even more the next time through a harvest (Fig. 2.9).

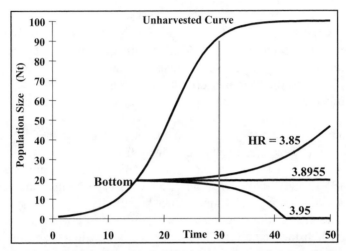

Figure 2.9 The sensitivity to changes in harvest rate of a population growing according to a discrete logistic population model with a regular and constant harvest being taken starting at a population size approximately 20% of the maximum population size. The harvest rate (HR) is indicated showing that taking 3.8955 population units per time period leads to a constant population size. A slight deviation in harvest rate to become smaller or larger leads to either rapid population growth or a rapid decline to extinction. See Example Box 2.4.

2.6 THE LOGISTIC MODEL IN FISHERIES

The models we have dealt with so far may be thought not to have much relevance to fisheries. However, if we extend the logistic model to include catch we obtain

$$B_{t+1} = B_t + rB_t \left(1 - \frac{B_t}{K}\right) - C_t \qquad (2.14)$$

where B_t is stock biomass at time t, r is the intrinsic rate of growth, K is the unfished or virgin biomass equivalent to the carrying capacity, and C_t is the catch level over time t. It is common practice to assume that catch is proportional to fishing effort and stock size (though this is only the case if the catchability coefficient, q, does not vary through time or with stock size). If we implement these further changes then Eq. 2.14 becomes the classic dynamic biomass model proposed by Schaefer (1954, 1957)

$$B_{t+1} = B_t + rB_t\left(1 - \frac{B_t}{K}\right) - qE_tB_t \qquad (2.15)$$

where E is the fishing effort and q is a parameter describing fishing gear efficiency (the catchability coefficient, in quantitative terms, is the proportion of the stock biomass B taken by one unit of effort, i.e., $C_t = qE_tB_t$). This was the general form of model used in our simple fishery simulation (Fig. 2.9) and we will see more of this and other slightly more complex models when we discuss surplus-production models.

Some key changes which have occurred recently in fisheries stock assessment methods have been a shift away from equilibrium analyses, and a move away from the traditional interpretation of maximum sustainable yield (MSY) towards investigating alternative harvesting strategies. Also, models are no longer constrained to express only linear or symmetrical responses to changes in density. There is no point in criticizing the logistic equation when the actual problem was that it was inappropriate to use a linear relation between density-dependent effects and population size. The changes in methods mentioned above should lead to the production of much better fisheries management advice. Unfortunately, it is also more likely to lead one to conclude that one's information is inadequate for the production of useful advice. Using the old equilibrium methods and inappropriate models, definite management advice could usually be produced. However, the equilibrium methods often produced completely inappropriate conclusions. It is far better to know that one's data is uninformative than to provide bad advice. At least then one can be honest about the foundations of management decisions.

2.7 AGE-STRUCTURED MODELS

2.7.1 Age-Structured and Exponential Growth Models

An age-structured model attempts to capture the composite behaviour of the cohorts making up a population, which entails following the development and changes in each cohort separately. This is an improvement over a simple whole-population model for the obvious reason that different aged animals tend to have different growth rates and be different sizes (weights). A larger individual will clearly contribute more biomass to a catch than a smaller one and generally will contribute a higher egg-production. If the necessary information is available, then an age-structured model has the potential to reflect natural population processes and the impacts of harvesting rather better than simpler models.

If the model concerns a 'good' biological population or stock then immigration and emigration will be, or is assumed to be, minimal. If this is the case then once a cohort has recruited, the numbers in that cohort or age-class can only decline. How this decline is modelled determines the design of the model. In the introduction to this chapter, it was pointed out that at least some age-structured models have a relationship with the exponential model of population growth. The explanation is quite simple when we remember that the exponential model of population growth can be used to model a declining population. An equation denoting the changing numbers within a cohort could be represented as

$$N_{t+1} = N_t e^{-Z} \qquad (2.16)$$

Where N_t is the number in the cohort at time t, Z is the total rate of instantaneous mortality (fishing + natural mortality), and e^{-z} is the proportional survivorship. The relationship with the exponential growth model (Eq. 2.2) is clear but note that the Z value is given as negative indicating that there can only be an exponential decline in cohort size through time (Fig. 2.10).

2.7.2 Annual vs. Instantaneous Mortality Rates

The use of the exponential term in Eq. 2.16 is fundamental to population modelling. At the start of this chapter (after Eq. 2.1) it was pointed out there was a possibility of confusion over rates of change referring to either proportional changes or absolute changes in population size. One place where this is often a problem and which leads us into an important foundation of fisheries science is the confusion over annual and instantaneous rates of change.

An important aspect of the relationship between fish stocks and fishing is captured by the notion of the mortality imposed by fishing. As in Section 2.5, in a model, fishing constitutes an extra source of mortality that can greatly influence the dynamics of both the natural population and the model. Before we can determine if a level of fishing is sustainable, we would, ideally, obtain an estimate of the level of mortality introduced by the fishing activities. An important distinction is the one between Annual Mortality (easily understood) and Instantaneous Mortality rate (less easy to grasp intuitively).

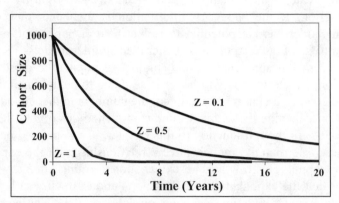

Figure 2.10 Cohort declines with different levels of total mortality (see Eq. 2.16). Because e^{-Z} has a negative exponent, it can never be greater than one and equates to the proportion of each cohort that survives over each time interval.

Fishing mortality is invariably referred to by the capital letter F, which refers to the 'instantaneous rate of fishing mortality'. If this is mistaken for the proportion of the fish stock caught annually (the annual mortality, often H the harvest rate) then confusion can arise, especially if F is greater than or equal to one (Fig. 2.10).

Most people are aware of the exponential growth properties of compound interest, which arise through increasing a starting amount by a constant proportion each time interval. Population or cohort mortality is similar to compound interest only the starting amount (the cohort size) is decreased by a constant proportion each period. In a biological population, this leads to an exponential decline in numbers.

It is also common knowledge that if interest is compounded at intervals shorter than a year then the overall effect is greater over a year than if it were merely compounded once. Hence we would expect that, given a particular proportional decrease, the effect if compounded daily, would be greater than if compounded weekly which would be greater than if compounded monthly (Table 2.2). Equally obviously, it is also a more gradual process. If this procedure of shortening the time over which the interest is compounded is taken to extremes and very tiny periods of time are used this would approximate the infinitesimals for which the exponential format of the equations operate (Table 2.2).

The exponential function acts to produce proportional changes out of events (such as a mortality) acting over very small time intervals on the same population. This is the same as compounding a negative interest rate very many times. The repeated application of a constant proportional decline (divided into however many time intervals are being considered) is what leads to the changes being able to be represented by the exponential equation. The instantaneous rate of fishing mortality is denoted F, and this is related or translated to the annual mortality due to fishing (the harvest rate, H) using

$$\text{Proportional Annual Mortality} = H = 1 - e^{-F}$$

$$F = -Ln(1-H)$$

(2.17)

Table 2.2 Outcome of applying a constant mortality rate apportioned among shorter and shorter time periods that add to one year. The Instantaneous mortality rate was 0.693147. Which when translated using Eq. 2.17 produces an annual mortality rate of 0.5 (i.e., 50% of the remaining cohort dies each year). $(1 - e^{-0.693147}) = (1 - 0.5) = 0.5$ the annual mortality rate derived from the instantaneous rate. To apply this rate twice in a year we divide 0.693147 by two, to apply it each week we divide by 52. Note how as the time interval becomes shorter the number remaining approaches the expected 500.

Time Period	Fraction of a Year (Number of Times applied)		Mortality rate applied each time period	Number Remaining after One Year
6 Months	0.5	(2)	0.34657	426.97
3 Months	0.25	(4)	0.17329	467.11
1 month	0.08333	(12)	0.05776	489.70
1 week	0.019231	(52)	0.01333	497.68
1 day	0.00274	(365)	0.00189	499.67
½ day	0.00137	(730)	0.00095	499.84
3 hours	0.000342	(2920)	0.00024	499.96
Infinitesimal	~0.0	(∞)	~ 0.0	500.00

A general symbol for the annual mortality would be useful but we cannot use lowercase f as this is one of the generally accepted symbols for fishing effort (and sometimes fecundity) and we cannot use M as this is generally used to designate instantaneous natural mortality. Some people use E to represent the Exploitation rate but that can also be confused with effort and also commonly refers only to recruited or legal sized fish. In this book, we will use the letter H to indicate annual fishing mortality, the harvest rate.

The important distinction to understand is the one between the F that everyone tends to talk about, the instantaneous rate, and the proportion of the stock being taken each year (H) as catch (Fig. 2.11).

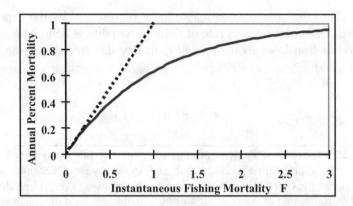

Figure 2.11 The relationship between instantaneous and annual fishing mortality. Note that at low levels of F the two are approximately equal (the curve approximates the straight dotted line which is a 1:1 reference line) but with F values greater than about 0.2 the equivalence disappears and much larger values of F are required to increase the annual proportion killed by fishing. The relationship is asymptotic with extremely high levels of instantaneous fishing mortality required to catch 100% of the population in a single year.

2.7.3 Selection of a Target Fishing Mortality

An important question to answer is what level of fishing mortality to impose on a stock and the manner in which the catch should be harvested. In fisheries management several standard reference or target levels of fishing mortality exist and a number of these derive from the analysis of yield-per-recruit. A yield-per-recruit analysis is the simplest form of age-structured population model. A discussion of these methods will thus provide a gentle introduction to age-structured modelling and suggest some fishing targets.

2.8 SIMPLE YIELD-PER-RECRUIT

2.8.1 Is there an Optimum Fishing Mortality Rate?

Most exploited aquatic populations are made up of a number of discrete cohorts of different ages. In polar, temperate, and sub-tropical regions, most commercial species produce one cohort each year. Even in the tropics, despite breeding seasons sometimes being extended throughout much of the year, similar cohort-based population structure can be discerned (even if it is just size-based, a form of ecological-cohort). Russell's (1931) equation of stock dynamics can be applied to each cohort separately in an age-structured population. The effects of age-structure on stock dynamics become especially important when individual growth and the time delays of ageing are taken into account.

The historical intuition, which some still hold today, that total yield from a fishery will always increase with increases in total effort, was shown to be incorrect before the 1930s. When trying to formalize this idea the first problem was to find a clear demonstration of this, at the time, counter-intuitive notion. Also, there is the implication that if maximizing effort does not necessarily maximize catch is there such a thing as an optimal fishing rate that would lead to a maximum yield? Russell (1942) provided a nice empirical demonstration that the optimal harvest was not necessarily taken by fishing as hard as possible (i.e., that an intermediate fishing mortality could give a bigger yield in an age-structured population).

Russell's (1942) example was of an hypothetical fish species for which the implications of fishing at two different levels were illustrated (Table 2.3). The major implication of having the high harvest rate of 80% mortality per annum (an $F = 1.609$) relative to a lower rate of 50% per annum (an $F = 0.693$) is that in this particular example, despite basically the same number of fish being taken, the total catch weight was over 60% greater with the lower F. This result stems simply from the lower catch rate leaving more fish in the sea for longer so that they grow to a heavier weight before being caught. At the lower harvest rate there is a greater biomass caught and the number of fish remaining in the sea after fishing is also greater. Hence, the lower fishing mortality yields a larger catch and is more risk averse. Even if the population were not in equilibrium the same principle holds. The benefit in yield remains obvious but obviously the potential stable population size cannot be determined or compared.

By carrying out these analyses for a wider range of different fishing mortality rates one can search for the fishing mortality rate which would be expected to produce the maximum yield from the fishery. This is commonly termed F_{max} (pronounced "F max") and was a common fisheries target mortality rate in the past (Fig. 2.12).

The yield drops off at very low levels of F (Fig. 2.12) because very few fish are caught. This type of analysis suggests that an obvious refinement to fishing would be to use selective fishing gear that only catches the larger individuals in which the gains through individual growth would be maximal. The tabular approach is too simple, however, because it ignores natural mortality and differing vulnerability to fishing gear of different sized/aged animals. If these are included, the problem would then be one of estimating the optimal size (or equivalent age class) at first capture for a given F.

One aspect of standard yield-per-recruit analysis that must be emphasized is that this analysis only takes into account growth and mortality; generally recruitment variation is not included. Because recruitment is not modelled explicitly but is assumed to be constant (hence yield-per-recruit) the standard yield-per-recruit analysis does not attend to the issue of whether the fishing rate predicted to produce the maximum yield is likely to be sustainable.

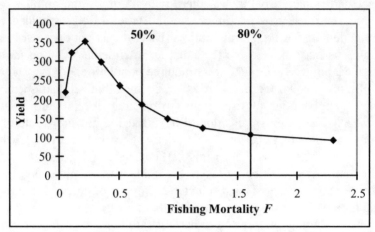

Figure 2.12 The equilibrium yield-per-recruit obtained from Russell's (1942) example with different annual harvest rates (5%, 10%, 20%, 30% up to 90%). Given zero natural mortality and assuming fish are all equally vulnerable to fishing from age class 1, then the optimal fishing mortality, the F_{max}, in terms of maximum yield, is $F = 0.223$ or 20% annual harvest rate (Table 2.3 and Example Box 2.5).

2.8.2 What is the Optimum Age or Size at First Capture?

If fishing gear can be made to be age (meaning average size) selective this opens the possibility of attempting to catch some cohorts while avoiding others (i.e., treating the cohorts differently). If individual growth is an important component of productivity, then we would wish to catch cohorts selectively and would also want to include the effects of natural mortality in the analysis. An extreme example will illustrate the significance of selecting the right age at first capture.

Certain arrow squid species (such as the *Notodarus* species found around New Zealand and South America) have a short lifecycle that lasts a year at most. In that year they hatch, grow, breed, and then die. Around New Zealand, arrow squid is often a 75,000 t fishery but just after the squid have spawned and the adults have died the total biomass of juvenile squid is likely to be in the order of only 5 to 15 t. They grow at a tremendous rate and losses to total biomass due to natural mortality tend to be trivial in comparison (though the numbers dying are not insignificant). This means that if the squid were to be caught early in their lifecycles they would not have grown as much as they could and the total catch would be reduced. The skill in managing such species is to wait long enough for them to grow to a size at which the total stock biomass will give a large, profitable catch but not to wait so long that they pass maturity, breed, and die (or even that there would be insufficient time to take the possible catch).

Table 2.3 Comparison of the effects of two fishing mortality rates on an hypothetical fishery in which natural mortality is ignored. Note the total number of fish caught in each case is effectively the same but the total yield is much greater with the lower fishing mortality, and more stock numbers remaining and presumably greater resilience. The # refers to numbers of fish (data from Russell 1942). Note that each cohort is exposed to the same level of fishing mortality and that this analysis assumes 1000 recruits each year so the system is in equilibrium. The population with the lower fishing mortality ends with a higher standing stock and the greatest yield. As an exercise, construct an Excel worksheet to duplicate this table (see Example Box 2.5).

Age	Mean Wt kg	Annual Mortality of 80% $F = 1.61$ $H = 0.8$			Annual Mortality of 50% $F = 0.69$ $H = 0.5$		
		Stock Size #	Catch #	Catch Wt kg	Stock Size #	Catch #	Catch Wt kg
1	-	1000.0	-	-	1000.0	-	-
2	0.082	200.0	800.0	65.6	500.0	500.0	41.0
3	0.175	40.0	160.0	28.0	250.0	250.0	43.8
4	0.283	8.0	32.0	9.1	125.0	125.0	35.4
5	0.400	1.6	6.4	2.6	62.5	62.5	25.0
6	0.523	0.3	1.3	0.7	31.3	31.3	16.3
7	-	0.1	0.3	0	15.6	15.6	-
8	-	0	0.1	0	7.8	7.8	-
9	-	0	0	0	3.9	3.9	-
10	-	0	0	0	2.0	2.0	-
11	-	0	0	0	1.0	1.0	-
Total Yield	-	**1250.0**	**1000.0**	**105.9**	**1999.0**	**999.0**	**161.5**

In Australia, a fine example of the value of fishing for the right-sized animals is seen in the banana prawns (*Penaeus merguiensis*) of the northern prawn fishery, where fishing is based principally in the Gulf of Carpentaria (Pownall, 1994). This species is found in large breeding aggregations (fishers refer to these as "boils") but like the squid they are short-lived and start life as very small animals. The best prices for prawns are obtained for the larger individuals so it would be bad economics to begin the harvest of the banana prawns too early.

Example Box 2.5 Simplified Yield-per-Recruit. This Excel sheet calculates the yield from two different schedules of natural and fishing mortality (C1C2) and (F1F2). The total annual mortality (in C3 and F3 for the two populations) is used to calculate the animals captured (in columns D and G), which value is subtracted from the population size to determine the remainder (cols C and F). By summing the respective columns (using sum(C6C16), etc) the standing stock and yield from each fishery can be determined. By saving as values (Edit/Paste Special/Values) in columns to the right, along with their respective mortality rates, it should be possible to build up the information required to duplicate Fig. 2.12. By setting the natural mortality to zero the contents of Table 2.3 can be duplicated. The age at first capture (tc) is used later.

	A	B	C	D	E	F	G	H
1	Annual F		0.8			0.5		
2	Annual M		0.1			0.1		
3	Total A		=C1+C2			=F1+F2		
4		tc	1				1	
5	Weight Kg	Age	N	Catch N	Catch Kg	N	Catch N	Catch Kg
6	0.042	0	1000.00			1000		
7	0.082	1	=C6-D7	=C6*C$3	=A7*D7	=F6-G7	=F6*F$3	=A7*G7
8	0.175	2	=C7-D8	=C7*C$3	=A8*D8	160.0	240.0	42.0
9	0.283	3	1.00	9.00	2.5	64.0	96.0	27.2
10	0.400	4	0.10	0.90	0.4	25.6	38.4	15.4
11	0.523	5	0.01	0.09	0	10.2	15.4	8.0
12	0.700	6	0	0.01	0	4.1	6.1	4.3
13	0.850	7	0	0	0	1.6	2.5	2.1
14	0.925	8	0	0	0	0.7	1	0.9
15	0.990	9	0	0	0	0.3	0.4	0.4
16	1.000	10	0	0	0	0.1	0.2	0.2
17		Totals	1111.10	1000.00	92.5	1666.6	999.9	149.6

The fishing season is very short with catch rates becoming unprofitable after only a few weeks (nowadays just three to four). Year to year variations in catch are considered to be the result of variation in recruitment and not in fishing effort so the assumption has been made that despite the high fishing mortality there will always be sufficient prawns to provide adequate recruitment in subsequent years. Unlike with New Zealand squid, where the fishing industry determine when they want to fish their quota, the opening date for the banana prawn season was set by an economic analysis and model of the fishery which was developed early on and is continually improved. An explanation of the current management follows.

> The northern prawn fishery is open from April until the end of November with a midseason closure in June and July. By opening the fishery in April new season banana prawns have had a chance to grow to a larger, more valuable size. Catchability of banana prawns is also higher in April as they congregate for spawning. Owing to these factors, effort is focused mainly on banana prawns following the opening of the fishery (Dann and Pascoe, 1994, p.11).

The Australian management of banana prawns is an improvement over the New Zealand squid fishery in that the optimum yield-per-recruit (optimum size of individual animal) has been determined and an effort made to ensure that the animals at least approach this size before they are fished. Where both are risky is in the assumption that all out fishing effort will not affect subsequent recruitment. Squid can certainly be recruitment overfished, as can tropical prawns (Garcia et al., 1989). However, in species with highly variable recruitment it is hard to detect recruitment failure until after it has happened.

2.8.3 From Empirical Table to Mathematical Model

Given the objective of maximizing yield (total catch), analytical methods were developed in the 1950s for calculating the optimum age or size at which to first begin capture for a particular fishing mortality. However, standard yield-per-recruit analyses do not attend to whether such catches would be sustainable (i.e., they ignore absolute recruitment). If a species grew to a large size before it was mature and suffered a significant natural mortality, the equilibrium yield-per-recruit analysis could lead to recommending that heavy fishing be imposed upon immature individuals. The development of the alternative technique of egg-per-recruit analyses, which take yield-per-recruit one step further, is an attempt to avoid the risk of recruitment overfishing.

The examples of squid and prawns above are exceptional because the gains through individual growth so outweigh the possible losses through natural mortality (e.g., predation, disease) that the benefits of waiting for the animals to grow are obvious (which is why they are used as examples). Russell (1942) demonstrated the value of searching for the optimum fishing mortality and his example was also clear. The benefits of avoiding the capture of fish in younger age classes by using selective gear can be demonstrated in a similar manner. Using the same hypothetical model presented by Russell (1942), we can arrange that no fishing mortality occurs until either the third or fifth year. If we also add a level of natural mortality to all years, so as to increase reality, we can derive a range of optimum yield curves (*cf.* Fig. 2.12) each for a different age at first vulnerability to fishing (Fig. 2.13).

2.8.4 The Model Structure and Assumptions

A yield-per-recruit analysis is carried out by constructing a model of the development of a cohort through time, which takes into account the growth, and mortality of individuals. For these simple calculations to be valid, a major assumption is made that the age-structure of a fish population has attained equilibrium with respect to the mortality imposed upon it. This would imply that recruitment is constant and what happens to one cohort as it ages is representative of what happens to all cohorts and so represents a cross-section of the entire population at any one time (Pitcher and Hart, 1982).

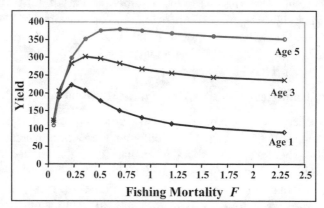

Figure 2.13 Yield-per-recruit from Russell's (1942) example with $M = 0.1054$ (annual = 10%) included. Calculations are repeated assuming fish are vulnerable to fishing from three different age classes onwards. The yield curves suggest that the optimal yield would be obtained with a fishing mortality of $F = 0.693$ or $H = 50\%$ with fish capture starting at age five (see Example Box 2.6).

The assumption of equilibrium is quite unrealistic because in many fisheries the age structure of the population changes markedly with time. However, this assumption simplifies the model considerably and is useful for introducing the ideas involved. It is a major weakness of the method, however, and implies that any results should be assumed to be highly uncertain and potentially risky. In our cohort model we assume that R individuals of age t_r are recruited (become vulnerable to possible capture) to the fishery each year. Hence, by definition, t_r is the minimum age at which fish could be targeted by the fishery. Note that there is a distinction between being vulnerable and being targeted. To add a little more realism we will also impose natural mortality to underlay the fishing mortality. To keep things simple to start with we will assume that once fish are recruited (at age t_r) they are subject to a constant rate of natural mortality, M (measured as an instantaneous rate). The mortality effects of fishing are then added to the model and this can be done in a number of ways. In our simplest of models we will arrange things so that we can set the age at first capture (t_c) to be greater than the age at recruitment. With this arrangement, because of targeting, fish are not vulnerable to harvesting before the age t_c but thereafter they are subject to a constant rate of fishing mortality, F (again, measured as an instantaneous rate). The model structure is completed by assuming that fish older than t_{max} years of age are no longer vulnerable to fishing so that only natural mortality remains. Note that the model assumes both knife-edge recruitment at age t_r and knife-edge selection at t_c, i.e., either none or all fish in an age-class are either recruited or not or are vulnerable to harvesting or not, and once vulnerable all age-classes are equally vulnerable (Fig. 2.14).

Example Box 2.6 In order to take account of the age of first capture, replace the contents of C7 in Example Box 2.5 with =if(B7>=C$4,C6-(C6*C$3),C6-(C6*C$2)). Copy this down to C16. This equation is now equivalent to Eq. 2.19. In D7 put =if(B7>=C$4,C6*C$3,0) and copy down. Do the equivalent changes in columns F and G. Now, by changing the ages of first capture in C4 and F4 it should be possible to develop the information needed to duplicate Fig. 2.13.

    There are many ways to implement the same model. The mortality rates could be put in columns as IF statements, turning on if the age were greater or equal to the age at first capture. One could use the formal equations as listed below. As an exercise, put together an alternative worksheet that implements the formal equations below.

2.8.5 The Model Equations

To generate yield-per-recruit information in a comparative manner for various ages at first capture (Fig. 2.13) we need to translate Russell's (1942) empirical example into our first version of an age-structured model. In the empirical depiction of the imposition of fishing mortality and growth (Table 2.3), all ages or cohorts experienced the same level of fishing mortality.

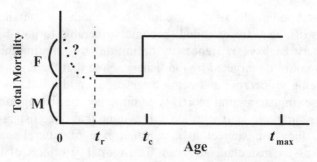

Figure 2.14 A diagrammatic representation of the assumptions made in simple yield-per-recruit analysis. Here t_r is the age at recruitment, t_c is the age at first possible capture, and t_{max} is the age at which fish cease to be vulnerable to fishing. The model does not consider the dynamics of individuals younger than t_r years of age but simply assumes that there are R recruits of this age entering the stock each breeding period. M is the constant rate of natural mortality, and F is the constant rate of fishing mortality (both F and M are instantaneous rates). Knife-edge selection is shown by the vertical rise in mortality at t_c.

Obviously as a year concludes, the members of one age-class progress to become the members of the next age-class. We require an equation with which we can follow the progression of a particular cohort of individuals as they proceed through the possible age-classes. The cohort begins at age t_r with a particular level of recruitment R and then, through time, the numbers will decline as natural and fishing mortality have their respective impacts. A common representation of the numbers N at t_r, N_0 is

$$N_{t_r} = N_0 = R \qquad (2.18)$$

We need to find a way to model natural mortality acting throughout the life of the animals in a cohort while fishing mortality only acts between ages t_c and t_{max}. As per the equation for exponential decline (Eq. 2.14) the numbers at any age $t + 1$ starting from a given time zero will be

$$N_{t+1} = N_t\, e^{-(M+F_i)} \qquad (2.19)$$

This is simply an exponential decline where the growth rate is negative and equal to M plus F (the exponential term is equivalent to 1-H). Note there are two sets of subscripts, those for time or age and those for whether fishing mortality is present. The value of F is only greater than zero for ages where i is greater than or equal to the age of first capture t_c, i.e., there may be some ages which are unaffected by fishing mortality. The normal steady decline brought about by natural mortality is interrupted and made more severe by the onset of fishing mortality (Fig. 2.15).

If this model is run, and the population is assumed to be in age-structured equilibrium, the curves in Fig. 2.15 also represent the age-structure of the population. Clearly, fishing mortality can have a major impact on the age-structure, effectively removing all older animals. This occurs even without size selectivity because it is an expression of the repeated application of a constant mortality rate. The older a fish is the greater the number of years the cohort will have been exposed to mortality.

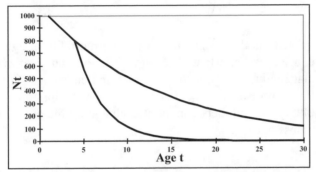

Figure 2.15 A comparison of a cohort starting with an R of 1000 at age 1 ($t_r = 1$), an age of first capture ($t_c = 5$), no t_{max}, an M of 0.075, and an F of 0.25. The upper line is what would happen if fishing mortality was zero for all ages. This exhibits a standard exponential decline of numbers through time. The lower curve is identical to the upper until age five when fishing mortality is added, whereupon the numbers drop far more precipitously. Without F there are still 0.1% of the initial cohort left by the age of 60 while with an F of 0.25 the same population level is reached by the age of 17. Fishing can obviously affect the age-structure of a stock.

But this only describes how the population changes in terms of numbers, whereas we need to know how much yield would be obtained under different regimes of age at first capture and different levels of F. The number dying at any time N_Z is simply the difference between the numbers at time $t+1$ and time t

$$N_Z = N_t - N_{t+1} \tag{2.20}$$

Remember that N_{t+1} will always be less than N_t. By substituting Eq. 2.19 into this, replacing the N_{t+1}, we obtain

$$N_Z = N_t - N_t e^{-(M+F_i)} = N_t \left(1 - e^{-(M+F_i)}\right) \tag{2.21}$$

where the term in large brackets is the proportion that dies from all causes. This is simply the complement of the survivorship $[e^{-(M+F)}]$. Equation 2.21 determines the number of individuals of age t that die as the cohort proceeds to time $t+1$.

The numbers that die due to fishing mortality (i.e., the number caught N_{ct}) is simply the fraction (F/Z) of the numbers that die, where Z is the total instantaneous mortality rate

$$N_{ct} = \left(\frac{F_i}{F_i + M} \right) N_t \left(1 - e^{-(M+F_i)} \right) \qquad (2.22)$$

Remember that F_i is the instantaneous fishing mortality at age i, which is zero before, age t_c and constant afterwards. The total numbers caught under any particular regime of age at first capture and constant fishing mortality is simply the sum of all the N_{ct}. In order to convert this to a yield, Y, we need to multiply the numbers caught at each age by the average weight w_t for each of those ages

$$Y = \sum_{t=t_c}^{t_{max}} w_t N_{ct} \qquad (2.23)$$

Finally, the yield in Eq. 2.23 is that expected for a recruitment of R or N_0. To generate the strict yield-per-recruit calculations, we need to divide the estimated yield, Y, by the initial recruitment to scale the calculations to units of yield-per-recruit (Eq. 2.24). A common alternative would be to carry out the calculations with an initial recruitment of one

$$\frac{Y}{R} = \sum_{t=t_c}^{t_{max}} w_t N_{ct} \qquad (2.24)$$

Using Eq. 2.24 the data required to generate Fig. 2.13 can be produced. This can then be used to compare different harvesting strategies for different species, which perhaps have different levels of natural mortality operating (Fig. 2.16).

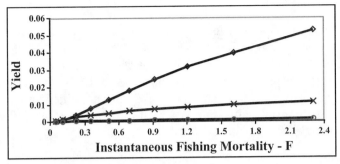

Figure 2.16 Annual natural mortality of 0.75 ($M = 1.386$) with a range of F fishing mortality rates measured over a year and three different ages at first capture. The order of the three age-at-first-capture lines is reversed relative to Fig. 2.13. Here the upper curve is the yield for a harvesting from year one, the middle line is for year three, and the lower line (almost coincident with the x-axis) is for harvesting at year five. The reversal of the lines arises because with high levels of M, most fish die before growing to a large size, so it is optimal for yield to catch them earlier. It may not be optimal in terms of allowing animals to reach maturity and breed, but that is a different objective for the fishery. Note the greatly reduced yield that has occurred because the weight at age data was not changed.

As natural mortality becomes very large, an analysis of yield-per-recruit on a time scale of years becomes less useful because it simply predicts that fishing mortality should become extremely large for the youngest age-class in the fishery (Fig. 2.16). In such circumstances (e.g., with prawns), data would be required from shorter time intervals such as months whereupon useful answers could again be determined.

2.8.6 Yield-Per-Recruit Management Targets

The expected outcomes from a yield-per-recruit analysis are a target fishing mortality (the mortality rate to aim for) and a target age at first capture. The age at first capture would be used to set regulations regarding gear type (e.g., mesh sizes, hook sizes, escapement holes, minimum sizes) while the target fishing mortality could be used to set a constant fishing rate harvesting strategy (one of the options possible when managing a fish stock). This all assumes the overall objective is to maximize yield from the fishery.

Remember that yield-per-recruit (YPR) analyses by themselves do not attend to the sustainability of the predicted optimal F values. The actual target mortality chosen should reflect this fact. An obvious target fishing mortality to choose might be taken to be the fishing mortality, F_{max}, which gives rise to the maximum yield (in Fig 2.13 this would be approximately $F_{max} = 0.69$, starting on age-class five, in Fig. 2.16, $F_{max} = 2.4$ or greater, starting on age class one). Empirical evidence, which has accrued since the inception of these methods, indicates that, in part because of uncertainties inherent in equilibrium YPR analyses, F_{max} tends to be too high and leads to stock declines. Instead of F_{max}, many fisheries around the world are now being managed using an $F_{0.1}$ (pronounced F zero point one) strategy (Hilborn and Walters, 1992). The value of $F_{0.1}$ is determined numerically by finding the fishing mortality rate at which the slope of the YPR curve is 10% of the slope at the origin (Fig. 2.17).

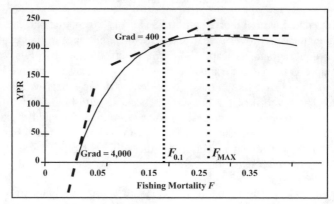

Figure 2.17 Illustration of the concepts of the reference fishing mortalities $F_{0.1}$, which is the mortality where the gradient of the yield-per-recruit (YPR) curve is 10% that at the origin, and F_{max}, which is the mortality that predicts the highest YPR (gradient = zero). In this case $F_{0.1} = 0.139$ (14.2% per annum) with a yield of approximately 213 units, while $F_{max} = 0.196$ (21.7% per annum) with a yield of 222 units. Therefore, for a 33% decrease in annual mortality there is a loss of only 4% from the yield.

It should be noted that the $F_{0.1}$ strategy is *ad hoc* and has no theoretical justification except that empirically it appears to be more conservative or risk averse to departures from the assumptions of the yield-per-recruit analyses. For a small loss in yield, one often gains a great deal in stock resilience to poor recruitment years and other sources of uncertainty. It may be the case that even the $F_{0.1}$ strategy is insufficiently conservative and this can only be determined by further experience with more fisheries. By being less than the F_{max} strategy, however, it is possible that the resulting strategy would be somewhat less economically optimum although this would depend upon the relationship between effort and fishing mortality. It should also be noted that F_{max} is not necessarily the same fishing mortality as would give rise to the MSY (F_{MSY}). They are different fishing target or reference points. Hilborn and Walters (1992) considered the introduction of the $F_{0.1}$ strategy to be remarkable.

$F_{0.1}$ policies may be one of the most significant changes in fisheries harvesting practice since the earlier widespread acceptance of MSY. They are significant not because of any theoretical breakthrough, or any intrinsic elegance, but simply because they provide a replacement for F_{max} and MSY and appear to often be robust (Hilborn and Walters, 1992, p. 461).

2.8.7 Uncertainties in Yield-Per-Recruit Analyses

Unfortunately, there are a number of limitations to the yield-per-recruit analyses that go toward adding appreciable uncertainty to the estimates of optimum age/size at first capture and optimum fishing mortality (estimates of all parameters can only be made with limited statistical precision). Using $F_{0.1}$ instead of F_{MAX} is an attempt to alleviate these uncertainties.

Yield-per-recruit analyses assume that the fishery concerned has reached an equilibrium with the given fishing mortality. This is a very severe limitation for some, especially intrinsically unstable fisheries (high recruitment variability). It also assumes that natural mortality and the growth characteristics of the population are constant with stock size. There is also the problem that management measures, which can be taken to implement the recommended age limits, can only be implemented using size-selective gear. Age classes are rarely uniform in size so the outcome will never be as optimistic as that predicted. With annual species, YPR recommendations can be enforced through using closed seasons.

Finally, of course, setting or estimating F is a very difficult process requiring an accurate estimate of the population size and good records of total commercial catch. One always ends with an estimate of F; there is no real way of eliminating the inherent uncertainty in any fisheries assessment parameter. One method of trying to avoid recruitment overfishing is to conduct egg-per-recruit analyses as well as the more traditional yield-per-recruit.

2.8.8 Types of Over-Fishing

While implementation of the predictions of the yield-per-recruit approach may be difficult, if the primary objective of management is to maximize yield of biomass from a fishery then this sort of analysis is fundamental. A good example of a fishery optimized for yield is the snapper (*Pagrus auratus*) fishery out of Shark Bay, Western Australia.

If one is dealing with a fishery in which most of the fish are being caught before the size at which they reach their optimum for yield, then "growth-overfishing" is said to be occurring. If a stock is being fished so hard that there are appreciable impacts upon subsequent recruitment, then there is said to be "recruitment overfishing".

There are sometimes good (or at least economic) reasons for catching a species at less than its optimum size/age. For example, if the optimal price is obtained for individuals smaller than the optimum for yield there may be a good reason to ignore the yield-per-recruit analysis. Perhaps in such cases it would be better to perform a dollar-per-recruit analysis although with the volatility of fish prices this would probably not be practical. On the other hand, fishing a stock so hard that recruitment overfishing is the result can lead to a vicious cycle of stock reduction until it collapses (i.e., recruitment overfishing is risk prone and not risk averse).

3

Model Parameter Estimation

3.1 MODELS AND DATA

3.1.1 Fitting Data to a Model

A mathematical model of a biological population is always a simulation of nature. These models can be purely simulations in which values for all model parameters are not estimated but are simply given plausible values. Such models can be used to illustrate the implications of a particular idea or model structure (e.g., Haddon, 1999). However, when models are descriptive and/or explanatory of a particular situation in nature, it is necessary to estimate values for at least some of the model parameters by optimizing the fit between the expectations from the model and data observed from nature. Parameter estimation is fundamental to the science of modelling populations and this chapter focuses on the different ways that data can be fitted to a model.

The design of a model, such that it adequately represents the structure of the modelled system, primarily relates to determining which variables are to be included (i.e., whether the model will include age-structure, relate to numbers or biomass, etc) and the relationships between them (linear, non-linear, etc.). Model specification is clearly a vital step in modelling any system and one tends to proceed by development from a simple model to a more complex representation.

Once the model has been designed, if it is not just a simulation model then it remains to be "fitted", or have its parameters "tuned" to the observable world, using whatever data is available from the system. Fitting a model to a data set implies estimating values for the model's parameters to optimize the agreement between the model's predictions and the fitted data from nature.

In formal terms there are three requirements for any attempt at fitting a model to observed data.

1. A formal mathematical model of a system, which is capable of generating predictions about the observable world.
2. Data from the system to be used when estimating the parameter values.
3. A criterion (sometimes called a merit or objective function) to judge the quality of fit between the model's predictions and the observed data.

An example, to keep in mind during the following discussion, could be a fishery stock assessment model that predicts changes in catch-rates and age-structure through time. Fitting the mathematical model to nature entails varying trial values of the model parameters until an optimum agreement is found between the model predictions of how catch-rates and catch-at-age will change through time and the observed time series of real information.

As described in Chapter 1, there is more to the process of modelling than the three formal requirements listed above. If this were not the case then one would invariably finish with a multi-parameter empirical model providing the best statistical fit but not being interpretable in a realistic way. If an explanatory model is wanted then we need a rather obvious but non-quantifiable extension to the third requirement of the quality of fit criterion. As well as optimising some statistic or criterion of best fit, the optimum model should also give rise to biologically sensible predictions. It can happen that a model fit that generates an optimal mathematical solution still predicts biological nonsense (e.g., predictions of an enormous initial population size and almost no productivity, with the history of fishing being one of the gradual erosion of the accumulated biomass). Not all such deviations from biological reality are so obvious, and we need to guard against a lack of realism in the outcomes of the fitting process. In short, we generally want our models to be simple but realistic. It comes down to how one selects which model to use and what is meant by "quality of fit".

3.1.2 Which Comes First, the Data or the Model?

We should always be asking the question, given a particular model, with particular parameter values, how well does it predict (how likely are) the observed data?

The process of model fitting has two parts. First, a set of one or more models is selected or designed and, secondly, values for the model's parameters are found that optimize the quality of fit according to the formal criterion selected by the modeller. Except for at least some of the variables used, model selection is independent of the data (the variables we observed, against which the model is to be compared, must obviously be included in the model). Thus, we always determine how well the data fit, or are reflected by, the selected model(s). Once we focus on a particular model, the optimum parameter values for that particular model are determined by the fixed set of observed data. In fact, what tends to happen is that one starts with a relatively simple representation of whatever is being modelled and fits that to available data, extending and articulating the model in steps.

The process of positing a model plus parameter values and comparing its implications against nature, reflects the hypothetico-deductive approach to scientific practice (Popper, 1963; Lakatos, 1970). However, remember, from Chapter 1, that fitting a model to data only tests how successfully it can describe the data, not how well it explains the data.

3.1.3 Quality of Fit vs. Parsimony vs. Reality

With a set of structurally different models, there are criteria other than just quality of numerical fit that should be used to determine which model should be preferred. For example, if we consider the difference between two models of stock production: the first assumes a symmetric production curve against stock size (linear density-dependence) while the second has no such restriction (Fig. 3.1).

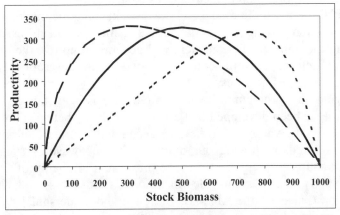

Figure 3.1 Three different biomass production curves (cf. Chapter 2). The equation used was: productivity = $(r/p)*B((1-(B/K)^p)$, where r is a growth rate, K is the equilibrium biomass, B is the stock biomass, and p is the coefficient of asymmetry. When p = 1 (solid line) the equation simplifies to the standard logistic Schaeffer production curve, symmetric about K/2. When p is less than 1, the production curve has a mode lower than K/2 (p = -0.25, dashed line) and when p is greater than 1 the mode is to the right of K/2 (p = 7, dotted line). Whenever p < 1 or > 1, the curve is asymmetric and needs to be scaled to have a similar absolute maximum productivity to the Schaeffer curve.

We would expect the asymmetric production curve to be an improvement over the symmetric in terms of both biological reality as well as mathematical generality (because it is biologically more flexible and because the second model contains the first as a special case when p = 1). In this case, there is an increase in realism and probably of quality of fit but there has also been an increase in model complexity. If the quality of fit between two models were equivalent, one would tend to use the most realistic or the simplest. The general question to be answered is: do the benefits derived from adding extra parameters to a model outweigh the losses and thereby justify their addition? There is no simple answer to this question because the realism of a model is not quantifiable.

Adding parameters to a realistic model may convert it into a multi-parameter empirical model (*cf*. Fig. 1.6, p. 14). Extra parameters are likely to improve the quality of fit of most models because they permit greater flexibility in the model outcomes. In the extreme, one could have the same number of parameters as one had data points and could obtain a perfect fit of the model to the data. Such a model would be both *ad hoc* and useless but it would certainly fit the data points well. Clearly, the quality of fit between a model and data is not everything.

Selecting an optimum model requires a balance between improving the quality of fit between the model and the data, keeping the model as simple as possible, and having the model reflect reality as closely as possible. Increasing the number of parameters will generally improve the quality of fit but will increase the complexity and may decrease the reality. The latter quality is the hardest to assess.

Ignoring the problem of whether a model is realistic, quantitative measures have been developed that assess the balance between the relative quality of fit (the variation in the data accounted for) and the number of parameters fitted. Following our intuitions, these measures suggest that parameter addition be rejected if the improvement in quality of fit is only minor. Such measures include likelihood ratio tests and Akaike's Information Criterion (the AIC, see Burnham and Anderson, 1998); these will be described later.

A further problem with selecting an optimum fit relates to data quality. If the available data is limited in quantity and/or quality (sadly this is common when modelling fisheries) the number of parameters that can be estimated adequately is also limited. The data might be said to be uninformative about those extra parameters. Adding extra parameters to a model that can already be fitted reasonably well to the data may make the fitting procedure highly unstable. A more complex model that provides a significantly better fit may be possible but the solution may be less precise or more biased. Punt (1990) appeared to have this problem with data on Cape hake (Haddon, 1998).

3.1.4 Uncertainty

Irrespective of how closely one can fit a model to a set of data there is no logical or automatic implication that the model chosen is necessarily the best representation of the system being modelled. In addition, there is no guarantee that the model will accurately predict what will happen in the system in the future, or even that the parameter estimates are the optimum values for the system. Very often, it is possible to obtain essentially equally good fits to one's data with sometimes widely different sets of parameters (roughly, the more parameters being estimated the more likely this is to occur). This is especially the case when there are strong correlations between parameters. The correlations act to offset the effects of changing one parameter as altering the values of others means that different sets of parameter values describe the available data almost equally well. The different sets may even produce similar predictions. Model selection is especially difficult under such circumstances. When the results of a model fitting exercise are this uncertain, it suggests that the data are insufficiently informative to distinguish between the alternative possibilities. Uncertainty is an unpleasant commonplace in stock assessment and how best to approach it is a growing and vital part of fisheries modelling (Francis, 1992; Punt and Hilborn, 1997).

It is undoubtedly valuable knowing how to estimate model parameters but the value of this is greatly increased if we also have some way of determining the confidence with which we can use the estimated parameter values and other model outputs. We need to be able to characterize the uncertainty inherent in our analyses. The uncertainty in any analysis stems both from the data, which will contain random variation from observation errors and variation due to random factors, and from model uncertainty, where the model fails to capture the processes and stochasticity in nature. How to estimate this uncertainty will be addressed after we have considered parameter estimation in detail.

3.1.5 Alternative Criteria of Goodness of Fit

There are three different criteria of model fit commonly used today. Most biologists new to modelling will only be familiar with the least squared residual error approach. We will consider and compare methodologies that use least squares, maximum likelihood, and, briefly, Bayesian statistics as their criterion of model fit. The choice of the criterion of fit is, in some ways, a controversial field (Dennis, 1996; and associated papers). Some researchers stoutly defend a maximum likelihood approach but others imply that if an assessment is not conducted using a Bayesian approach it must be next to useless. It is hoped that this chapter will illustrate that much of this controversy is misplaced and that choice of the criterion of fit to be used in a modelling process should depend upon the objectives of the modelling process. The method to use should be whichever is most convenient and can do the job required without distorting the results.

3.2 LEAST-SQUARED RESIDUALS

3.2.1 Introduction

The most commonly used criterion of fit is still the one known as *least squares*. It is so-called because it involves a search for parameter values that minimize the sum of the squared differences between the observed data (e.g., time series of catch-rates or catch-at-age) and the predictions from the combination of model and particular parameter values. Typical fisheries data would never exactly fit a proposed model, even if that model were correct. The differences between the models' predictions and the data observations are known as residual errors (sometimes called "noise" on the signal). A statement made about a model's "error structure" is making a claim about the expected statistical distribution of the residual errors around each of the predicted observations. It is important to understand that when using the least-squared residual criterion of model fit, the residual errors are assumed to be normally distributed. The term, $N(\mu,\sigma^2)$, is the standard nomenclature for such normal, random residual errors, and should be read as implying a mean of μ, and a given variance, σ^2. In general terms

$$\textbf{Observed } \textit{value} = \textbf{Predicted } \textit{value} + \boldsymbol{\varepsilon} \qquad (3.1)$$
$$\textbf{Observed - Predicted} = \boldsymbol{\varepsilon}$$

where ε (epsilon) is the residual or random error. Strictly, the phrase used should always be "residual error", but it is very common to use just one or the other term (residual and/or error) interchangeably instead of the full phrase. Note that the residual error term is added and not multiplied to the predicted (fitted or expected) value (Eq. 3.1; Fig. 3.2). It is common to use the two terms (residual and/or error) to describe one concept and three terms (predicted, fitted, and expected) to describe another concept. By pointing out the effective equivalence within these sets of terms, it is hoped that possible confusion might be avoided.

Normal residuals can be either positive or negative (Fig. 3.2). Thus, it would not suit our purpose just to search for parameter values that would minimize the sum of the residuals, as the negative values would cancel some of the positive in an *ad hoc* manner. However, it would be a real option to minimize the sum of the absolute values of the residuals, although this is not particularly convenient mathematically. Despite this problem, methods have been devised for some statistical procedures (Birkes and Dodge, 1993; see Example Box 3.1).

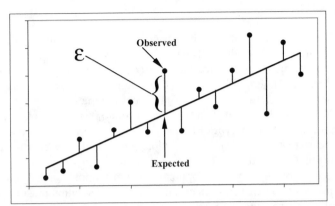

Figure 3.2 View of the residual errors for a linear relation between two hypothetical variables. The error terms represented are those for the regression of Y on X; where the observed and expected values are on the Y-axis (Y = a + bX + ε) for given values of the X variable. The residuals for a regression of X on Y would be horizontal. The residual errors (the ε's) are the differences between the expected values of the Y variable, given by the regression line, and the observed values. Clearly, for a well-fitted line they will be both positive and negative in value.

Squaring each residual error removes the problem of negative residuals and is mathematically very convenient. Note that squaring each residual will give extra weight to larger residuals and less weight if a residual is less than one (because squaring a fraction produces a smaller number), while with absolute residuals each value would have equal weight. Such differences in emphasis and method are why one obtains a different "optimal" fit depending on which criterion is used (Fig. 3.3; Example Box 3.1).

The objective when using the least-squares criterion is to minimize the sum of the residual errors squared. From this we gain

$$SSR = \sum \left(Observed - Expected \right)^2 \qquad (3.2)$$

Where *SSR* is the sum of the squared residuals, i.e., $SSR = \Sigma[Y - (a+bX)]^2$.

It should be remembered that there is an analytical solution for the simple least squares solution of a linear regression, which can be found in any statistical text (e.g., Snedecor and Cochran, 1989; Sokal and Rohlf, 1995).

Example Box 3.1 Fitting a straight line using the sum of absolute residuals (SAR) and the sum of squared residuals (SSQ). Size relates to drained body weight and Eggs is the fecundity in 1000s for the crab *Ovalipes catharus* (data is a subset from Haddon, 1994). PredSSQ and PredSAR are the predicted values of eggs using the parameters defined for the two different criteria of fit respectively. SSQ represents the squared residuals and SAR the absolute residuals, Totals, represents the respective sums of both types of residual errors. The intercepts and gradients are clearly different (Fig. 3.3). The solutions are found in each case using the Solver. The solution for the SAR criterion is highly unstable; try different starting points and compare the apparently optimal solution found. Compare this with the relative stability of the least squares criterion. The SSQ solution must pass through the mean of each variable. The SAR solution must pass through two of the points. Plot columns A to D as in Fig. 3.3. Which line fit looks best to your eyes?

	A	B	C	D	E	F
2		**Intercept**	21.50974	17.07990	**Totals**	
3		**Gradient**	2.59998	2.37019	62764.18	=sum(F5:F21)
4	**Size**	**Eggs**	**PredSSQ**	**PredSAR**	**SSQ**	**SAR**
5	20.71	89.35	=C$2+A5*C$3	=D$2+A5*D$3	=(B5-C5)^2	=abs(B5-D5)
6	30.35	82.399	=C$2+A6*C$3	=D$2+A6*D$3	=(B6-C6)^2	=abs(B6-D6)
7	37.04	166.97	=C$2+A7*C$3	=D$2+A7*D$3	=(B7-C7)^2	=abs(B7-D7)
8	39.50	98.324	*Copy*	*down*	*to*	*Row 21*
9	55.60	135.427	166.07	148.86	938.92	13.44
10	67.90	240.713	198.05	178.02	1820.25	62.70
11	69.46	181.713	202.10	181.71	415.81	0.00
12	84.12	193.161	240.22	216.46	2214.58	23.30
13	94.31	310.425	266.71	240.61	1910.64	69.81
14	108.47	213.247	303.53	274.17	8151.00	60.93
15	125.54	411.056	347.91	314.63	3987.22	96.42
16	132.7	366.567	366.53	331.60	0.0016	34.96
17	137.31	298.439	378.51	342.53	6411.90	44.09
18	141.34	529.351	388.99	352.08	19700.85	177.27
19	178.60	440.394	485.87	440.40	2067.76	0.00
20	224.31	683.008	604.71	548.74	6130.29	134.27
21	229.89	545.681	619.22	561.96	5407.95	16.28

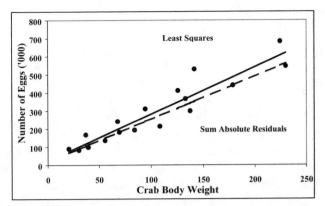

Figure 3.3 Number of eggs vs. mass for the swimming crab *Ovalipes catharus* (data is a subset from Haddon, 1994). The upper line is the best fitting least squares linear regression (eggs = 25.51 + 2.599*Wt) while the lower dashed line is the best fitting least sum of absolute residuals (eggs = 17.08 + 2.37*Wt). The lines differ because the error structures used to fit the lines are different. Moreover, because the form of the residual errors used differs, the lines cannot be directly compared statistically. Both are optimal fits according to their respective criteria (see Example Box 3.1). Most people tend to prefer the line that keeps the data around it in a symmetrical fashion.

In the analytical solution is

$$b = \frac{\sum_{i=1}^{n}(X_i - \overline{X})(Y_i - \overline{Y})}{\sum_{i=1}^{n}(X_i - \overline{X})^2} \quad \text{and} \quad a = \overline{Y} - b\overline{X} \quad (3.3)$$

where b is the gradient, a is the intercept, X_i represents the n independent observations, and Y_i the n dependent observations. The X and Y with the bars (e.g., $\overline{X}$), of course, represent the mean values of the observations concerned. It is invariably more efficient to use an analytical solution to a problem and one should do so wherever it is possible.

A major assumption of least squares methodology is that the residual error terms are distributed as a normal distribution about the predicted variable with equal variance for all values of the observed variable; that is the σ^2 in the $N(0, \sigma^2)$ is a constant. If data are transformed in any way, the transformation effects on the residuals may violate this assumption. Conversely, a transformation may standardize the residual variances if they vary in a systematic way. As always, a consideration or visualization of the form of the residuals is vital.

The Sum of Absolute Residuals (SAR) is intuitively attractive in that it gives all data point's equal weight. However, at least in the linear model, it is a fact that the best fitting line must always pass literally through two of the points (Birkes and Dodge, 1993; see Fig. 3.3). This has the disadvantage of sometimes forcing the residuals to be asymmetric.

It should be noted that the standard least squares linear regression is a regression of Y on X. All this means is that the X-values are assumed to be measurable with no errors and to be independent of the Y-values while the Y-values are assumed to be dependent upon the "given" X values. In the fitting process, this means the residuals would be vertical (Fig. 3.2). Of course, in most biological processes it would not be possible to measure the so-called independent variable without error. In an ideal world, it would perhaps be best to have residual errors that were perpendicular to the expected line (i.e., neither Y on X or X on Y) as in functional regression or principle components analysis. Ricker (1973) and McArdle (1990) discuss this problem at length. Generally, the bias that using Y on X might introduce is likely to be small although it would become greater as the variability about the predicted curve becomes greater.

3.2.2 Selection of Residual Error Structure

A common alternative to the least squares approach would be to use *Maximum Likelihood* as the criterion of quality of fit. Parameters are selected which maximize the probability density or likelihood that the observed values (the data) would have occurred given the particular model and the set of parameters selected.

One great advantage of the maximum likelihood approach is that it forces one to be explicit about the statistical form of the expected residual errors. That is, whether they are normal and additive, as with the regression examples we have seen ($Y = a + bX + \varepsilon$), or lognormal and multiplicative (as in $Y = aX^b e^\varepsilon$), or follow some other distribution (more on this later). One should also consciously select the form of the errors when using the least squares criterion. Whichever error structure is selected, with least squares it is necessary to devise a linearizing transformation to convert the selected error structure into normally distributed residuals. Thus, with lognormal errors, a log-transformation will permit linear, least square methods (e.g., $y = ax^b e^\varepsilon$ becomes $Ln(y) = Ln(a) + bLn(x) + \varepsilon$). If no normalizing transformation is possible, or if it fails to linearizes the modelled relation, then it becomes necessary to use non-linear methods to fit the models to data. In such cases, if one still used normal residuals this would greatly influence the optimum fit. If the residuals required are not normal and there is no normalizing transformation, then one has no choice but to use maximum likelihood methods. Generally, it must be remembered that with least squares, the selection of the residual error structure is implicit and at worst, it is *ad hoc*.

We will consider maximum likelihood methods in detail after we have introduced non-linear parameter estimation methods and expanded on the method of least squares.

3.3 NON-LINEAR ESTIMATION

3.3.1 Parameter Estimation Techniques

Our examples have so far been limited to the estimation of parameters for simple linear models. Using least squares, it would be simpler to use the analytical solution for linear regressions (Eq. 3.3) described in any statistical textbook rather than invoking the solver facility built into Excel. Of course, most fisheries models are far more complex, involving many more parameters and non-linear relationships between the variables. There are usually no analytical solutions or linearizing transformations available for these more difficult to fit models. Numerical methods are needed when fitting these multi-parameter, non-linear models to data. In order to understand the strengths and weaknesses of such numerical methods it is necessary to have some understanding of the algorithms or strategies used in the more commonly available methods used for fitting complex models. When no analytical solution exists for fitting a model to data, we need to search for the optimum parameter values. We can define three types of search: graphical searches, directed searches, and heuristic searches. To understand non-linear parameter estimation we will first consider the graphical approach to searching for optimal parameter values.

3.3.2 Graphical Searches for Optimal Parameter Values

Consider the simple problem of fitting a curve to the relationship between carapace width and eggs carried per batch in the crab *Ovalipes catharus* (Fig. 3.4; Haddon, 1994). The values exhibited by the data points suggest the exponential relationship

$$\text{Eggs} = a.e^{b.\text{CWidth}} e^{\varepsilon} \tag{3.4}$$

where "Eggs" refers to fecundity, "Cwidth" refers to the carapace width, and a and b are the parameters to be estimated. Because the residual errors in Eq. 3.4 are lognormal, it can be linearized by natural logarithmic transformation

$$\text{Ln(Eggs)} = \text{Ln}(a) + b.\text{Cwidth} + \varepsilon \tag{3.5}$$

Equation 3.5 has the form of a linear regression, which can be solved analytically, leading to Ln(a) = 2.7696 and b = 0.0347. By back-transforming the Ln(a), i.e., Exp(2.7696), we obtain a = 15.953, which produces a satisfactory fitted line (Fig. 3.4; see Example Box 3.2).

An alternative approach would be to carry out a grid search across plausible parameter values and plotting the resulting sum of squared residuals as a contour plot (Fig. 3.5). This defines a valley or pit in the sum of squares surface of possible values and the optimal parameter values relate to the bottom of the pit. Obviously, the graphical search grid must bracket the minimum sum of squares.

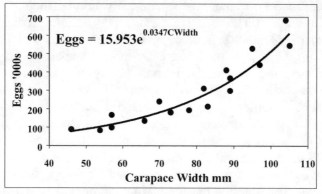

Figure 3.4 The exponential relationship between number of eggs in an egg-mass and carapace width for the New Zealand crab *Ovalipes catharus*. Data are a sub-set from Haddon (1994). The exponential relationship describes over 90% of the variation in the data. The solid line illustrates the given optimal solution (see Example Box 3.2).

The contour plot makes it possible to constrain the parameter values being used in the search for the bottom of the least squares pit. To extend the analogy with a pit, this is a downhill search for the bottom or minimum depth. Each of the contours represents combinations of parameter values that give rise to the same value of sum of squared residuals. The contours are of the sums of squares so they represent combinations of parameters giving fits of equal quality.

Instead of a simple trial and error search for the minimum, it is possible to use this analogy of a downhill search and, using information gained from previous trials, conduct an informed trial and error search. The contour map indicates visually how to improve the parameter estimation. This entails moving the trial parameters to form combinations that are most likely to lead to a maximal reduction in the sum of squares (the bottom of the pit). This informed trial and error is continued until there is no detectible improvement to the sum of squares. Of course, it would only be effective if the range of trial values used actually brackets the optimal solution.

As we have seen from the contours, the same sum of squares can be obtained from different combinations of parameters. If we contemplate the sixth significant digit in the parameter estimates then the optimum fit would not be graphed as a point but as a very small contour circle. If there is no analytical solution and the model has to be fitted using some numerical method then the optimal fit is a compromise between the time taken to find the solution and the accuracy of the fit. In Example Box 3.2, the impact of different trial values of the two parameters on the predicted values and graph (Fig 3.4) can be determined.

The graphical search is informative in a number of ways. The shape of the contour plot provides information about relationships between the parameters of the model. If the parameters were completely independent of one another (the ideal but rare case), then the contours would form perfect circles. If the contours are not circles but ovals then parameter correlation is indicated (e.g., Fig. 3.5).

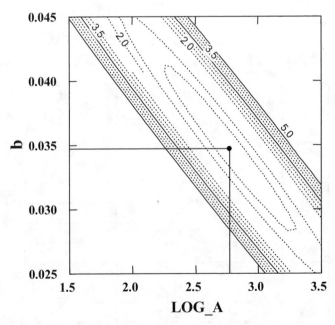

Figure 3.5 Graphical search for the optimal parameter values for Eq. 3.5 given the data in Example Box 3.2. Different combinations of the two parameters produce different values of the sum of squared residuals. Plotting these as contours makes it possible to home in on the optimum values. The optimal solution is indicated (at 0.0347 and 2.769).

Example Box 3.2 Fitting an exponential curve to eggs at carapace width data. Width is in mm and Eggs is fecundity in 1000s for *Ovalipes catharus* (*cf.* Fig. 3.4; data is from Haddon, 1994). The *Pred* column is the predicted values of eggs. SSQ represents the sum of squared residuals (in C1). The solution is found using the Solver, minimizing C1 by changing C2 and C3; an exact solution can be found using the array function Linest, but that would require a separate column of transformed egg numbers (try this, don't forget to use <ctrl><shift><enter> to enter the array function). Note the log transformation of the number of eggs in column D (see Eq. 3.5). By plotting column B against A as separate points, and then column C against column A as a connected line you should be able to mimic Fig. 3.4. Use this sheet to investigate the relative precision of different solutions. The values in columns E and F represent different detailed trial values of Ln(a) and b with their respective sum of squared residuals. Paste the values of your trials values and SSQ into spare cells in columns E and F, i.e., copy B1:C3 to keep a record of your trials. Do different parameters that give the same SSQ alter the graph visually? How precise should one try to be when using biological data of limited original accuracy?

	A	B	C	D	E	F
1		SSQ	=sum(D5:D21)		Trial Values	
2		Ln(a)	2.7696		1) SSQ	0.630620
3		b	0.034734		1) Ln(a)	2.7696
4	Width	Eggs	Pred	ResidSQ	1) b	0.034734
5	46	89.35	=C$2+A5*C$3	=(Ln(B5)-C5)^2		
6	53.8	82.399	=C$2+A6*C$3	=(Ln(B6)-C6)^2	2) SSQ	0.630620
7	57	166.97	=C$2+A7*C$3	=(Ln(B7)-C7)^2	2) Ln(a)	2.77
8	57	98.324	*Copy down*	*To row 21*	2) b	0.03473
9	66	135.427	5.062	0.024		
10	70	240.713	5.201	0.080		
11	73	181.713	5.305	0.011		
12	78	193.161	5.479	0.046		
13	82	310.425	5.618	0.014		
14	83	213.247	5.653	0.084		
15	88	411.056	5.826	0.037		
16	89	366.567	5.861	0.002		
17	89	298.439	5.861	0.026		
18	95	529.351	6.069	0.041		
19	97	440.394	6.139	0.003		
20	104	683.008	6.382	0.021		
21	105	545.681	6.417	0.013		

3.3.3 Parameter Correlation and Confounding Effects

The resolution of the contours shown in Fig 3.5 is sufficient to indicate approximately, where the "best" fit solution would lie. The fact that the sum of squared residual values forms a diagonal trough indicates that there is a negative correlation between the two parameters. This means we cannot readily distinguish between different combinations of the two parameters. What it implies is that if we were to force a particular value of *a* on the model we could still recover a reasonable fit because the value of *b* could be altered accordingly to produce a similar curve. The plot provides us with a visual indication of the quality of fit and the confidence we can have that our best fit provides a good indication of the actual situation.

Very strong parameter correlation is a problem that occurs with many fisheries statistics. For example, the total mortality for a fish population is a combination of natural and fishing mortalities and there is always difficulty in separating these different phenomena. Parameter correlations imply that the so-called independent variables (X-axis variables) cannot be independently determined and in fact, are often strongly correlated. When purportedly independent variables are correlated, it becomes impossible to determine which "independent" variable is most closely related to changes in the so-called dependent variable. The effects of however many independent variables are correlated are said to be confounded.

Sadly, the graphical search strategy is of limited use for models with more than three parameters. In the example of the fecundity vs. carapace width there were only two parameters and the sum of squared residuals can be visualized as a third axis described by the *a* and *b* parameters to form a third dimension to generate a surface. With a three parameter model the criterion of fit values would be described as a four dimensional volume or three surfaces. With n parameters (more than three), the criterion of fit would be described by an n+1 dimensional hyper-volume. Even with a hyper-volume the analogy of moving downhill or over the surface of the criterion of fit, is still a useful intuition to hold in mind.

The basic idea and strategy of the downhill search lies behind very many of the numerical methods for fitting non-linear models to observed data, even with n parameters. With the least squares criterion, the mathematical problem is one of minimization of the sum of the squared residuals. To move beyond the graphical search, some other more efficient numerical method is required. The graphical approach is essentially a grid search that could be pursued with more and more precision once the minimum values had been bracketed. Automatic minimization routines do something similar without the need for the grid or human intervention. There are numerous algorithms but here we will only describe the directed and the heuristic search.

3.3.4 Automated Directed Searches

A common approach, known as the Levenberg-Marquardt algorithm (after the two inventors), uses the downhill analogy directly. It requires a single set of starting parameter values to begin the search. These guessed parameters specify a particular point on the sum of squared residuals (SSR) surface (which has the same number of dimensions as there are parameters, and so may be a hypervolume).

Despite the number of dimensions, the analogy remains of a surface having a steepest path downwards to a minimum. By automatically making tiny increments to each of the parameter values in turn, the algorithm can estimate the gradient of the surface along each of the parameter dimensions (a form of numerical partial differentiation). The algorithm then alters the parameters in the direction that should lead to the steepest decline down the SSR surface. Given this new point on the surface the relative gradients are considered once again from the new viewpoint and the parameters incremented again in whichever direction will lead to the greatest decline in the SSR. This is continued until a minimum is reached or further benefits are minimal (Press *et al.*, 1989). The algorithm automatically directs the new trial values along a path that should maximize the decline in the criterion of fit.

An obvious problem with this search strategy is the possibility of local minima on the SSR surface that could be confused for the global minima by a non-linear-fitting algorithm (Fig. 3.6). An equally obvious test of the generality of any solution is to start the search with initial parameter guesses that widely bracket the final estimated values. They should all converge on the same answers if they are to be considered as adequate estimates.

A further problem, which is becoming less and less important, is that a large model with very many parameters may take a very long time to converge on a possible solution for all parameters (Schnute *et al.*, 1998). More likely is that the non-linear solver will fail through instability of some form.

The solver in Excel97 and Excel 2000 is a great improvement over earlier versions and is useful for many smaller problems. For larger problems, it may be necessary to resort to custom computer programs or to use a meta-programming language such as AD-Model Builder (Schnute *et al.*, 1998).

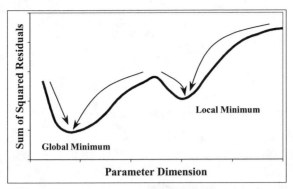

Figure 3.6 Schematic representation of a single parameter dimension with a local minimum that could confuse an automatic minimization routine. Once in the vicinity of the local minimum every change in the parameter value makes the SSR value get larger and many routines would understandably stop and claim to have found a minimum.

3.3.5 Automated Heuristic Searches

A robust alternative to the Levenberg-Marquardt algorithm is the Simplex algorithm. This is an n+1 dimensional bracketing search method (Nelder and Mead, 1965) where there are n parameters to be estimated. This method requires n+1 sets of initial trial values that attempt to bracket the optimal solution or at least define the direction in which to move the set of trial values over the surface of the criterion of fit. By comparing the relative fit of the different sets of trial parameter values, the n+1 trials permit the direction that should lead to maximum improvement in fit to be determined approximately. N+1 sets of trial values are required so it becomes possible for the overall set to bracket the optimum combination. The n+1 set of trial values move over/through the n-dimensional hyper-volume, always in the approximate direction that should improve the value of the criterion of fit. Some speak of the n+1 dimensional search object "crawling" over the fitting criterion surface towards the minimum. The analogy with an amoeba flowing through n-dimensional space has been used (Press *et al.*, 1989). This continues until the only way the set of trial values can improve the fit is for the n+1 parameter combinations to contract towards each other (i.e., they really bracket the optimum solution and contract towards it). The simplex algorithm is very robust but can be slow, depending on the complexity of the model and how close the starting values are to the optimum.

Whichever approach is used one should get to know the limitations of any particular non-linear solver that might be used, and always attempt to find the same solution from a number of different sets of initial parameter guesses.

3.4 LIKELIHOOD

3.4.1 Maximum Likelihood Criterion of Fit

Maximum likelihood, used as the criterion of quality of fit, is usually characterized as the determination or search for the set of model parameters that maximize the probability that the observed values (the data) would have occurred given the particular model and the set of parameters selected (Neter *et al.*, 1996; Press *et al.*, 1989). Using maximum likelihood requires the model to be defined so that it specifies probabilities or likelihoods for each of the observations (the available data) as a function of the parameter values and other variables in the model. To obtain the likelihoods, one needs to define how the residual errors are distributed about the expected values derived from the model. To understand the idea of maximum likelihood parameter estimation we need first to formalize the relatively familiar idea of using probability distributions. Many different statistical probability distributions are used to describe different residual error structures. We will consider the normal, the lognormal, the binomial, the poisson, the gamma, and the multinomial distributions as well as their uses in fisheries modelling.

3.4.2 The Normal Distribution

Most biological scientists would have some understanding of a claim that a set of observations is expected to exhibit a normal distribution about their expected value (the mean). What this implies is that the observed values of a variable X are expected to be distributed symmetrically about their mean and that large deviations from the mean would occur less often than small deviations. The expected relative rates of occurrence of different sized deviations from the mean are described by the probability density function (pdf) for the normal distribution (Fig. 3.7). Relative frequency histograms count the relative occurrence of individuals in a population that are found in defined classes of the variable under study (e.g., body weight). A pdf for the normal distribution describes, in effect, the expected relative frequency (probability density) curve generated for a continuous variate instead of for discrete classes of the variate. The pdf for the normal curve has two parameters, the mean or expectation of the distribution (μ) and its standard deviation (σ). Once they are set for a variable X, substituting different values of the variable into the equation generates the well-known normal curve (Fig. 3.7; Eq 3.6; Example Box 3.3).

For any given value of the variable X, the value of this function (Eq. 3.6) defines its probability density. To someone unfamiliar with it, this equation may look rather daunting but it can be implemented easily in most spreadsheet programs (see Example Box 3.3) or on a hand calculator. The right hand version of the pdf (Eq. 3.6) demonstrates that each observation is being converted into a residual and is then standardized by dividing by the standard deviation ($[(X - \mu)/\sigma]$, i.e., observation X minus the mean, the expected value μ, divided by the standard deviation). In other words, with the normal distribution, we are determining the probability density of particular values considered in terms of how many standard deviations these values are from the mean (*cf.* Fig. 3.9).

$$\text{Probability Density} = \frac{1}{\sigma\sqrt{2\pi}} e^{\left(\frac{-(X-\mu)^2}{2\sigma^2}\right)} = \frac{1}{\sigma\sqrt{2\pi}} e^{\left(-\frac{1}{2}\left(\frac{X-\mu}{\sigma}\right)^2\right)} \tag{3.6}$$

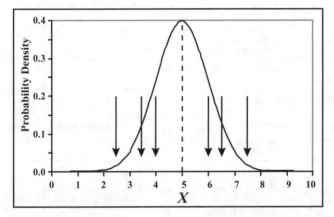

Figure 3.7 The arrows represent a set of 6 fictional observations, 2.5, 3.5, 4.0, 6.0, 6.5, and 7.5, with a mean of 5.0 and standard deviation of approximately 1.95. The probability density function (pdf) of a normal distribution with a mean of 5.0 and standard deviation of only 1.0 (Eq. 3.6), from which they are hypothesized to have been sampled, is superimposed about them. Note the symmetry. From Eq 3.6, with a mean and standard deviation of one, the pdf value of X at 2.5 and 7.5 is 0.018, at 3.5 and 6.5 it is 0.130, and at 4 and 6 it is 0.242. These values are the probability densities of the respective X values given the selected parameter values. See Example Box 3.3.

Example Box 3.3 Properties of the Normal probability density function. Cell B1 is named *Mean*, and B2 as *StDev*. In cell E1 put the equation =1/(StDev*sqrt(2*PI())), and name the cell Const. In cell B5 (and copy down to row 105, where X = +5 in column A), put Eq. 3.6: =Const*exp(-((A5-Mean)^2)/(2*StDev^2))). Alternatively, you could use the Excel function, Normdist(A5,Mean, StDev, false), which provides the same answer without having to bother with the Const. Check the help for a description of this function. In C5 (and copy down), put the function Normdist(A5,Mean, StDev, true), to obtain the cumulative probability density. The last number in column C should be very close to 1. The sum in B3 should be close to 10, which reflects the increments of 0.1 that were used to step through from –5 to 5. If you reconstruct the worksheet to increment the values in column A by 0.05 instead of 0.1, don't forget to modify cell B3. What happens to the value in B3 when you make this change? Standardize the separate probability densities by dividing each one by their sum, as in column D. If these are then cumulated they will also sum to one. Column C doesn't quite reach one because there is a small but finite probability of values being greater than 5 given a mean of 0 and standard deviation of 1. Plot columns B and C against column A to see the standard normal curve and the cumulative normal curve (*cf.* Fig. 3.9). Alter the values in B1 and B2 and see the impact on the curves. See the text for the difference between probability density (column B) and probability. Obtain the six probability densities for the particular X values in Fig. 3.7.

	A	B	C	D	E
1	**Mean**	0		$1/\sigma \sqrt{2\pi}$	0.39894228
2	**StDev**	1			
3	**Sum**	**=Sum(B5:B105)**			
4	**X**	**Prob. Density**	**Cumulative Prob. Density**	**Standardized**	**Cumulative**
5	-5	1.4867E-06	2.871E-07	=B5/$B$3	=D5
6	-4.9	2.4390E-06	4.799E-07	=B6/$B$3	=D6+E5
7	-4.8	3.9613E-06	7.944E-07	3.9613E-07	7.887E-07
8	-4.7	6.3698E-06	1.302E-06	6.3698E-07	1.426E-06
9	.	*Copy down*	*Copy down*	*Copy down*	*Copy down*
.	.	.	.	.	.
.	.	.	.	.	.
102	4.7	6.3698E-06	0.9999987	6.3698E-07	0.9999992
103	4.8	3.9613E-06	0.9999992	3.9613E-07	0.9999996
104	4.9	2.4390E-06	0.9999995	2.4390E-07	0.9999999
105	5.0	1.4867E-06	0.9999997	1.4867E-07	1

3.4.3 Probability Density

Note the use of the phrase "probability density" instead of probability. In terms of probability, the sum of the probabilities of the full set of possible outcomes for a particular event must equal one. Thus, when tossing an unbiased coin the possible outcomes are a head or a tail, each with a probability of 0.5, and these add to one for any particular coin-tossing event (a discrete variable). With a continuous variable, an event would be the making of a single observation X. However, speaking mathematically, and ignoring such things as the limits of measurement, with a truly continuous variable there are an infinity of potential observed values of X within the range of what is possible. The sum of the probabilities of all possible events (i.e., X between $-\infty$ to $+\infty$) must equal one and so, with an infinity of possibilities, the probability of any particular value (e.g., exactly 2.5) would be infinitesimally small. To obtain literal probabilities for continuous variates we need to quantify an area under the pdf. Thus, we can have a probability for a range of a continuous variate X but not for a particular value.

Clearly there is a difference between probability density and probability, because, in the sample of six observations (Fig. 3.7), the probability of observing an X value of exactly 3.5 is infinitesimal but using the given parameter values, the probability density of exactly 3.5 is 0.130. To grasp the ideas behind using likelihoods it is necessary to understand this difference between probability density and probability (see Example Boxes 3.3 and 3.4).

The graph of the pdf values (Fig. 3.7) is analogous to a histogram of expected relative proportions for a continuous variate. This captures our intuitions with regard to the relative chance of obtaining different values of the continuous variable X. Thus, with our example (Fig. 3.7) and the given parameter values, the pdf value for an observation of 4 (pdf = 0.242) is 13 times greater than (more likely than) the pdf value for an observation of 2.5 (pdf = 0.018). We can use this characterization to determine whether the observed data is consistent with the hypothesized normal pdf curve from which they are assumed to be sampled.

The term "density" is used as an analogy to express the weight or mass of probability above particular values of the variable X. Of course, when the range over which each density operates is infinitesimal the overall probability would also be infinitesimal. However, consider what would be the case if, as an approximation, one were to assume the same probability density operated over a small range of the variable of interest (Fig. 3.8).

Using the approximation that the probability density of $X = 3.5$ represents the probability density of values of X close to 3.5, then the probability (as contrasted with probability density) of observing a value between 3.45 and 3.55 is the area $\alpha + \beta + \gamma = 0.129518 \times 0.1 = 0.0129518$. This is only an approximation to the real area under the pdf curve between those ranges (which is $\alpha + \beta + \delta = 0.0129585$). The difference, or error, is $\delta - \gamma$, and this would become smaller and smaller as the range over which the approximation was applied became smaller. The limit would be where the range was infinitesimal whereupon the solution would be exact but also infinitesimal.

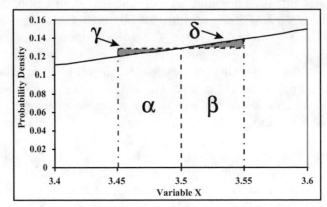

Figure 3.8 Probability density for a normally distributed variate X having a mean of 5.0 and a standard deviation of 1.0 (as in Fig.3.7). The pdf value at $X = 3.5$ is 0.129518. The true area under the curve between 3.45 and 3.55 is $\alpha + \beta + \delta$, an approximation to this is $\alpha + \beta + \gamma$ (see Example Box 3.4).

Instead of summing under the curve in a series of discrete, small steps (which provides only an approximation as in Fig. 3.8 and Example Box 3.4) it is better to sum the area under the pdf curve in an infinitesimal way using integration (as with the Excel function normdist, with cumulate set to true). The assertion that the sum of the probabilities of the set of all possible outcomes for any given observation cumulates to one can be expressed by integrating Eq. 3.6 between the maximum and minimum possible values for the variable X (Eq. 3.7).

$$\text{cdf} = 1 = \int_{X=-\infty}^{+\infty} \frac{1}{\sigma\sqrt{2\pi}} e^{\left(\frac{-(X-\mu)^2}{2\sigma^2}\right)} \tag{3.7}$$

where cdf refers to the cumulative distribution function. This provides an exact solution to the area under the pdf and translates the probability densities into true probabilities (Fig. 3.9).

Example Box 3.4 Estimating probabilities under the normal distribution. The calculations involved reflect those seen in Fig. 3.8. P(3.45-3.55) refers to the probability of the variable lying between 3.45 and 3.55 and equals $\alpha + \beta + \delta$. The terms α and $(\beta + \gamma)$ are derived from the normdist function (see help for details). The approximate probability can be derived from the single value probability density for 3.5 multiplied by the area concerned, and the difference between the strict probability and the approximate (i.e. cells B6 and D6) can be seen. Note we need to use six decimal places to detect the difference. The use of the probability density in the approximation only becomes valid once we multiply it by the range over which it is intended to apply and thereby generate an area under the normal curve.

	A	B	C	D
1	3.45	=normdist(A1,5,1,true)		
2	3.5	=normdist(A2,5,1,true)	=normdist(A2,5,1,false)	
3	3.55	=normdist(A3,5,1,true)	$\alpha + \gamma = \beta$	=C2*0.05
4	α	=B2-B1		
5	$\beta+\delta$	=B3-B2		
6	P(3.45-3.55)	=B4+B5	**Approx. P(3.45-3.55)**	=2*D3
7	δ	=B5-D3	$\alpha+\beta+\delta=P(3.45-3.55)$	=B4+D3+B7
8	γ	=D3-B4		

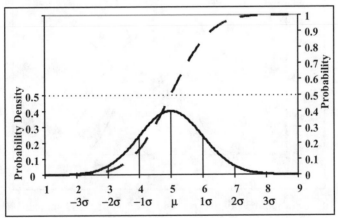

Figure 3.9 The relationship between the probability density function (solid line) and its integral (or the cumulative distribution function) for the normal distribution (in this instance with a mean of 5.0 and a standard deviation of 1.0). Integration is equivalent to summing the area under the pdf so we see that it sums to one across all possibilities. We can also see that the symmetry of the normal pdf leads to an observation being the mean or less than the mean having a probability of 0.5.

Integration of the pdf can be performed across more ranges of the variate to produce the probability of observing values within a given range (Example Box 3.4). For example, if we integrated the pdf between X values of five and four we would be calculating the probability of making an observation with a value between four and five (Fig. 3.10). To do this does not require an exercise in calculus each time as there are published tables of the cumulative normal frequency distribution (e.g., Table A3 in Snedecor and Cochran, 1967). Using such tables or the functions in Excel, one can determine the cdf value for 5.0 and subtract from it the cdf value for 4.0 to produce the area under the pdf between the limits of 4 and 5 (Fig. 3.10).

The values we obtain by substituting real values into Eq. 3.6 are known as probability densities, and these are infinitesimally summed or integrated between given limits to produce probabilities, as in Eq. 3.7. Thus, in terms of true probability, with respect to continuous variates, pdf values can only be interpreted directly when referred to in the context of a range of the possible values (e.g., Fig. 3.10). Later, when we consider discrete variates (e.g., the binomial distribution - heads or tails, tagged or untagged), the probability density functions actually generate discrete probabilities. With discrete variate pdf's, probability density and probability are the same because the number of possible outcomes to any particular event are limited by definition.

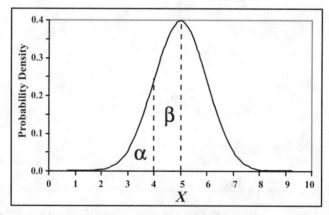

Figure 3.10 A normal distribution pdf with a mean of 5 and standard deviation of 1. The total area under this curve from -∞ to +∞ sums to one. As the curve is symmetrical we can therefore conclude that the probability of observing a value of less than or equal to 5 would be 0.5. Therefore, the area $\alpha + \beta = 0.5$. To obtain the probability of a value falling between 4 and 5 we need the probability of obtaining a value of four or less (i.e., $\alpha = 0.158655$) and this must be subtracted from the cdf for the value of five or less to leave the probability in which we are interested i.e., β (= 0.341345).

3.4.4 Likelihood Definition

We have seen how the true probability of individual observations is infinitesimal with continuous variates. However, we have also repeated a statement found in many statistical texts about Likelihood being the probability of the observed data given a set of parameters. This apparent anomaly requires clarification. With any model of a process there will be parameter sets for which the model predictions are obviously inconsistent with the available data and are thus "unlikely". Conversely, there will be other parameter sets, which produce predictions closely consistent with the data and these we feel are far more "likely". Maximum likelihood methods are attempting to find the parameter set that is most likely in this sense of the word "likely". Press *et al.* (1989) present a clear statement of the problem of applying likelihood methods to parameter estimation.

> It is not meaningful to ask the question, 'What is the probability that a particular set of fitted parameters $a_1 \ldots a_M$ is correct?' The reason is that there is no statistical universe of models from which the parameters are drawn. There is just one model, the correct one, and a statistical universe of data sets that are drawn from it! (Press *et al.*, 1989, p549).

What they are claiming is that a set of observations only constitutes a sample of what is possible. If we were able to take another sample we would expect to obtain similar, but not exactly the same values. This is because our sampling is based upon the assumption that there is some underlying explanatory model for the behaviour of the system and this constrains the measurable variate's behaviour to follow the predicted model output plus the residual terms defined by the pdf of the errors. If this is the case, then, as stated, there is only one correct model and only one set of correct parameters. This means that each different set of parameter values is a separate hypothesis and not a sample from some statistical distribution of parameter values. Press *et al* go further and state

> ... we identify the probability of the data given the parameters (which is a mathematically computable number), as the *likelihood* of the parameters given the data. This identification is entirely based upon intuition. It has no formal mathematical basis in and of itself; ... (Press *et al.*, 1989, p. 550).

This appears to be a very weak foundation upon which to base the serious business of model parameter estimation. However, one will often see either a definition or an implied definition of Likelihood that is very exact

$$L(\theta) = \prod_{i=1}^{n} pdf(X_i \mid \theta) \tag{3.8}$$

which is read, the likelihood of the parameter(s) θ is the product of the pdf values for each of the n observations X_i given the parameter(s) θ. It is common knowledge that when events are independent it is necessary to multiply their separate probabilities to obtain the overall probability (e.g., the probability of three heads in a row is 0.5 x 0.5 x 0.5 = 0.125). With continuous variates, the same process is involved so we use the product (capital Pi, Π) of the separate probability densities and not the sum. The product of all the separate pdf values when the parameters are set equal to the hypothesized values is called the likelihood value and is usually designated $L(\theta)$ (Neter *et al.*, 1996). Thus, in our earlier example (Fig. 3.7) we could assume a standard deviation of one and then search for the value of the mean that would lead to the maximum likelihood. We have observations at 2.5, 3.5, 4.0, 6.0, 6.5, and 7.5. If we were to assume a range of different values for the mean, we would find that the respective pdf values for each observation would alter so that their product, the Likelihood for the particular guessed value of the mean would also vary. Clearly to search for the most likely value of the mean we need to trial different values and search for that value which maximizes the product of the pdf values (Example Box 3.5).

Probability densities only relate to probabilities in the context of a range of values so it appears that the definition of Likelihood we have been using is too strong. As we noted earlier if the observed variate is a continuous variable then the probability of each particular value would simply be infinitesimal. The common argument is that what we are really talking about is the product of the probability densities for our particular data values but over a tiny range around this value (Edwards, 1972; Press *et al.*, 1989). In practice, we just calculate the probability density for precise values and not a range of values. Edwards (1972) suggests we are effectively using a constant range of one about each data value, but this is hardly what would be called a tiny range.

There is no such conceptual problem with probability density functions for discrete statistical distributions (e.g., binomial) because the probability densities are exact probabilities. It is only with continuous pdfs that a problem of interpretation arises. While this may seem pedantic and only related to definitions, the source of likelihoods for continuous variate has concerned mathematical statisticians for decades and a variety of solutions have been proposed (Edwards, 1972).

Example Box 3.5 Maximum likelihood (ML) search for an estimate of the mean value of a set of observations. Columns B to D are example trials (Fig. 3.11). Row 9 contains the product of each set of likelihoods. Using the solver, maximize the value in E9 by changing cells E1:E2. The StDev of the data, =StDev(A3:A8) = 1.94936, which is larger than the estimate in E2. The ML estimate of the standard deviation is =$\Sigma(x-\mu)^2/n$, the divisor is n and not n-1. In a separate cell calculate the usual StDev and beside it put =sqrt((E2*E2*6)/5). Are they the same? This difference between the usual population estimate of the standard deviation and the ML estimation of the same statistic becomes important when we want to use ML to fit data to a model and need to estimate the unbiased variance.

	A	B	C	D	E
1	Mean	4.5	5	5.25	5
2	Obs \ StDev	1	1	1	1.7795
3	2.5	0.05399	0.01753	0.00909	=normdist($A3,E$1,E$2,false)
4	3.5	0.24191	0.12952	0.08628	=normdist($A4,E$1,E$2,false)
5	4	0.35207	0.24197	0.18265	=normdist($A5,E$1,E$2,false)
6	6	0.12952	0.24197	0.30114	Copy down to Row 8
7	6.5	0.05399	0.12952	0.18265	0.15715
8	7.5	0.00443	0.01753	0.03174	0.08357
9	Π(Likelihood)	0.1425 x 10^{-6}	0.3108 x 10^{-6}	0.2502 x 10^{-6}	=product(E3:E8)

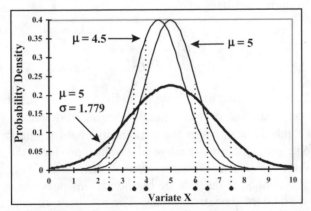

Figure 3.11 Different hypothesized normal distributions superimposed upon the six data points from our example. Shifting the mean away from 5.0 leads to a decrease in the overall likelihood. However, increasing the standard deviation from 1.0 to 1.779 increased the likelihood to its maximum. This implies the best fitting normal distribution is one with a mean of 5.0 and standard deviation of 1.779. Note that increasing the standard deviation widens the spread of the distribution, and hence increasing the probability densities of the observations at 2.5, 7.5, 3.5, and 6.5, while decreasing the pdf values of the central observations 4.0 and 6.0.

3.4.5 Maximum Likelihood Criterion

Earlier we estimated parameters using the minimal sum of the squared residuals as a criterion of good fit. With maximum likelihood as the criterion of optimal fit, parameter estimation is a matter of finding the set of parameters for which the observed data are most likely. In order to apply the maximum likelihood method to parameter estimation we need two things:

1. A list of hypotheses to be considered with the model under consideration (i.e., what combinations of parameters we are going to trial).
2. A function required to calculate the probability density/likelihood of the observed data if the hypotheses were true.

It is usual to search over ranges of parameter values, focussing with more detail on combinations with the largest likelihoods just as with the non-linear parameter searches already described and the likelihood search in Example Box 3.5.

3.4.6 Likelihoods with the Normal Probability Distribution

As already shown, probability densities for the normal distribution are calculated from the familiar

$$L\{X \mid \mu, \sigma\} = \frac{1}{\sigma\sqrt{2\pi}} e^{\left(\frac{-(X-\mu)^2}{2\sigma^2}\right)} \tag{3.9}$$

where $L\{X|\mu,\sigma\}$ is the likelihood of any individual observation X, given μ the population mean and σ the population standard deviation (that is, the likelihood of the data X, given the hypothesis about μ and σ; Fig 3.12). Using this equation with a wide range of X values produces the bell-shaped curve everyone associates with the normal distribution (Fig. 3.12), the value of the likelihood is determined by the various possible combinations of mean and standard deviation values. The y-axis represents the relative likelihood of observing each specific X value if one is sampling a given population. This probability distribution can be used to estimate likelihoods wherever the error terms or residuals in a model are normally distributed.

As an example, we can consider the relationship between the caudal peduncle width and standard length of female orange roughy from the Lord Howe Rise. This linear relationship (Fig. 3.13) can be described by a very simple model, a one parameter linear regression (where C is the caudal peduncle width and S is the standard length)

$$C_i = bS_i + \varepsilon_i \qquad (3.10)$$

The residuals (ε_i) are assumed to be normally distributed with a mean of zero and variance σ^2, so if we use maximum likelihood methods there are two parameters to estimate: b and σ. The probability density of any particular observation C_i, given values for b, σ, and the data S_i, is $L\{C_i|b, \sigma, S_i\}$ and can be obtained

$$L\{C_i \mid b, \sigma, S_i\} = \frac{1}{\sigma\sqrt{2\pi}} e^{\left(\frac{-(C_i - bS_i)^2}{2\sigma^2}\right)} \qquad (3.11)$$

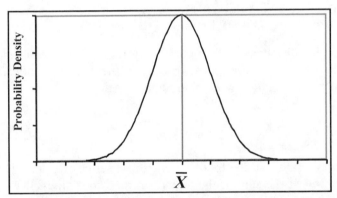

Figure 3.12 Likelihood distribution for the standard normal function (mean = 0 and variance = 1). Note the symmetrical distribution of values around the mean. The tick marks on the X axis are in units of standard deviations above and below the mean. The height on the likelihood curve indicates the relative likelihood of randomly obtaining a particular X value. A value of X approximately 2 standard deviations from the mean is approximately 1/20th as likely to occur as the mean value. Likelihoods have meaning only relative to one another.

The total probability density of all n observations given a particular pair of values for the parameters b and σ is just the product of the probability density for each of the n separate observations. Probability densities are calculated for independent events and, as such, have to be multiplied together and not added; hence, we do not use Σ the sum but instead use Π (capital Pi) the product

$$L\{C \mid b, \sigma, S\} = \prod_{i=1}^{n} \frac{1}{\sigma\sqrt{2\pi}} e^{\left(\frac{-(C_i - bS_i)^2}{2\sigma^2}\right)} \qquad (3.12)$$

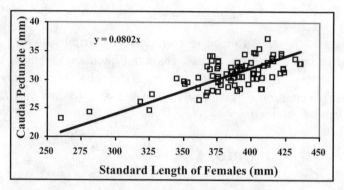

Figure 3.13 Relation between caudal peduncle and standard length for female orange roughy from Lord Howe Rise. The regression line has an $r^2 = 0.3855$, $F = 52.1$, $P < 0.0001$, df = 83. Data described in Haddon and Willis (1995).

As probability densities are commonly rather small numbers, the risk of rounding errors becomes great if many of them are multiplied together. So, given a function $f(X)$, we should remember that

$$\prod_{i=1}^{n} f(X_i) = e^{\sum Ln(f(X_i))}$$ (3.13)

which simply says that the Π, or product, of a series of values, is the same as the antilog of the sum of the logs of those same values. If we omit the antilog we would be dealing with log-likelihoods. Using this latter approach tiny numbers and the potential for rounding errors may be avoided. Eq 3.12 would take the form

$$LL\{C \mid b, \sigma, S\} = \sum_{i=1}^{n} Ln \left[\frac{1}{\sigma\sqrt{2\pi}} e^{\left(\frac{-(C_i - bS_i)^2}{2\sigma^2} \right)} \right]$$ (3.14)

Equation 3.14 can be simplified by expanding the logarithm and removing the terms that stay constant from the summation term as the parameters change.

or
$$LL = nLn \left(\frac{1}{\sqrt{2\pi}\hat{\sigma}} \right) + \frac{1}{2\hat{\sigma}^2} \sum_{i=1}^{n} \left[-\left[(C_i - bS_i)^2 \right] \right]$$ (3.15)

where σ-hat

$$\hat{\sigma}^2 = \frac{\sum_{i=1}^{n}(C_i - bS_i)^2}{n} \tag{3.16}$$

is the variance estimated from the data (hence the hat). By expanding Eq. 3.15, using Eq. 3.16, we can produce a simplification that makes for easier calculation. The summation term in Eq. 3.15 is cancelled by the inverse σ^2 from Eq 3.16 leaving -n/2. Further simplification is possible

$$LL = nLn\left(\left(\sqrt{2\pi}\hat{\sigma}\right)^{-1}\right) - \frac{n}{2} \tag{3.17}$$

giving

$$LL = -n\left(Ln\left(\sqrt{2\pi}\right) + Ln(\hat{\sigma})\right) - \frac{n}{2} \tag{3.18}$$

and finally

$$LL = -\frac{n}{2}\left[Ln(2\pi) + 2Ln(\hat{\sigma}) + 1\right] \tag{3.19}$$

The objective when fitting data to a model is to maximize the log-likelihood and any of Eqs. 3.14, 3.17, 3.18, or 3.19 could be used. Among people working with non-linear models, it appears to be a tradition to minimize instead of maximizing a criterion of fit. In practice, all this means is that one minimizes the negative log-likelihood (i.e., for normal errors remove the negative symbol from Eqs. 3.18 and 3.19).

The log-likelihood, LL{C|b, σ, S}, can be back-transformed to the likelihood

$$L\{C\,|,b,\sigma,S\} = e^{LL\{C|b,\sigma,S\}} \tag{3.20}$$

The optimum combination of b and σ can then be found by searching for the maximum of either the likelihood L{C|b,σ,S} or the log-likelihood (see Fig. 3.14), or the minimum of the negative log-likelihood. In Excel, with this problem, the result obtained was the same irrespective of whether Eq. 3.14 or Eq. 3.19 was used. This occurred despite there being 94 pairs of data points. Generally, however, with large samples, Eq. 3.19 would be the safest option for avoiding rounding errors and the machine-limits to numbers (the likelihoods can become very, very small during the search).

There is no correlation between the parameter b and the standard deviation σ, as evidenced by the circular contours of the log-likelihood surface (Fig. 3.14). The maximum likelihood produces the same estimate of *b* as the minimum sum of squared residual estimate (*b* = 0.080155).

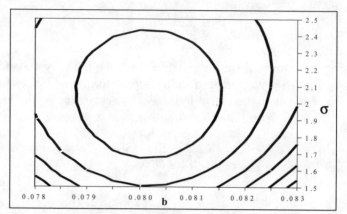

Figure 3.14 Circular maximum likelihood contours for the female orange roughy from the Lord Howe Rise morphometric data (caudal peduncle v standard length) given different values of the single parameter regression parameter *b* and the standard deviation of the residuals about the regression σ. Optimum fit at *b* = 0.08016 and σ = 1.990.

3.4.7 Equivalence with Least Squares

For linear and non-linear models, having normally distributed residuals with constant variance, fitting the model by maximum likelihood is equivalent to ordinary least squares. This can be illustrated with a straight line, normal regression model

$$y_i = \alpha + \beta x_i + \varepsilon_i \tag{3.21}$$

where the random deviations ε_i are independent values from a normal distribution with mean zero and variance σ^2 [$N(0, \sigma^2)$]. In other words, for a given value of the independent variable x_i, the response variable y_i follows a normal distribution with mean $\alpha + \beta x_i$ and variance σ^2, independently of the other responses. The probability distribution for the set of observations $y_1, y_2, ... y_n$, is formed from the product of the corresponding normal pdf values (likelihoods), so that the likelihood of the parameters α and β is given by

$$L\{y \mid \alpha, \beta, \sigma\} = \prod \frac{1}{\sigma\sqrt{2\pi}} e^{\left(\frac{-(y_i - \alpha - \beta x_i)^2}{2\sigma^2}\right)} \tag{3.22}$$

To fit by maximum likelihood we need to find the parameter values (α, β, and σ) that maximize the total likelihood (Eq. 3.22). Equivalently, we could maximize the sum of the logarithm of the likelihoods (or minimize the sum of the negative logarithms!)

$$LL\{y\mid\alpha,\beta,\sigma\}=nLn\left(\sigma\sqrt{2\pi}\right)+\frac{1}{2\sigma^{2}}\sum_{i=1}^{n}-\left(y_{i}-\alpha-\beta x_{i}\right)^{2} \qquad (3.23)$$

In Eq. 3.23 the only part of the equation that is not constant as parameters alter is the summation term (the other terms being constant can validly be removed from the summation). The summation term is equivalent to the sum of squared residuals used in standard linear regression. Therefore, fitting by maximum likelihood will produce an equivalent result to fitting the line which minimizes the sum of the squared deviations of the observations about the fitted line, i.e. fitting by ordinary least squares. This is the case for all models as long as their residual errors are normal, additive, and with constant variance.

3.4.8 Fitting a Curve Using Normal Likelihoods

The most commonly used equation used to describe growth in fisheries modelling is the von Bertalanffy growth curve

$$\hat{L}_{t}=L_{\infty}\left(1-e^{-K(t-t0)}\right)+\varepsilon \qquad (3.24)$$

where L_t is the expected size at age t, L_∞ is the average maximum size, K is a growth rate parameter, $t0$ is the hypothetical age at zero length, and ε is the normal error term (see Chapter 8). Kimura (1980) provided a set of data relating to the growth of male Pacific hake. To fit the von Bertalanffy curve using normal likelihoods, the observations X, are the observed length at age, and the expected values are derived from Eq. 3.24 (Example Box 3.6).

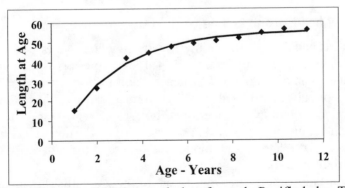

Figure 3.15 Kimura's (1980) growth data for male Pacific hake. The solid line represents the expected values (Example Box 3.6) using additive, normal random residual errors.

Example Box 3.6 Fitting a von Bertalanffy growth curve using normal likelihoods. Data on male Pacific hake from Kimura (1980). In column C put Eq. 3.24 =$B$1*(1-exp(-$B$2*(A6-$B$3))) and copy down to row 16. Sum column D in D3 (=sum(D6:D16). We use the equivalent of Eq. 3.16 in D2. The values in E1 and E3 will differ unless you put =D2 in B4. Why is that the case? Plot columns B as points and C as a line, against A (Fig. 3.15). Note that altering the value for σ has no effect on the location or shape of the curve, it only affects the relative likelihoods. Find the optimum fit by using the solver to minimize the negative log-likelihood in either E1 or E3. Alternatively, solve by minimizing the sum of the squared residuals in D3. Do the results obtained with maximum likelihood differ from those obtained by least squares?

	A	B	C	D	E
1	L_∞	55.000			=(11/2)*(Ln(2*PI())+2*Ln(D2)+1)
2	K	0.350		=sqrt(D3/11)	
3	t0	0.000		43.3607	=-sum(E6:E16)
4	σ	1.000		SSQ	**Maximum Likelihood**
5	Years	Length	Expect	Resid^2	**Ln(Likelihood)**
6	1	15.4	16.242	=(B6-C6)^2	=Ln(normdist(B6,C6,$B$4,false)
7	2	26.93	27.688	=(B7-C7)^2	=Ln(normdist(B7,C7,$B$4,false)
8	3.3	42.23	37.672	=(B8-C8)^2	-11.3074
9	4.3	44.59	42.789	3.243425	-2.5407
10	5.3	47.63	46.395	1.525007	-1.6814
11	6.3	49.67	48.936	0.538431	-1.1882
12	7.3	50.87	50.727	0.020470	-0.9292
13	8.3	52.3	51.989	0.096835	-0.9674
14	9.3	54.77	52.878	3.579456	-2.7087
15	10.3	56.43	53.505	8.557435	-5.1977
16	11.3	55.88	53.946	3.739299	-2.7886

3.4.9 Likelihoods from the Log-Normal Distribution

In fisheries models, the probability density function that is perhaps most commonly used to describe untransformed residual errors is the lognormal distribution. The distributions of catches, and of efforts, across a fleet are often lognormally distributed while catch-rates can usually be described using lognormal multiplicative errors (Fig. 3.16). Events that relate to each other in an additive manner tend to be described by the normal distribution, while those that relate in a multiplicative way tend to be described by the lognormal distribution.

Natural logarithmic transformation of lognormally distributed data generates a normal distribution (multiplications are converted to additions), thus, obviously, the lognormal distribution is closely related to the normal distribution. The probability density function for lognormal residual errors is (Hastings and Peacock, 1975)

$$L(x_i) = \frac{1}{x_i\sqrt{2\pi}\sigma}e^{\left[\frac{-\left(Ln(x_i)-Ln(m)\right)^2}{2\sigma^2}\right]} \tag{3.25}$$

where $L(x_i)$ is the likelihood of the data point x_i in question, m is the median of the variable, $m = e^\mu$ and $\mu = Ln(m)$, where μ is the mean of $Ln(x_i)$, and σ is the standard deviation of $Ln(x_i)$. Equation 3.25 is equivalent to the pdf for normal distributions, with the data log-transformed and divided by each x_i value.

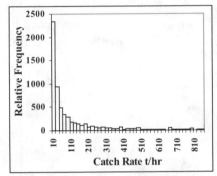 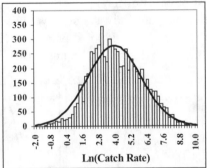

Figure 3.16 The left hand panel shows the raw data from south coast catches of Blue Warehou (*Seriolella brama*) from the Australian Fisheries Management Authority's trawl database for the Australian South East Fishery. That this is log-normally distributed is illustrated in the right hand panel, where the distribution is approximately normal with a mean (μ) of 3.93 and standard deviation (σ) of 2.0) after log-transformation.

The relationship between the normal and lognormal probability density functions can be seen in Eq 3.25. To use the normal pdf to generate pdf values for the lognormal distribution one would log transform the observed and expected values of the variable x, find the normal likelihood value and then, strictly, divide this by the untransformed observed x value. This last step, of dividing by the observed x value, has no effect on the optimisation of model fitting and is usually omitted. Strictly, however, if one wanted to plot up the pdf of a lognormal distribution Eq. 3.25 should be used to the full (Example Box 3.7; Fig. 3.17). There is the potential for confusion when describing the parameters of the lognormal distribution so care must be taken when doing so. The location parameter (position along the x-axis) is the median ($m > 0$) of the x-variate and not the mean, but the shape parameter σ ($\sigma > 0$) is the standard deviation of the log of the x-variate. Given a continuous variate x, the main parameters of the lognormal distribution are (Hastings and Peacock, 1975):

1. The median $\qquad\qquad\qquad\qquad m = e^{\mu}$ $\qquad\qquad\qquad$ (3.26)
    where μ is the mean of Ln(x).

2. The mode of the lognormal pdf $m / e^{\sigma^2} = e^{(\mu - \sigma^2)}$ $\qquad$ (3.27)
    where σ is the standard deviation of Ln(x).

3. The mean of the lognormal pdf $me^{(\sigma^2/2)} = e^{(\mu + \sigma^2/2)}$ $\qquad$ (3.28)

The lognormal distribution is always skewed to the right (Fig. 3.17). The most obvious difference between the lognormal and normal curves is that the lognormal is always positive. In addition, as the mode is determined by both the median and the σ parameter (Eq. 3.27), both can affect the location of the mode (Fig. 3.17; Example Box 3.7).

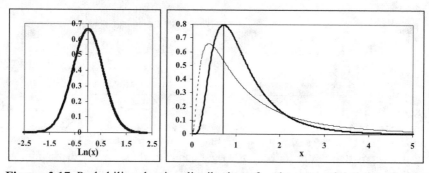

Figure 3.17 Probability density distributions for the same data expressed as a lognormal distribution in the right hand panel and as a normal distribution in the left hand panel after the x-variate was log-transformed. The thick line in both panels relates to data having a median (m) of 1 (i.e., $\mu = 0$) and a σ of 0.6. The thin vertical line in the right hand graph indicates the mode of $0.6977 = 1/e^{0.36}$. The thin curve in the right hand graph relates to parameters m = 1 and σ = 1. The x-axes in both graphs have been truncated to areas where the probability density or likelihood values were significantly greater than zero (see Example Box 3.7). As the variance increases for a given median, the mode moves towards zero and the curve skews further to the right.

The likelihood equation can be simplified as with the normal likelihoods. In fact, given the appropriate transformation one can use the equivalent of Eq. 3.19

$$LL = -\frac{n}{2}\left[Ln(2\pi)+2Ln(\hat{\sigma})+1\right]+\sum_{i=1}^{n}\frac{1}{x_i} \qquad (3.29)$$

where

$$\hat{\sigma}^2 = \sum_{i=1}^{n}\frac{\left(Ln(x_i)-Ln(\hat{x}_i)\right)}{n} \qquad (3.30)$$

note the maximum likelihood version of σ^2 using n instead of n-1 (Example Box 3.5).

Example Box 3.7 Relationship between the normal pdf and the lognormal pdf. Column C is filled from C8:C1007 with numbers starting from 0.01 down to 10 in steps of 0.01. Column A is the natural log of column C, and column B is the normal likelihood (probability density) =normdist(A8,$D$4,$D$5,false). Column D is the lognormal likelihood (Eq. 3.24) which can also be found by dividing the normal likelihood in column B by the value of the observed data x in column C (i.e. =B8/C8 in D8). Try this approach and compare it with the outcome of Eq. 3.24. Column E uses the Excel function (e.g., =LogNormDist(C8, $D$4,$D$5) in E8) which provides the cumulative distribution by default. Notice that this function uses μ instead of the median m. Plot column B against A as one graph and then column D against C, and column B against C as continuous lines. The first two should mimic Fig. 3.17. Modify cells D3 and D5 and observe the effects on the shape and location of the different curves. Note the effect of division by x (column D). Note the value in D6. Why does it have this value (cf. Example Box 3.3)? Compare the value of D2 with the mode on the graph of column D; try putting 0.45 in D5, what is the impact on the sum in D6.

	A	B	C	D	E
1	Mean of the pdf			=D3*Exp((D5^2)/2)	
2	Mode of the pdf			=D3/(Exp(D5^2))	
3	Median m, of the pdf			1	
4	Ln(m) = μ =Avg(Ln(x))			=Ln(D3)	
5	σ = StDev(Ln(x))			1	
6				=sum(D8:D1007)	
7	Ln(x)	Normdist(μ,σ)	X	NormDist/x	Lognormdist
8	=Ln(C8)	=normdist(A8,d4,d5,false)	0.01	=B8/C8	2.06279E-06
9	=Ln(C9)	=normdist(A9,d4,d5,false)	0.02	=B9/C9	4.57817E-05
10	-3.50656	0.000853	0.03	0.0284287	0.00022702
11	-3.21888	0.002244	0.04	0.0560987	0.00064353
12	-2.99573	0.004489	0.05	0.0897783	0.00136900
13	Copy these columns down to row 1007		0.06	Copy down to Row 1007	

3.4.10 Fitting a Curve Using Log-Normal Likelihoods

Fitting a curve using lognormal residual errors is very similar to using normal random likelihoods. Recruitment in fisheries is notoriously variable with occasional very large year classes occurring in some fisheries. Stock recruitment relationships are generally taken to exhibit lognormal residual errors. Numerous stock recruitment relationships have been described (see Chapter 9) but here we will restrict ourselves to the most commonly used equation, that by Beverton and Holt (1957). As an example, we will attempt to fit a stock recruitment curve to some real data. Hilborn and Walters (1992) indicate that there can be more than one form of the Beverton and Holt stock recruitment equation but all would be expected to have lognormal residual errors. We will use the following version

$$R_i = \frac{aS_i}{b + S_i} e^{N(0,\sigma^2)} \tag{3.31}$$

where R_i is the recruitment in year i, S_i is the spawning stock size that gave rise to R_i, a is the asymptotic maximum recruitment level, and b is the spawning stock size that gives rise to 50% of the maximum recruitment. The residual errors are lognormal with a μ of 0 and variance σ^2 (Example Box 3.8; Fig. 3.18). Penn and Caputi (1986) provide data on the stock recruitment of Exmouth tiger prawns. They used a Ricker stock-recruitment relationship but we will use the Beverton and Holt relationship (Eq. 3.31).

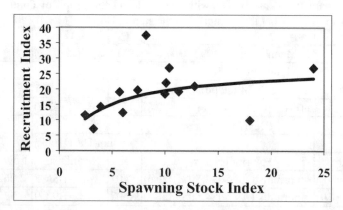

Figure 3.18 A Beverton and Holt stock recruitment relationship (Eq. 3.31) fitted to data from Penn and Caputi (1986) on Exmouth Gulf tiger prawns (*Penaeus semisulcatus*). The outliers to this relationship were thought to be brought about by extreme environmental conditions (see Chapter 9).

The estimated parameters obtained by using lognormal likelihoods are identical to those obtained through using either the minimum sum of squared residuals (on log-transformed data) or through using normal likelihoods on log-transformed data (i.e., omitting the division by the observed recruitment as in Eq. 3.25; see Example Box 3.8).

Example Box 3.8 Fitting a Beverton and Holt stock recruitment relationship using lognormal residual errors. The predicted recruitment values are in column C as =($B$1*A5)/($B$2+A5) and copied down. In column D put the log-likelihood =Ln(normdist(Ln(B5),Ln(C5),$B$3,false)). The natural logarithmic-transformations are important; without them one would be using normal random errors and not lognormal. Finally, in E1, put the negative of Eq. 3.29 =(C1/2)*(Ln(2*pi())+2*Ln(E2)+1). Plot column B against A (as points) and C against A as a line (*cf.* Fig. 3.18). Using the solver and minimizing D1 by varying cells B1:B3 can find the optimum line. If E1 is minimized instead, only cells B1:B2 need be varied (σ is estimated directly). If you want D1 and E1 to be the same then put =E2 into B3. To generate the strict lognormal likelihoods the normal likelihoods need to be divided by each spawning value i.e., =Ln(normdist(Ln(B5),Ln(C5),$B$3,false)/A5). While this alters the likelihoods generated, compare the results obtained when using this version to those obtained using the simpler version. Were any differences found? Try minimizing the sum of squared residuals in cell E3, are the results different from those obtained using maximum likelihood?

	A	B	C	D	E
1	a	27.366	=count(E5:E18)	=-sum(D5:D18)	5.2425
2	b	4.0049			=sqrt(E3/C1)
3	σ	0.3519			=sum(E5:E18)
4	Spawn	Recruit	Expect	LL	resid2
5	2.4	11.6	10.25	0.0642	=(Ln(B5)-Ln(C5))^2
6	3.2	7.1	12.15	-1.0416	=(Ln(B6)-Ln(C6))^2
7	3.9	14.3	13.50	0.1122	=(Ln(B7)-Ln(C7))^2
8	5.7	19.1	16.07	0.0053	0.0298
9	6	12.4	16.41	-0.1917	0.0786
10	7.4	19.7	17.76	0.082	0.0108
11	8.2	37.5	18.39	-1.9258	0.5080
12	10	18.5	19.54	0.1134	0.0030
13	10.1	22.1	19.60	0.0671	0.0145
14	10.4	26.9	19.76	-0.259	0.0952
15	11.3	19.2	20.21	0.115	0.0026
16	12.8	21	20.84	0.1253	0.0001
17	18	9.9	22.39	-2.5626	0.6657
18	24	26.8	23.45	0.0537	0.0178

3.4.11 Likelihoods with the Binomial Distribution

Very many texts introduce ideas relating to maximum likelihood estimation with a worked example using the Binomial distribution. This seems to occur for two reasons. The first is that the binomial distribution can relate to very simple real examples, such as tagging/recapture experiments, where single parameters are to be estimated and only single likelihoods need be considered (i.e., no products of multiple likelihoods are required). The second is that the values calculated for this discrete distribution are true probabilities and not just probability densities. Thus, the complication of identifying likelihoods that are different from true probabilities is avoided. Remember that this distinction is not necessary with discrete probability distributions where the probability density function can only generate values (probabilities) for the possible discrete events. In fisheries stock assessment most analytical situations would need to use continuous probability density functions (pdfs), but there are situations where discrete pdfs, such as the binomial distribution, are necessary.

In situations where a study is determining whether an observation is true or false (a so-called Bernoulli trial; e.g., a captured fish either has or does not have a tag) and the probability of success is the parameter p, then it would generally be best to use the Binomial distribution. The Binomial probability density function generates true probabilities and is characterized by two parameters, n the number of trials and p the probability of success in a trial (an event proving to be true)

$$P\{m\,|\,n,p\} = \left[\frac{n!}{m!(n-m)!}\right] p^m \left(1-p\right)^{(n-m)} \tag{3.32}$$

which is read as the probability of m events proving to be true out of n trials (a sample of n) where p is the probability of an event being true (see Example Box 3.9). The term (1-p) is often written as q, that is (1-p)=q. The term in the square brackets in Eq. 3.32 is the number of combinations that can be formed from n items taken m at a time, and is sometimes written as

$$\binom{n}{m} = \frac{n!}{m!(n-m)!} \tag{3.33}$$

It is always the case that n >= m, one cannot have more successes than trials. The distribution can vary in shape from highly skewed to approximately normal (Fig. 3.19; Example Box 3.9).

Tagging programmes, designed to estimate the size of a population, provide a common example of where a Binomial distribution would be used in fisheries. Thus, if one has tagged a known number of animals, n_1, and later obtained a further sample of n animals from the population it is possible to estimate the total population, assuming all of the assumptions of this form of tagging manipulation have been kept. The observation is of m, that is, how many in a sample of n are tagged (i.e., tagged = true, untagged = false).

Example Box 3.9 Examining the properties of the Binomial probability density function (Eq. 3.32). The parameters of the Binomial are n, the number of trials/observations, and p, the probability of a successful trial or observation. Set up a worksheet as below, where the column of m values (the number of observed successes) stretches from 0 to n down column A. Down column B insert and copy down =binomdist(A4, $B$1, $B$2, false) to obtain the likelihoods or probabilities directly (look up this function in the Help). Down column C insert and copy down = fact($B$1)/(fact(A4)*fact($B$1-A4)) to obtain Eq. 3.33. Put =($B$2^A4)*(1-$B$2)^($B$1-A4) into D4 and copy down, and then put =C4*D4 in E4 and copy down to obtain the likelihoods again. By plotting Column B against Column A as a simple scatterplot you should be able to mimic Fig. 3.19. Vary the value of p and observe how the distribution changes. Vary n (adjusting the length of the columns of numbers to match the n value) and see how above a value of 170 the binomdist() function operates beyond that of the fact() function. Look ahead to Equation 3.41 (the log-likelihood) and implement that on this worksheet to see how it handles the larger values of n.

	A	B	C	D	E
1	n	20			
2	p	0.25			
3	m	Binomdist	FactTerm	p(1-p)term	Likelihood
4	0	0.003171	1	0.003171	0.00317
5	1	0.021141	20	0.001057	0.02114
6	2	0.066948	190	0.000352	0.06695
7	3	0.133896	1140	0.000117	0.13390
8	4	0.189685	4845	3.92E-05	0.18969
	Extend down to equal n	Copy down to match Column A			

A real example of such a study was made on the New Zealand fur seal pup (*Arctocephalus forsteri*) population on the Open Bay Islands off the west coast of the South Island of New Zealand (Greaves, 1992). The New Zealand fur seal appears to be recovering after having been badly over-exploited in the 19th century with new haul-out sites starting to be found in the South and North Island. The fishery officially closed in 1894 with complete protection within the New Zealand Exclusive Economic Zone beginning in 1978 (Greaves, 1992). In cooperation with the New Zealand Department of Conservation, Greaves journeyed to and spent a week on one of these offshore islands. She marked 151 fur seal pups by clipping away a small patch of guard hairs on their heads, and then conducted a number of colony walk-throughs to resight tagged animals (Greaves, 1992). Each of these walk-throughs constituted a further sample of varying sizes, and differing numbers of animals were found tagged in each sample (Table 3.1). The question is, what is the size of the fur seal pup population (X) on the Island?

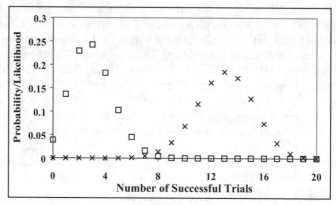

Figure 3.19 Two examples of the Binomial distribution. The left hand set of squares are for n = 20 and p = 0.15, while the right hand set of crosses are for n = 20 and p = 0.65. Note that zero may have a discrete probability, and that large p values tend to generate discrete, approximately normal distributions.

All the usual assumptions for tagging experiments are assumed to apply i.e., we are dealing with a closed population - no immigration or emigration, with no natural or tagging mortality over the period of the experiment, no tags are lost and tagging does not affect the recapture probability of the animals. Greaves (1992) estimated all of these effects and accounted for them in her analysis. Having tagged and resighted tags in a new sample, a deterministic answer can be found with the Peterson estimator (Caughley, 1977; Seber, 1982)

$$\frac{n_1}{X} = \frac{m}{n} \qquad \therefore \qquad X = \frac{n_1 n}{m} \tag{3.34}$$

where n_1 is the number of tags in the population, n is the subsequent sample size, m is the number of tags recaptured, and X is the population size. An alternative estimate adjusts the counts on the second sample to allow for the fact that in such cases we are dealing with discrete events. This is Bailey's adjustment (Caughley, 1977)

$$X = \frac{n_1 (n+1)}{m+1} \qquad (3.35)$$

Like all good estimators, it is possible to estimate the standard error of this estimate and thereby generate 95% confidence intervals around the estimated population size

$$\text{StErr} = \sqrt{\frac{n_1^2 (n+1)(n-m)}{(m+1)^2 (m+2)}} \qquad (3.36)$$

Instead of using the deterministic equations, a good alternative would be to use maximum likelihood methods to estimate the population size X, using the Binomial probability density function. We will continue to refer to these values as likelihoods even though, in this case, they are also true probabilities.

Table 3.1 Counts of New Zealand fur seal pups made by Greaves (1992) on Open Bay Island, West Coast, South Island New Zealand. She had tagged 151 animals (i.e., n_1 = 151). The column labeled n is the subsequent sample size, m is the number of tags re-sighted, X is the population size, and 95%L and 95%U are the lower and upper 95% confidence intervals, respectively, calculated as ± 1.96 times StErr, the standard error. Calculations as per Eqs. 3.33 and 3.34. The first two rows are individual independent samples, while the last row is the average of six separate counts (data from Greaves, 1992). The average counts lead to population estimates that are intermediate in value.

m	n	X	95%U	95%L	StErr
32	222	1020	704	1337	161.53
31	181	859	593	1125	135.72
29	185	936	634	1238	153.99

We are only estimating a single parameter, X, the population size, and this entails searching for the population size that is most likely given the data. With the binomial distribution, $P\{m|n,p\}$, Eq. 3.32 provides the probability of observing m tagged individuals from a sample of n from a population with proportion p tagged (Snedecor and Cochran, 1967; Hastings and Peacock, 1975). If one implemented this exact equation in a spreadsheet, adjusting the data to match Eq. 3.35, we would quickly meet the limitations of computer arithmetic. For example, in Excel, one can use the =Fact() function to calculate the factorial of numbers up to 170, but beyond that leads to a numerical overflow as the result is too large to be represented as a normal real number. Fortunately, Excel provides a =binomdist() function that can operate with much larger numbers. It seems likely that the binomial probabilities have been implemented as log-likelihoods for the calculation and then back-transformed. Given

$$P\{m \mid n, p\} = \left[\frac{n!}{m!(n-m)!} \right] p^{m} (1-p)^{(n-m)} \tag{3.37}$$

log-transforming the component terms

$$Ln\left(p^{m} (1-p)^{(n-m)} \right) = mLn(p) + (n-m)Ln(1-p) \tag{3.38}$$

and

$$Ln\left[\frac{n!}{m!(n-m)!} \right] = Ln(n!) - \left(Ln(m!) + Ln((n-m)!) \right) \tag{3.39}$$

noting that

$$Ln(n!) = \sum_{i=1}^{n} Ln(i) \tag{3.40}$$

we obtain the log-likelihood

$$LL\{m \mid n, p\} = \sum_{i=1}^{n} Ln(i) - \left(\sum_{i=1}^{m} Ln(i) + \sum_{i=1}^{n-m} Ln(i) \right) \tag{3.41}$$
$$+ mLn(p) + (n-m)Ln(1-p)$$

The proportion p of fur seal pups that are marked is, in this case, $p = 151/X$, and, with the first example in Table 3.1, n_1 is 151, and with the Bailey correction, n is 222+1, and m is 32+1 (see Eq. 3.35). The maximum likelihood estimate of the actual population size X is determined by searching for the value of X for which $P\{m|n,p\}$ is maximized. Making the likelihood equation (3.32) explicit we can write

$$L\{data \mid X\} = \left[\frac{223!}{33!(223-33)!} \right] \left(\frac{151}{X} \right)^{20} \left(1 - \frac{151}{X} \right)^{(223-33)} \tag{3.42}$$

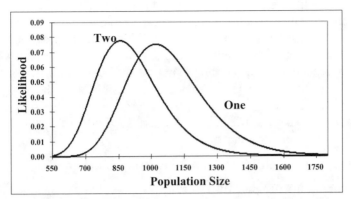

Figure 3.20 Likelihood distribution against possible population size X for two estimates of fur seal pup population size made through a tagging experiment. The right hand curve is that from 151 pups tagged, with a subsequent sample of 222 pups found to contain 32 tagged animals (with a mode of 1020). The left hand curve is for the same 151 tagged pups, with a subsequent sample of 181 pups containing 31 tagged animals (with a mode of 859). The modes are the same as determined by Eq. 3.35 (Example Box 3.10).

Using Eq. 3.37, and setting population size X to different values between 550 and 1800, in steps of 1 (1251 hypotheses), we gain a set of likelihoods to be plotted against their respective X values (outer curve in Fig. 3.20). This can be repeated for the alternative set of observations from Table 3.1 (see Fig. 3.20; Example Box 3.10).

3.4.12 Multiple Observation

When one has multiple surveys, observations, or samples, or different types of data, and these are independent of one another, it is possible to combine the estimates to improve the overall estimate. Just as with probabilities the likelihood of a set of independent observations is the product of the likelihoods of the particular observations (Eq. 3.43).

$$L\{O_1, O_2, ..., O_n\} = L\{O_1\} \times L\{O_2\} \times ...... \times L\{O_n\} \qquad (3.43)$$

This can be illustrated with the New Zealand fur seal pup-tagging example (Example Box 3.10). The two independent re-sampling events, listed in Table 3.1, can be combined to improve the overall estimate. Instead of just taking the average of the observations and putting those values through the deterministic equations (Table 3.1) the independent likelihood analyses can be combined using Eq. 3.43. Thus, for each trial value of X, the separate likelihoods for each observation can be multiplied to give a joint likelihood (Fig. 3.21; Example Box 3.10).

In this instance, where there were six separate sets of observations made, these could all, in theory, be combined to improve the overall estimate. In practice, it might be argued that all six samples were not strictly independent and so should not really be combined. As with all analyses, as long as the procedures are defensible then the analyses can proceed (i.e., in this case it could be argued that the samples were independent – taken sufficiently far apart so there was no learning the locations of tagged pups, etc.).

3.4.13 Percentile Confidence Intervals Using Likelihoods

A consideration of the distribution of likelihoods against different hypothesized values of the parameters of interest (e.g., Fig. 3.21) demonstrates that as one moves away from the optimum parameter value the likelihoods decline in a smooth fashion. Such plots of likelihood against the parameter of interest are known as likelihood profiles. Approximate confidence intervals around parameters may be obtained by standardizing the likelihoods so they add to one (i.e., sum all the separate likelihoods and divide each one by the total), and then finding the particular parameter values that encompass the confidence intervals required (usually the central 95% of the total distribution). This approaches validity only if the increments on the parameter axis are relatively small so that the likelihood distribution approximates a continuous probability distribution (in the example of the fur seal pups the population size, X, can only increase in steps of whole units). This is an empirical likelihood profile method. In addition, the likelihoods at the extremes of the parameter values considered should be negligible.

The standardized likelihoods should cumulate to one and the 95% confidence intervals may be determined by searching for the 2.5 and 97.5 percentiles of the distribution (Example Box 3.10). An important difference between likelihood profile confidence intervals and those generated using 1.96 x the standard error is that the latter are forced by their nature to be symmetrical. This is not necessarily the case when using likelihood profile confidence intervals (Fig. 3.21).

3.4.14 Likelihood Profile Confidence Intervals

The uncertainty associated with a parameter estimate can be quantified by generating confidence intervals around the parameter of interest. There are a number of ways of generating confidence intervals around parameters when estimating those parameters with the maximum likelihood criterion. The previous section dealt with using empirical likelihood profiles to generate percentile confidence intervals. An alternative is to use a deterministic approximation.

Example Box 3.10 Using the Binomial distribution to estimate population size and confidence intervals. Column A must to be extended in steps of 1 down to row 1261 so that the 1251 hypothesized population sizes can have their relative likelihood calculated. Put =binomdist(B$7,B$6,B$8/$A11,false) in column B and copy across into column C (note the $ symbols and their order). The joint likelihoods are simply the separate values multiplied in column D and these have been standardized to sum to one by dividing through by the sum of column D. Plot columns B and C against column A to obtain a graph akin to Fig. 3.20. Vary the parameters in cells B6:C8 and consider how the likelihood profiles change. In the column next to the standardized likelihoods set out their cumulative distribution (i.e., in F11 put =E11, in F12 put =F11+E12, and copy down). Then search for the rows in which the values 0.025 and 0.975 occur. What do the population sizes at those cumulative likelihoods represent?

	A	B	C	D	E
1	**Experiment**	**One**	**Two**		
2	**Deterministic**	1020.39	858.81		
3	**Determ StErr**	161.530	135.722		
4	**Upper 95%**	1336.99	=C2+1.96*C3		
5	**Lower 95%**	703.80	=C2-1.96*C3		
6	**Sample n**	223	182		
7	**Tags found m**	33	32		
8	**Tagged p**	151	151		
9	**Σ Likelihoods**	31.8265	=sum(c11:c1261)	1.1890	1
10	**Pop. Size**	**One**	**Two**	**Joint**	**Std Joint**
11	550	2.91E-06	0.000589988	=B11*C11	=D11/$D$9
12	551	3.12E-06	0.000617058	=B12*C12	=D12/$D$9
13	552	3.35E-06	0.000645152	=B13*C13	=D13/$D$9
14	553	3.59E-06	0.000674299	Copy down	Copy down
15	Extend down	4.12E-06	0.000735869	3.03E-09	2.55E-09
16	to 1800 or to	4.41E-06	0.000768352	3.39E-09	2.85E-09
17	Row 1261	4.72E-06	0.000802007	3.78E-09	3.18E-09

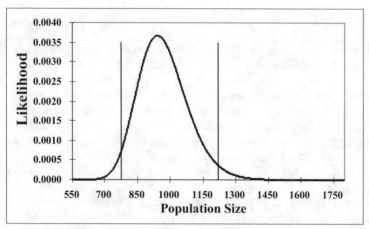

Figure 3.21 The combined likelihoods vs. possible population size for the two sets of observations made on the tagged population of New Zealand fur seal pups. The two sets of likelihood values from Fig. 3.20 were multiplied together at each hypothesized instance of population size to produce this composite likelihood distribution. Note the changed scale of the Likelihood axis. Also shown are the 95% percentile confidence intervals derived from the empirical likelihood profile. Note how the upper and lower intervals at not equal as would be the case if we used the usual standard error approach (Example Box 3.10).

Venzon and Moolgavkar (1988) describe a method of obtaining what they term "approximate 1 - α profile-likelihood-based confidence intervals." This simple procedure starts with finding the set of model parameters that generate the maximum likelihood for the data. The method then relies on the fact that likelihood ratio tests asymptotically approach the χ^2 distribution as the sample size gets larger. This means that with the usual extent of real fisheries data this method is only approximate. Likelihood ratio tests are, exactly as their name suggests, a ratio of two likelihoods or, if dealing with log-likelihoods the subtraction of one from another, the two are equivalent

$$\frac{L(\theta)_{Max}}{L(\theta)} = e^{\left(LL(\theta)_{Max} - LL(\theta)\right)} \tag{3.44}$$

where $L(\theta)$ is the likelihood of the θ parameter, and the *Max* subscript denotes the maximum likelihood (assuming all other parameters are also optimally fitted), and $LL(\theta)$ is the equivalent log-likelihood. The expected log-likelihoods for the actual confidence intervals for a single parameter, assuming all others remain at the optimum, are given by the following

(Venzon and Moolgavkar, 1988)

$$2\times\left[LL(\theta)_{Max} - LL(\theta)\right] \le \chi^2_{1,1-\alpha}$$

$$LL(\theta) = LL(\theta)_{Max} - \frac{\chi^2_{1,1-\alpha}}{2}$$

(3.45)

where $\chi^2_{1,1-\alpha}$ is the $(1-\alpha)^{th}$ quantile of the χ^2 distribution with 1 degree of freedom (e.g., for 95% confidence intervals $\alpha = 0.95$ and $1-\alpha = 0.05$, $\chi^2_{1,1-\alpha} = 3.84$). For a single parameter θ_i, the approximate 95% confidence intervals are therefore those values of θ_i for which two times the difference between the corresponding log-likelihood and the overall optimal log-likelihood is less than or equal to 3.84 ($\chi^2_{1,0.05} = 3.84$). Alternatively, one can search for the θ_i that generates a log-likelihood equal to the maximum log-likelihood minus half the required χ^2 value (i.e., $LL(\theta)_{Max} - 1.92$; see Eq. 3.45; Example Box 3.11).

As there is often parameter correlation within fisheries models (and many other disciplines) the assumption that keeping other parameters at the value for which the overall optimal fit was obtained will introduce bias such that the interval estimates would tend to be smaller than they should be (Fig 3.22).

The likelihood profile method may be applied to more than one parameter to avoid the problem of parameter correlation. The search for values that satisfy Eq 3.45 (e.g., with χ^2 set at 5.99 for two parameter, 2 degrees of freedom) is somewhat more complicated but Excel's solver could be set to search for the requisite values if it were started in the correct directions.

Clearly, when considering more than one parameter the confidence intervals become wider (Fig. 3.22). This only occurs if there is any correlation (covariance) between the parameters. If there were no parameter correlation (*cf*. Fig. 3.14) then there would be no interference between the parameters and the confidence intervals would be unbiased.

An alternative approach to producing confidence intervals is to use bootstrapping methods, which have the advantage of automatically accounting for changes in all parameters at once. However, this has the disadvantage of being a computer intensive method, although with improvements in computer speeds this is now less of a problem (see Chapter 6).

Example Box 3.11 Approximate likelihood profile confidence intervals. Extend Example Box 3.10 as below. Instead of enumerating all possible population sizes X, and calculating the binomial likelihood for each one use the solver to maximize G4 (the binomial likelihood) by changing cell F4, the hypothesized population size. One can either maximize the likelihood in G4 or the log-likelihood in H4. Given the optimum parameter estimate then we can calculate the likelihood expected at the 95% confidence intervals for X the population size in cell H7. In this instance the optimum LL is –2.5897 and subtracting 1.92 from that gives – 4.5104. Run the solver again only this time set the target cell to H4, and have the solver vary cell F4 until the target cell has a value of –4.5104. If the search finds the upper confidence interval you will need to set the starting value of X below the optimum in order that the search will move to the lower confidence interval. Compare the results with the percentile confidence intervals produced in Example Box 3.10. The values in G8 and H8 relate to using simple likelihoods instead of log-likelihoods. Write out the equations and reconcile them with Eq. 3.45.

	F	G	H
3	X	Likelihood	LL
4	1020.394	=binomdist($B$7,$B$6,$B$8/F4,false)	=Ln(G4)
5			
6	d of f	Chi2	Required LL
7	1	=CHIINV(0.05,F7)	=H4-G7/2
8		=G4/sqrt(exp(G7))	=exp(H7)

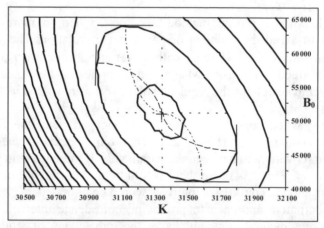

Figure 3.22 Maximum likelihood contours for two parameters from a surplus-production model of northern Australian tiger prawns from the Gulf of Carpentaria Chapter 10). With no parameter correlations the contours would be perfect circles (*cf.* Fig 3.14). The second central circle represents the 95% confidence intervals. The vertical and horizontal dotted lines inside this circle indicate the likelihood profile confidence intervals for single parameters (holding the other parameters constant) while the S-shaped dashed lines illustrate the true two-parameter confidence intervals that are wider than the single parameter intervals.

3.4.15 Likelihoods from the Poisson Distribution

The Poisson distribution is another discrete statistical distribution whose probability density function generates actual probabilities. As with the binomial distribution, however, we will continue to refer to these as likelihoods.

The Poisson distribution is often used in ecology to represent random events. Most commonly, it will be used to describe the spatial distribution of organisms (if the mean density is roughly the same as the variance of the density, this is taken to indicate a random distribution) (Seber, 1973). It reflects one of the properties of the Poisson distribution, which is that the expectation of the distribution (its mean) is the same as its variance (Hastings and Peacock, 1975). A variate X describes the number of events per sample and this can only assume values from 0 and upwards. To be distributed in a Poisson fashion the variable should have two properties: it should be rare, that is, its mean value should be small relative to the number of events possible, and each event must be independent of the other events – that is, each event must be random with respect to each other. The Poisson probability density function has the following form

$$P(X) = \frac{\mu^{X}}{e^{\mu} X!} \tag{3.46}$$

where X is the observed number of events and μ is the expected or mean number of events. As with the Binomial distribution the Poisson contains a factorial term, so the possibility of very small and very large numbers is a problem when computing the values of the distribution. The log-transformation solves this problem

$$L\{X \mid \mu\} = \frac{\mu^{X}}{e^{\mu} X!} \tag{3.47}$$

Expanding the terms

$$LL\{X \mid \mu\} = Ln(\mu^{X}) - Ln(e^{\mu} X!) \tag{3.48}$$

and simplifying

$$LL\{X \mid \mu\} = X.Ln(\mu) - \mu - \sum_{i=1}^{X} Ln(i) \tag{3.49}$$

Hilborn and Walters (1992) present an hypothetical example about the analysis of a tagging-multiple-recapture tagging study, where the Poisson distribution could be used in a fisheries context. We will use a similar example to illustrate the importance of selecting the correct probability density function to represent the residual errors for a modelled situation.

Example Box 3.12 The properties of the Poisson distribution. We can compare the Excel Poisson function with the result of the log-likelihood calculation. Name cell B1 as mu. In column A put potential values of X from 0 to 50 down to Row 53. In column B put Eq 3.49 and copy down, back transform it in column C to obtain the Likelihood or probability of the X value. In column D use the Excel Poisson function for comparison. Plot column C against A as points to obtain a graph like Fig. 3.23. Vary the value of μ and observe the effect on the distribution of values. How does the spread or variance of the distribution change as μ increases?

	A	B	C	D
1	μ	4		
2	X	Log-Likelihood	P(X)	P(X)
3	0	=(A3*Ln(mu))-(mu+Ln(fact (A3)))	=exp(B3)	=poisson(A3,mu,false)
4	1	=(A4*Ln(mu))-(mu+Ln(fact (A4)))	=exp(B4)	=poisson(A4,mu,false)
5	2	-1.9206	0.1465	0.1465
6	3	-1.6329	0.1954	0.1954
7	4	-1.6329	0.1954	0.1954
8	5	-1.8560	0.1563	0.1563
9	6	-2.2615	0.1042	0.1042
10		Extend and copy these columns down to row 53 (where X = 50)		

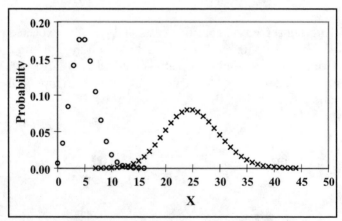

Figure 3.23 Two examples of Poisson distributions. The left hand distribution has a mean of 5, and the right hand distribution a mean of 25. Note the top two values in each case (i.e., 4 and 5, and 24 and 25) have the same probabilities. The larger the value of μ the closer the approximation is to a discrete normal distribution (Example Box 3.12).

The objective of the tagging-multiple-recapture analysis here is to estimate the constant rate of total mortality consistent with the rate of tag returns. Such surveys are a form of simplified Schnable census (Seber, 1973). It is simplified because while it is based around a multiple-recapture tagging study there is only a single tagging event. By tagging a known number of animals and tracking the numbers being returned in a set of equally spaced time periods (could be weeks, months, or years), an estimate of the total mortality experienced by the animals concerned can be determined. We will assume 200 animals were tagged and 57, 40, 28, 20, 14, 10, and 6 fish were recaptured. In total, 25 tags were therefore not retaken. The instantaneous total rate of mortality (the parameter to be estimated) is assumed to apply at a constant rate through the sampling period. As time passes the number of fish alive in a population will be a function of the starting number and the number dying. As we saw in Chapter 2, this relationship can be represented as

$$N_t = N_0 e^{-Zt} \quad \text{or} \quad N_{t+1} = N_t e^{-Z} \tag{3.50}$$

where N_t is the number of fish alive at time t, N_0 is the number of fish alive at the start of observations, and Z is the instantaneous rate of total mortality. This is a straightforward exponential decline in numbers with time where the numbers at time t are being multiplied by the survivorship to give the numbers at time t+1. If tagging has no effect upon catchability, this relationship (Eq. 3.50) can be used to determine the expected rate of return of the tags.

The expected number of tags captured at time t (C_t) given N_t and Z, is simply the difference between the expected number of tagged fish alive at time t and at time t + 1

$$C_t = N_t - N_{t+1} = N_t - N_t e^{-Z} = N_t \left(1 - e^{-Z}\right) = \mu \tag{3.51}$$

Thus, μ is simply the number of tags multiplied by the complement of the survivorship between periods. The Poisson distribution can give the probability of capturing X tags, given μ. For example, with 200 tags in the population and a $(1-e^{-Z})$ of 0.05 we have $\mu = 200 \times 0.05 = 20$, we would expect to recapture 20 tags. However, we can calculate the likelihood of only capturing 18 tags

$$L\{18 \mid \mu = 20\} = \frac{20^{18}}{e^{20}18!} = 0.08439 \tag{3.52}$$

By plotting likelihood/probability against hypothesized fishing mortality values a maximum can be seen to occur at approximately $F=0.37$ (Fig. 3.24).

Example Box 3.13 A comparison of normal random residual errors and Poisson distributed residuals. Data from a hypothetical tagging program. The Tags column records the number of tags returned over equal periods of time. Name cell C1 as Z, and cell F1 as N0. Fill columns D and E as shown. In C8 put the equivalent to Eq. 3.49 =(B8*Ln(D8))-(D8+Ln(fact(B8))) and copy down. This contains the factorial function limited to a maximum X of 170. One could always use the Excel function: =Poisson(X, μ, false) to obtain the probabilities directly. Cell C3 is the negative sum of the log-likelihoods (=-sum(C8:C14)), and F3 the sum of the squared residuals (=sum(F8:F14)). Use the solver to minimize the negative log-likelihoods and the squared residuals in turn and determine whether the results are the same. Plot column F against column D to observe the impact on the squared residuals of using the different residual structures. To observe the residuals proper create a new column with =(B8-D8) in it and plot that against column D. When the least squares method is used the residuals are symmetrically and approximately equally arranged about the expected values. With the Poisson residuals there is an obvious trend with the expected value. Can you explain this in terms of the properties of the two statistical distributions? Put 400 in F1 and double each of the observed data values. Does the optimal solution change? What would be the advantage of tagging a larger sample at the start of the survey (*cf.* Fig. 3.24)?

	A	B	C	D	E	F
1		Z	0.3479		N_0	200
2						
3		ΣLL	17.103		SSQ	9.6696
4						
5						
6	t	Tags	P(Tags)	$C_t = \mu$	Nt	$(\mu - Tags)^2$
7	0				=N0*exp(-(Z*A7))	
8	1	57	-2.9686	=E7*(1-exp(-Z))	=N0*exp(-(Z*A8))	=(B8-D8)^2
9	2	40	-2.79279	=E8*(1-exp(-Z))	=N0*exp(-(Z*A9))	=(B9-D9)^2
10	3	28	-2.61751	29.30	70.4367	1.7026
11	4	20	-2.43277	20.69	49.7419	0.4828
12	5	14	-2.25752	14.61	35.1273	0.3777
13	6	10	-2.0836	10.32	24.8067	0.1028
14	7	6	-1.94994	7.29	17.5183	1.6599

It might seem to be a viable alternative to use a simple least squared residuals approach to match the number of tags observed against the predicted number of tags from Eq. 3.51. Thus, we would be assuming normal random residual errors with a constant variance and minimizing the sum of squared residuals should provide us with an optimum agreement between the observed number of tags and the expected

$$SSQ = \sum \left(T_i - \left[N_t \left(1 - e^{-Z} \right) \right] \right)^2 \qquad (3.53)$$

where T_i is the number of tags returned in period i. The difference between the two model fits relates to the different properties of the pdfs. With normal errors, the variance of the residuals is constant but with Poisson errors, the variance increases with the expected number of tags. If it seems more likely that the variance will increase with number of tags returned, then clearly the Poisson is to be preferred. This must be decided independently of the quality of fit, because both analyses provide optimum parameter estimations (Example Box 3.13).

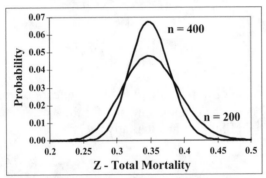

Figure 3.24 Hypothesized total mortality vs. the likelihood for the imaginary example of the results of a tagging experiment. The total mortality consistent with the tagging results using Poisson likelihoods, was approximately 0.348 (Example Box 3.13). The two curves refer to different tag numbers but with equivalent relative numbers of returns.

3.4.16 Likelihoods from the Gamma Distribution

The Gamma distribution is less well known to most ecologists than the statistical distributions we have considered in previous sections. Nevertheless, the Gamma distribution is becoming more commonly used in fisheries modelling, especially in the context of length-based population modelling (Sullivan et al., 1990; Sullivan, 1992). The probability density function for the Gamma distribution has two parameters, a scale parameter, b (b > 0; an alternative sometimes used is λ, where $\lambda = 1/b$), and a shape parameter c (c > 0). The distribution extends over the range of $0 \leq x \leq \infty$, where x is the variate of interest. The expectation or mean of the distribution, $E(x)$, relates the two parameters, b and c, Thus

$$E(x) = bc \qquad \text{or} \qquad c = \frac{E(x)}{b} \qquad (3.54)$$

The variance of the distribution is b^2c, and, for values of $c > 1$, the mode is calculated as $b(c-1)$, which is thus less than the expectation or mean (Fig. 3.25).

A typical use of this distribution would be to describe the relative likelihood of each of a range of sizes to which a particular sized animal might grow. The probability density function for determining the likelihoods for the gamma distribution is (Hastings and Peacock, 1975)

$$L\{x \mid c, b\} = \frac{\left(\dfrac{x}{b}\right)^{(c-1)} e^{\frac{-x}{b}}}{b\Gamma(c)} \qquad (3.55)$$

where x is the value of the variate, b is the scale parameter, c is the shape parameter and $\Gamma(c)$ is the Gamma function for the c parameter. Some books (and the Excel help file) give an alternative version

$$L\{x \mid c, b\} = \frac{x^{(c-1)} e^{-x/b}}{b^c \Gamma(c)} \qquad (3.56)$$

but these are equivalent algebraically. Where the shape parameter, c, takes on integer values the distribution is also known as the Erlang distribution (Hastings and Peacock, 1975)

$$L\{x \mid c, b\} = \frac{\left(\dfrac{x}{b}\right)^{(c-1)} e^{\frac{-x}{b}}}{b(c-1)!} \qquad (3.57)$$

where the Gamma function is replaced by Factorial($c-1$).

The Gamma distribution is extremely flexible, ranging in shape from effectively an inverse curve through a right hand skewed curve, to approximately normal in shape (Fig. 3.25; Example Box 3.14). Its flexibility makes it a very useful function for simulations (see Chapter 7).

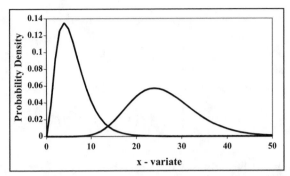

Figure 3.25 Two examples of the Gamma distribution with different parameter combinations. Both curves have a scale parameter, $b = 2$. The left hand curve has an expected value of 6 (giving a shape parameter $c=3$), while the right hand curve has an expectation of 26 ($c = 13$). Note the modes are at $b(c-1)$ and not at the expectations of the distributions (Example Box 3.14).

Example Box 3.14 Likelihoods from the Gamma and Erlang distributions. Continue the series of x values down to 50 (row 55), copying the respective column down. The Excel function in column B can give simple likelihoods or give the cumulative distribution function by using the "true" parameter instead of the "false". In this way, true probabilities can be derived. The Erlang distribution is identical to the Gamma distribution when c is an integer put =(((A5/\$B\$1)^(\$B\$2-1))*exp(-A5/\$B\$1)) /(\$B\$1*fact(\$B\$2-1)) into C5 and copy down. Plot column B against A to mimic Fig. 3.25. Vary b and $E(x)$ and determine the affect upon the curve. When c is not an integer, the Fact function in the Erlang distribution truncates c to give an integer.

	A	B	C	D
1	**Shape b**	2	**Mode**	=B1*(B2-1)
2	**Scale c**	4	**Variance**	=(B1^2)*B2
3	**E(x)**	8		
4	**x**	**Gamma**	**Erlang**	
5	0.1	=gammadist(A5,\$B\$2,\$B\$1,false)	9.91E-06	
6	1	=gammadist(A6,\$B\$2,\$B\$1,false)	0.006318	
7	2	0.030657	0.030657	
8	3	0.062755	0.062755	
9	4	0.090224	0.090224	
10	Extend or copy down to row 55 where x = 50			

It is possible that one might fit the Gamma function to tagging data in order to provide a probabilistic description of how growth proceeds in a species (but one would need a great deal of tagging data (Punt *et al.*, 1997).

In these instances of fitting the Gamma distribution, the presence of the Gamma function in the equation implies that it will be liable to numerical overflow errors. It would always be risk averse to work with log-likelihoods instead of likelihoods

$$LL\{x\,|\,c,b\} = \left((c-1)Ln\left(\frac{x}{b}\right) - \frac{x}{b}\right) - \left(Ln(b) + Ln(\Gamma(c))\right) \qquad (3.58)$$

This may appear to be rather of little assistance as we are still left with the trouble of calculating the natural log of the Gamma function. Surprisingly, however, this is relatively simple to do. Press *et al.*, (1989) provide an excellent algorithm for those who wish to write their own procedure, and there is even a GammaLn function in Excel.

An example of using the Gamma distribution in a real fisheries situation will be produced when we consider growth and its representation. It will be demonstrated when it is used to create the growth transition matrices used in length-based models.

3.4.17 Likelihoods from the Multinomial Distribution

We use the binomial distribution when we have situations where there can be two possible outcomes to an observation (true/false, tagged/untagged). However, there are many situations where there are going to be more than two possible outcomes to any observation and in these situations, we should use the multinomial distribution. In this multivariate sense, the multinomial distribution is an extension of the binomial distribution. The multinomial is another discrete distribution that provides distinct probabilities and not just likelihoods.

With the binomial distribution we used P(m|n,p) to denote the likelihoods. With the multinomial, this needs to be extended so that instead of just two outcomes (one probability p) we have a probability for each of k possible outcomes (p_k) in our n trials. The probability density function for the multinomial distribution is (Hastings and Peacock, 1975)

$$P\{x_i\,|\,n, p_1, p_2, ..., p_k\} = n!\prod_{i=1}^{k}\frac{p_i^{x_i}}{x_i!} \qquad (3.59)$$

where x_i is the number of times an event of type i occurs in n trials, n is the sample size or number of trials, and the p_i are the separate probabilities for each of the k types of events possible. The expectation of each type of event is $E(x_i)=np_i$, where n is the sample size and p_i the probability of event type *i*. Because of the presence of factorial terms that may lead to numerical overflow problems, a log-transformed version of Eq. 3.59 tends to be used

$$LL\{x_i \mid n, p_1, p_2, ..., p_k\} = \sum_{j=1}^{n} Ln(j) + \sum_{i=1}^{k} \left[x_i Ln(p_i) - \sum_{j=1}^{x} Ln(j) \right] \quad (3.60)$$

In real situations the factorial terms will be constant and are usually omitted from the calculations, thus

$$LL\{x_i \mid n, p_1, p_2, ..., p_k\} = \sum_{i=1}^{k} \left[x_i Ln(p_i) \right] \quad (3.61)$$

Examples may provide a better indication of the use of this distribution. Whenever we are considering situations where probabilities or proportions of different events or categories are being combined we should use the multinomial. This might happen in a catch-at-age stock assessment model that uses the proportional catch-at-age. The model will predict the relative abundance of each age class and these will be combined to produce sets of expected catch-at-age. The comparison of the observed with the expected proportions is often best done using the Multinomial distribution. Another common use is in the decomposition of length frequency data into constituent modes (e.g., MacDonald and Pitcher, 1979; Fournier and Breen, 1983). We will consider the latter use and develop an example box to illustrate the ideas.

In November 1992, sampling began of juvenile abalone on Hope Island, near Hobart, Tasmania, with the aim of investigating juvenile abalone growth through the analysis of modal progression (Fig. 3.26).

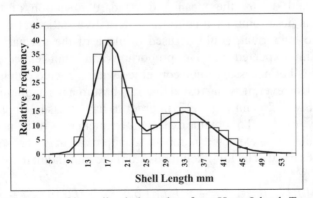

Figure 3.26 A subset of juvenile abalone data from Hope Island, Tasmania, taken in November 1992. Two modes are obvious and the solid line is the maximum likelihood best-fit combination of two normal distributions (Example Box 3.15).

The main assumption behind the analysis of length frequency information is that observable modes in the data relate to distinct cohorts or settlements. Commonly a normal distribution is used to describe the expected relative frequency of each cohort (Fig. 3.27), and these are combined to generate the expected relative frequency in each size class.

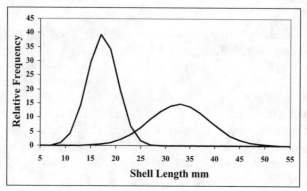

Figure 3.27 Two normal distributions fitted to abalone data using the multinomial distribution to combine the relative proportions (*cf.* Fig. 2.26; Example Box 3.15).

The normal probability density function is commonly used to generate the expected relative proportions of each of the k observed length categories, for each of the n age classes

$$p_{L_k} = \frac{1}{\sigma_n \sqrt{2\pi}} e^{\frac{-(L_k - \mu_n)^2}{2\sigma_n}} \tag{3.62}$$

where μ_a and σ_a are the mean and standard deviation of the normal distributions describing each cohort a. Alternative statistical distributions, such as the lognormal, could be used in place of the normal. As we are dealing with expected relative proportions and not expected relative numbers it is not necessary to be concerned with expected numbers in each size class. However, it is suggested that it is best to have a graphical output, similar to Fig. 2.26 and Fig. 2.27, in order to have a visual appreciation of the quality of fit. To obtain expected frequencies, it is necessary to constrain the total expected numbers to approximately the same as the numbers observed. The log-likelihoods from the multinomial are

$$LL\{L \mid \mu_n, \sigma_n\} = -\sum_{i=1}^{k} L_i Ln(\hat{p}_i) = -\sum_{i=1}^{k} L_i Ln\left(\frac{\hat{L}_i}{\sum \hat{L}_i}\right) \tag{3.63}$$

where the μ_n and σ_n are the mean and standard deviations of the n cohorts being considered. There are k length classes and L_i is the observed frequency of length class i, while p_i-hat is the expected proportion of length class i from the combined normal distributions. Being the negative log-likelihood the objective would be to minimize Eq. 3.63 to find the optimum combination of the n normal distributions (cohorts) (Example Box 3.15).

Example Box 3.15 Using the multinomial to fit multiple normal distributions to length-frequency data. Extend the series in column A in steps of 2 down to 55 in row 32. Extend the observed frequencies such that, as shown, in row 15 at Mid_L of 21 obs = 23, then Mid_L of 23 has 13, then extending downwards, 7, 10, 14, 11, 16, 11, 11, 9, 8, 5, 2, and then zero for Mid_L of 47 to 55. Put =normdist($A7,B$1,B$2,false)*B$3 into B7, and copy it into C7, and down to row 32. Row 5 contains the sum of each of the columns, as in B5: =sum(B7:B32). Plot column E as a histogram against column A. Add column D to this plot as a line-chart (*cf.* Fig. 3.26). The values in B1:C3 are initial guesses, the quality of which can be assessed visually by a consideration of the plotted observed vs. expected. Cell D2 contains (-sum(G7:G32))+(E5-D5)^2. The second term is merely a penalty term designed to force the sum of the expected frequencies to equal the observed total. This has no effect on the relative proportions of each cohort. Use the solver to minimize D2 by changing cells B1:C3. Try different starting points. Set up a new column that provides normal residuals, i.e., =(E7-D7)^2, sum and minimize them (use the same penalty term to limit the relative frequencies). Are the optimum answers different? Compare the graphical images. Which do you prefer, that from multinomial or from normal random residuals?

	A	B	C	D	E	F	G
1	Mean	17.3319	32.7933				
2	Variance	3.0378	6.0196	706.2115	ML		
3	RelNos	302.1299	221.8833				
4							
5	Sum	151.064	110.936	262	262		
6	Mid-L	Cohort1	Cohort2	Expt	Obs	P_L	LL
7	5	0.01047	0.00035	=B7+C7	0	=D7/D$5	=E7*Ln(F7)
8	7	0.12208	0.00152	=B8+C8	0	=D8/D$5	=E8*Ln(F8)
9	9	0.92257	0.00595	0.92853	0	0.00354	0.00000
10	11	4.51975	0.02095	4.54070	6	0.01733	-24.33158
11	13	14.35417	0.06602	14.42019	12	0.05504	-34.79658
12	15	29.55238	0.18628	29.73866	35	0.11351	-76.15637
13	17	39.44181	0.47067	39.91249	40	0.15234	-75.26620
14	19	34.12491	1.06494	35.18984	29	0.13431	-58.22002
15	21	19.13975	2.15770	21.29745	23	0.08129	-57.72441
16	23	6.95907	3.91485	10.87392	13	0.04150	-41.36570

3.5 BAYES' THEOREM

3.5.1 Introduction

There has been a recent expansion in the use of Bayesian statistics in fisheries science (Chen and Fournier, 1999; McAllister and Ianelli, 1997; Punt and Hilborn, 1997; see Dennis, 1996 for an opposing view). An excellent book relating to the use of these methods was produced by Gelman *et al.*, (1995). Here we are not going to attempt a review of the methodology as it is used in fisheries [a detailed introduction can be found in Punt and Hilborn (1997)]. Instead, we will concentrate upon the foundation of Bayesian methods as used in fisheries and draw some comparisons with maximum likelihood methods.

Bayes' theorem is based around a manipulation of conditional probabilities. Thus, if an event, labelled A follows a number of possible events B_i, then we can develop Bayes' theorem by considering the probability of observing a particular B_i given that A has occurred

$$P(B_i \mid A) = \frac{P(A \& B_i)}{P(A)} \qquad (3.64)$$

In an analogous fashion, we can consider the conditional probability of the event A given a particular B_i

$$P(A \mid B_i) = \frac{P(A \& B_i)}{P(B_i)} \qquad (3.65)$$

re-arranging Eq. 3.65

$$P(A \mid B_i) P(B_i) = P(A \& B_i) \qquad (3.66)$$

Substituting Eq. 3.66 into Eq. 3.64 we obtain the basis of Bayes' theorem

$$P(B_i \mid A) = \frac{P(A \mid B_i) P(B_i)}{P(A)} \qquad (3.67)$$

If we translate the A as the data observed from nature and the B_i as the separate hypotheses (as models plus parameter values) we can derive the form of Bayes' theorem as it is used in fisheries. The $P(A|B_i)$ is just the likelihood of the data A given the hypothesis B_i. The $P(B_i)$ is the probability of the hypothesis before any analysis or consideration of the data. This is known as the prior probability of the hypothesis B_i. The $P(A)$ is simply the combined probability of all the combinations of data and hypotheses

$$P(A) = \sum_{i=1}^{n} P(A \mid B_i) P(B_i)$$ (3.68)

3.5.2 Bayes' Theorem

As stated above, Bayes' theorem relates to conditional probabilities (Gelman *et al.*, 1995) so that when we are attempting to determine which of a series of n discrete hypotheses is most probable we use

$$P\{H_i \mid data\} = \frac{L\{data \mid H_i\} P\{H_i\}}{\sum_{i=1}^{n} \left[L\{data \mid H_i\} P\{H_i\} \right]}$$ (3.69)

where H_i refers to hypothesis *i* out of the *n* being considered (an hypothesis would be a particular model with a particular set of parameter values) and the data are just the data to which the model is being fitted. $P\{H_i|data\}$, is the **posterior** probability of the hypothesis (a strict probability between 0 and 1) given the data (and any prior information). $P\{H_i\}$ is the **prior** probability of the hypothesis before the observed data are considered; once again this is a strict probability where the sum of the priors for all hypotheses being considered must sum to one). Finally, $L\{data|H_i\}$ is the likelihood of the data given hypothesis i, just as previously discussed in the maximum likelihood section (analogous to Eqs. 3.9 and 3.32). If the parameters are continuous variates (e.g., L_∞ and K from the von Bertalanffy curve), alternative hypotheses have to be described using a vector of continuous parameters instead of a list of discrete parameter sets, and the Bayesian conditional probability becomes continuous

$$P\{H_i \mid data\} = \frac{L\{data \mid H_i\} P\{H_i\}}{\int L\{data \mid H_i\} P\{H_i\} dH_i}$$ (3.70)

In fisheries, to use Bayes' theorem to generate the required posterior distribution we need three things:

1)   A list of hypotheses to be considered with the model under consideration (i.e., the combinations of parameters and models we are going to trial).
2)   A likelihood function required to calculate the probability density of the observed data given each hypothesis *i*, $L\{data|H_i\}$.
3)   A prior probability for each hypothesis, normalized so that the sum

of all prior probabilities is equal to 1.0.

Apart from the requirement for a set of prior probabilities, this is identical to the requirements for determining the maximum likelihood. The introduction of prior probabilities is, however, a big difference, and is something we will focus on in our discussion.

If there are many parameters being estimated in the model the integration involved in determining the posterior probability in a particular problem can involve an enormous amount of computer time. There are a number of techniques used to determine the Bayesian posterior distribution and Gelman *et al*. (1995) introduce the more commonly used approaches.

3.5.3 Prior Probabilities

There are no constraints placed on how prior probabilities are determined. One may already have good estimates of a model's parameters from previous work on the same or a different stock of the same species, or at least have useful constraints on parameters (such as negative growth not being possible or survivorship > 1 being impossible). If there is insufficient information to produce informative prior probabilities then commonly a set of uniform or non-informative priors are adopted in which all hypotheses being considered are assigned equal prior probabilities. This has the effect of assigning each hypothesis an equal weight before analysis. Of course, if a particular hypothesis is not considered in the analysis this is same as assigning that hypothesis (model plus particular parameters) a weighting or prior probability of zero.

One reason why the idea of using prior probabilities is so attractive is that it is counterintuitive to think of all possible parameter values being equally likely. Any experience in fisheries and biology provides one with prior knowledge about the natural constraints on living organisms. Thus, for example, even before thorough sampling it should have been expected that a deep-water (>800m depth) fish species, like orange roughy (*Hoplostethus atlanticus*), would likely be long lived and slow growing. This characterization is a reflection of the implications of living in a low temperature and low productivity environment. One of the great advantages of the Bayesian approach is that it permits one to move away from this counterintuitive assumption of all possibilities being equally likely. One can attempt to capture the relative likelihood of different values for the various parameters in a model in a prior distribution. In this way, prior knowledge can be directly included in analyses.

Where this use of prior information can lead to controversy is when moves are made to include opinions. The potential exists for canvassing a gathering of stakeholders in a fishery for their belief on the state of such parameters as current biomass (perhaps relative to five years previously). Such a committee-based prior probability distribution for a parameter could be included into a Bayesian analysis as easily as could the results of a previous assessment. There is often debate about whether priors from such disparate sources should be equally acceptable in a formal analysis. In a discussion on the problem of justifying the origin of priors, Punt and Hilborn (1997, p. 43) state:

> We therefore strongly recommend that whenever a Bayesian assessment is conducted, considerable care should be taken to document fully the basis for the various prior distributions. ... Care should be taken when selecting the functional form for a prior because poor choices can lead to incorrect inferences. We have also noticed a tendency to underestimate uncertainty, and hence to specify unrealistically informative priors – this tendency should be explicitly acknowledged and avoided.

The debate over the validity of using informative priors has been such that Walters and Ludwig (1994) recommended that noninformative priors be the used as a default in Bayesian stock assessments. However, besides disagreeing with Walters and Ludwig, Punt and Hilborn (1997) highlighted a problem with our ability to generate noninformative priors (Box and Tiao, 1973). The problem with generating noninformative priors is that they are sensitive to the particular measurement system. Thus, a prior that is uniform on a linear scale will not appear linear on a log scale (Fig. 3.28).

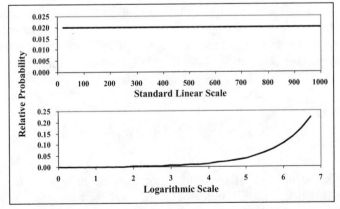

Figure 3.28 The same data plotted on a linear scale (upper panel) and a natural logarithmic scale (lower panel). The uniform distribution on the linear scale is distorted when perceived in logarithmic space. Note the vertical scale in the log panel is an order of magnitude greater than on the linear scale.

As fisheries models tend to be full of non-linear relationships, the use of noninformative priors is controversial because a prior that is non-informative with respect to some parameters will most likely be informative towards others. While such influences may be unintentional, they cannot be ignored.

3.5.4 An Example of a Useful Informative Prior.

We already have experience with a form of Bayesian analysis when we were considering Binomial likelihoods (Example Box 3.10; see Example Box 3.16). We took two likelihood profiles of a fur seal pup population estimate, each made from 1251 separate likelihoods, multiplied the respective likelihoods for each of 1251 hypothesized population sizes, and standardized the results to sum to one in order to obtain approximate percentile confidence intervals. This algorithm may be represented as

$$P\{H_i \mid data\} = \frac{L_1\{data \mid H_i\}L_2\{data \mid H_i\}}{\sum_{i=1}^{1251}\left[L_1\{data \mid H_i\}L_2\{data \mid H_i\}\right]} \tag{3.71}$$

where L_i refers to the likelihoods for the ith population estimation and the H_i refer to the 1251 separate hypothesized population sizes. If we had standardized the first population estimate's likelihood profile to sum to one we could then have treated that as a prior on the population size. This could then have been used in Eq. 3.69, along with the separate likelihoods for the second population estimate, and a formal posterior distribution for the overall population estimate been derived. Although these two algorithms sound rather different, they are in fact, equivalent, as can be seen algebraically. Treating the first estimate as a prior entails standardizing each hypothesized population size's likelihood to sum to one

$$P\{H_i\} = \frac{L\{data \mid H_i\}}{\sum_{i=1}^{n}L\{data \mid H_i\}} \tag{3.72}$$

where n is the number of separate hypotheses being compared (1251 in Example Box 10). The denominator in Eq. 3.72 is, of course, a constant. Thus, expanding Eq. 3.69 using Eq. 3.72 we obtain

$$P\{H_i \mid data\} = \cfrac{L_2\{data \mid H_i\} \cfrac{L_1\{data \mid H_i\}}{\sum_{i=1}^{n} L_1\{data \mid H_i\}}}{\left[\sum_{i=1}^{n}\left[L_2\{data \mid H_i\} \cfrac{L_1\{data \mid H_i\}}{\sum_{i=1}^{n} L_1\{data \mid H_i\}}\right]\right]} \qquad (3.73)$$

As the sum of separate likelihoods term is a constant, it can be moved out of the denominator's overall summation term and we can shift the numerator's inverse sum of likelihoods below the divisor. When the two are brought together, they can thus cancel out

$$= \cfrac{L_2\{data \mid H_i\} L_1\{data \mid H_i\}}{\cfrac{\sum_{i=1}^{n} L_1\{data \mid H_i\}}{\sum_{i=1}^{n} L_1\{data \mid H_i\}} \sum_{i=1}^{n}\left[L_2\{data \mid H_i\} L_1\{data \mid H_i\}\right]} \qquad (3.74)$$

which is identical to Eq. 3.71. Thus, the combination of two likelihood profiles from two separate assessments can be considered as an acceptable form of Bayesian analysis; that is, the origin of the prior distribution is not problematical.

Example Box 3.16 Bayesian posteriors are equal to standardized likelihoods when the prior probability is uniform. Recover Example Box 3.10 and extend it to include a column (perhaps in G or H) in which the likelihoods for experiment one are standardized. Thus, in G11 put = B11/$B$9, and copy down. Use this as a prior by generating another column of numbers, so in H11 put =G11*C11, and copy down. Sum that column into H9. Finally, in column I (cell I11) put =H11/$H$9, and copy down. Column I would then contain the posterior distribution according to Eq. 3.64. Plot this against column A and compare the curve with that found in column E (the standardized joint distribution). How similar are to two curves?

With discrete statistical distributions, there is no problem with combining the separate likelihoods because they are probabilities and not just likelihoods. With the continuous variables found in most fisheries models, numerical methods would be needed to conduct the integration required and this means that the answers would only be approximate. Nevertheless, the approximation would be acceptable assuming the integral step was sufficiently fine.

3.5.5 Non-Informative Priors

If there is no earlier assessment and noninformative prior probabilities are to be used then an even simpler adjustment can be made to Bayes' theorem. In the case of truly non-informative priors, the prior probability for each hypothesis would be equal or constant for each hypothesis, and thus the prior probability term could be removed from the summation term in the denominator

$$P\{H_i \mid data\} = \frac{P\{H_i\}P\{data \mid H_i\}}{P\{H_i\}\sum_{i=1}^{n} P\{data \mid H_i\}} \tag{3.75}$$

whereupon the prior terms can be cancelled and we are left simply with the standardized likelihoods (each separate likelihood divided by the sum of the likelihoods). Thus, using truly non-informative priors is the same as not using priors at all. The simplest approach then would be to omit them from the analysis altogether.

However, this raises the importance of the problem of the apparent impossibility of generating priors that are non-informative over all parameters and model outputs. For a particular model, if generating a truly non-informative prior across all parameters and model outputs requires a great deal of work or it even appears to be impossible, then trying to use non-informative priors would actually risk generating a biased analysis. If these interactions between prior distributions for different parameters and the effects of non-linearity within the model act to distort the priors so that they are no longer truly non-informative then we are left in a quandary. What this would imply is that even if we restrict ourselves to a straightforward maximum likelihood analysis as in Eq. 3.70 (equivalent to using uniform priors in the linear scale), then we are really imputing informative priors in some parts of the model. This unwitting imputation of potential bias is startling and raises many questions about the optimum approach to any non-linear analyses. It appears to imply that it is impossible to conduct a standardization of a full set of likelihoods without implying an unknown set of informative prior probabilities.

In opposition to this idea of the automatic imputation of informative priors, it appears that if priors are simply omitted from the analysis, then, from Eq. 3.70, the omitted priors must have been, by definition, non-informative (on all scales) and somehow equal on all scales. This is a matter for more formal investigation but its importance is crucial for the wide acceptance and further developments in the use of Bayesian analysis in fisheries modelling.

3.6 CONCLUDING REMARKS

Fisheries modellers can be an argumentative crowd, and each seems to have developed a set of methods that they favour. As seen above, there does not appear to be a criterion of quality of model fit that has no associated problems. Claims made that identify a particular strategy of analysis as being the optimum or best practice should therefore be contemplated with some doubt. One can use either maximum likelihood methods or Bayesian methods to generate assessments giving similar forms of output. At least some of the expressed enthusiasm for Bayesian methods appears to be excessive.

The optimum method to use in any situation depends largely on the objectives of the analysis. If all one wants to do is to find the optimum fit to a model then it does not really matter whether one uses least squares, maximum likelihood or Bayesian methods. Sometimes it can be easier to fit a model using least squares and then progress to using likelihoods or Bayesian methods to create confidence intervals and risk assessments.

Confidence intervals around model parameters and outputs can be generated using traditional asymptotic methods (guaranteed symmetric and, with strongly non-linear models, only roughly approximate), using likelihood profiles or by integrating Bayesian posteriors (the two are obviously strongly related), or one can use bootstrapping or Monte Carlo techniques.

It is not the case that the more detailed areas of risk assessment are only possible using Bayesian methods. Bootstrapping and Monte Carlo methods provide the necessary tools with which to conduct such work. The primary concern should be to define the objective of the analysis. It is bad practice to fit a model and not give some idea of the uncertainty surrounding each parameter and the sensitivity of the model's dynamics of the various input parameters.

Because there is no clear winner among the methodological approaches, then, if one has the time, it is a reasonable idea to use more than one approach (especially a comparison of likelihood profiles, Bayesian posteriors, and bootstrapping). If significant differences are found then it would be well to investigate the reasons behind them. If different procedures suggest significantly different answers, it could be that too much is being asked of the available data and different analyses would be warranted.

4

Computer Intensive Methods

4.1 INTRODUCTION

The phrase "computer intensive methods" is used to refer to a number of methods that include randomization tests (= permutation tests), jackknife techniques, bootstrap techniques, and Monte Carlo simulations and tests. In this chapter we will briefly introduce all four methods. Fuller descriptions and examples will be presented in following chapters.

"Computer intensive" can sound daunting but only if it is confused with meaning intensive computer programming (though some facility with writing relatively simple macros or computer programs is necessary). The word "intensive" refers to the use of available computing power. Such statistics and methods are not new, randomization tests were discussed by Fisher (1936), but have recently become much more widespread and popular as a direct result of the easy availability of powerful computers (Good, 1994; Manly, 1997).

The history of statistics is relatively short and has a number of phases. Statistics as a discipline is thought to have begun during or after the wish to understand the theory behind gambling. This led to the first developments in probability theory. However, the first developments in statistics of importance to our work here were related to parametric statistics. These statistics are completely dependent upon known probability distributions (such as the normal or binomial distributions, as in Chapter 3) and when they are used they require that the populations being sampled adhere to assumptions peculiar to the distributions concerned. Parametric statistics form the core of "traditional" statistics (ANOVA, regression, sample comparison, experimental design, etc.).

The second phase of particular interest to us, was the production of non-parametric statistics. Ronald A. Fisher (1936) first mooted the idea of testing hypotheses concerning observed patterns between groups of individuals by using a randomization test. These tests compare an observed pattern among groups with those possible by randomly allocating the available data across the groups of interest. Of course, with any reasonable amount of data such an approach was simply not practical at a time (1930s) when all calculations had to be done by hand or on a mechanical hand-calculator. Nevertheless, as a theoretical insight it eventually led to the development of a wide range of non-parametric test statistics. Almost all common non-parametric statistics have their basis in replacing the observed test data with ranks and computing test statistics based upon permutation tests (Good, 1994). Using ranks removes the unique or idiosyncratic aspects of each situation so that the same probability distribution of possible arrangements can be applied to every hypothesis test. The term "non-parametric" implies that the parametric form of the underlying population distribution need not be specified exactly.

Most recently there has been a growth of interest in general randomization tests, plus the development of the newer methods of jackknife and bootstrap statistics. All of these require far more calculations than classical methods, but with the rise of personal computing power, these statistical approaches may supersede the older methods that depend on parametric statistical distributions. They will constitute an important part of the future of statistical analysis, especially of biological data, which, typically, is non-normal.

4.2 RESAMPLING

The notion of resampling a population is fundamental to many statistics. If a set of data is normally distributed, then, given its mean and standard deviation, we can estimate the standard error of the mean analytically:

$$\text{StErr} = \frac{\sigma}{\sqrt{n}} \tag{4.1}$$

where n is the number of observations and σ is the standard deviation.

An alternative to the usual analytical methods would be to obtain a number of independent samples from the original population (i.e., resample the population a number of times). The standard deviation of the multiple sample means would also provide an estimate of the standard error. Thus, the notion of resampling should not be considered as remarkable. This is fortunate because resampling of one form or another is the foundation of all the computer intensive methods to be discussed. While resampling is not unusual, what is resampled, and how it is resampled, differs between the methods.

In the following sections and examples we will be considering the range of available computer intensive methods. Before this, we need to introduce a distinction concerning the approach used when making observations (sampling) that will be important in what follows. "Sampling without replacement" means making an observation from a population by removing individuals. Further observations in a sample could obviously not contain those particular individuals again. This contrasts with "sampling with replacement", where an observation is made non-destructively, i.e., without removing individuals. Clearly, when sampling with replacement, subsequent observations could be repeats of observations already made.

4.3   RANDOMIZATION TESTS

Randomization tests are used to test the null hypothesis that an observed pattern is typical of a random event. The null hypothesis is suggesting that each group of observations merely represents a random sample from a single population. The observed pattern among groups must be characterized by a test statistic (e.g., a mean difference between two groups). To test whether the observed pattern is significantly different from a random pattern the data from all groups are first combined. The observed test statistic is then compared with that obtained by randomly reallocating individuals from the combined data back to the groups to be compared (i.e., resampling without replacement). Of course, it is of no value doing this only once. The random resampling without replacement into the groups and recalculating the test statistic, must be repeated many times (1000+ being typical). The observed test statistic is compared with the empirical distribution of that test statistic given the data available. If the pattern observed is not different from random, then the observed test statistic value will be typical of those generated by random groupings. A significant difference is indicated by the observed test statistic lying beyond or at the extremes of the empirical distribution obtained from randomizing the data among the groups.

Permutation tests are good for testing hypotheses but standard errors cannot be calculated using this approach, and confidence intervals on parameter estimates can only be fitted very inefficiently.

Fisher (1936) first provided a theoretical description of randomization tests. Reviews of randomization techniques, with contrasting views on some methods, are given by Manly (1997) and Edgington (1995). There is a large, and expanding volume of literature on randomization tests providing both examples and theoretical developments. Some controversy has occurred but this has not prevented continued development (Basu, 1980, plus associated comments; e.g., Hinkley, 1980;

Kempthorne, 1980; Lane, 1980; Lindley, 1980). A detailed review and bibliography relating to randomization tests, is given by Good (1994).

4.4   JACKKNIFE METHODS

The name is reported as coming from considering this statistical method as a flexible tool rather like a Swiss-Army pocketknife. By considering known sub-sets of the available data one can produce estimates of standard errors as well as detect whether any parameter estimates are biased. Sub-setting the data involves estimating the statistics of interest on all possible combinations of the available data minus one data point: in a data set of n values there will be n subsets of $(n-1)$ data points and these are used to calculate the jackknife replicates. Using the jackknife replicates one can calculate what are known as "pseudo-values" for the statistic of interest. The difference between the original sample mean and the mean of the n pseudo-values provides the estimate of bias. The value of this methodology comes when one is not estimating the sample mean (which is an unbiased estimate) but some other parameter. The pseudo-values can also be used to calculate jackknife estimates of the parameter of interest and its standard error. Confidence intervals can be fitted using this standard error estimate but there is a problem deciding how many degrees of freedom to use so this approach is no longer recommended.

The jackknife methodology was first discussed by Quenouille (1956) who recommended the approach as a method for removing bias from parameter estimates. Tukey gave a paper to a conference of the American Institute of Statistics at Ames in Iowa. He introduced the notion of using the jackknife approach to produce parameter estimates with estimates of standard errors. Only the abstract of the conference talk was printed (Tukey, 1958) but that and his talk were enough to set off a number of developments (Hinkley, 1983). Jackknifing is discussed in Chapter 6.

4.5   BOOTSTRAPPING METHODS

Data sampled from a population are treated as being (assumed to be) representative of that population and the underlying probability density distribution of expected sample values. Given an original sample of n observations, bootstrap samples would be random samples of n observations taken from the original sample with replacement. Bootstrap samples, (i.e., random resampling from the sample data values with replacement) are assumed to approximate the distribution of values that would have arisen from repeatedly sampling the original sampled population. Each of these bootstrapped samples is treated as an independent random sample from the original population. This approach appears counter-intuitive to some but can be used to fit standard errors, confidence intervals, and to test hypotheses. The name "Bootstrap" is reported to derive from the story of the *Adventures of Baron Munchausen* in which the Baron escaped drowning by picking himself up by his own bootstraps and thereby escaping from a well (Efron and Tibshirani, 1993).

Efron (1979) first suggested bootstrapping as a practical procedure. He states (Efron and LePage, 1992) that development of the bootstrap began as an attempt to better understand the jackknife but quickly developed beyond the potential of the jackknife. The bootstrap could be applied to problems beyond those of bias and standard errors, in particular it was an approach which could provide better confidence intervals than those from the jackknife. Bickel and Freedman (1981) provided a demonstration of the asymptotic consistency of the bootstrap (convergent behaviour as the number of bootstrap samples increased). Given this demonstration, the bootstrap approach has been applied to numerous standard applications such as multiple regression (Freedman, 1981) and stratified sampling (Bickel and Freedman, 1984, who found a limitation). Efron eventually converted the material he had been teaching to senior level students at Stanford into a general summary of progress to date (Efron and Tibshirani, 1993).

4.6   MONTE CARLO METHODS

Monte Carlo simulations are carried out with a mathematical model of a situation plus a series of model parameters. Some of the parameters will not be known perfectly (i.e., there is uncertainty) so that instead of a particular value one would have a probability density distribution from which the parameter values are derived (e.g., normal, lognormal, hypergeometric, etc.). Each run of a Monte Carlo simulation involves randomly selecting a value for the variable, parameter, or data values, from the known distribution(s) and then determining the model's output. This process can be repeated many times to test hypotheses or determine confidence intervals. Such resampling from a theoretical distribution of values is effectively sampling with replacement (one could, in theory, obtain the same value more than once). The probability density distribution can never be exhausted of values.

Monte Carlo testing is often about comparing what was actually observed in a system with what one obtains from a model of the system. It involves an assessment of the properties of the system. In an hypothesis testing situation, if any of the hypotheses included in the model are incorrect, then the model output would not be expected to be consistent with the available observations.

Monte Carlo simulations are also the basis of risk assessment methods in fisheries by projecting the expected path of a fishery when it is exposed to a particular harvest strategy (e.g., a constant fishing mortality rate or constant catch level). When more than one of a model's parameters are each free to vary over a range of values, then the model output also becomes variable. If the model is run enough times one would expect to be able to generate a frequency distribution of possible outcomes (perhaps the biomass remaining in a stock after some years of exploitation at a given Total Allowable Catch or TAC). From this distribution one could derive the likelihood of various outcomes (e.g., stock collapse - defined in a particular way, or, current biomass falling below defined levels). In New Zealand, for example, such a model was used to determine the impact on orange roughy (*Haplostethus atlanticus*) stocks of different projected catch rates over the next 20 years (Francis, 1992). Thereby, the option of reducing the commercial catch slowly was demonstrated to be a more "risky" option for the stock than a rapid decline in catch rates.

4.7   RELATIONSHIPS BETWEEN METHODS

All the computer intensive methods we are going to consider may be viewed as different forms of random resampling where the observed sample data or its properties are taken to represent the expected range of possible data from the sampled population. So, instead of sampling from a theoretical probability distribution one can resample from the empirical distribution that is represented by the values in the sample (Table 4.1). Alternatively, one can sample from a parametric statistical distribution whose parameters are estimated from the original sample.

Randomization tests can be considered to be a special case of Monte Carlo testing where the original sample data is re-sampled without replacement so that each run uses all the available data. The only thing that changes is the assortment of data between groups.

Jackknife analysis is also a special case of Monte Carlo sampling where the available data forms the empirical distribution sampled. In this case it is systematically sub-sampled without replacement. It is the fact that values are omitted systematically, which leads to there being a fixed number (n-1) of jackknife replicates.

Non-parametric bootstrapping is another special case of the Monte Carlo process where the observed sample takes the place of a parametric probability distribution, or even the original population. In this case the situation is much more akin to parametric Monte Carlo sampling. The observed sample is sampled repeatedly, with replacement, just as if it were a continuous probability distribution.

Table 4.1 Relationships between computer intensive methods and their strategies for resampling from probability distributions. PDF refers to a probability density function. Parametric statistics are included for completeness but are not usually considered computer intensive. The non-parametric Monte Carlo and bootstrap methods are equivalent.

Method of Resampling	Computer Intensive Method
Resampling a theoretical PDF (e.g., t-distribution, χ^2 distribution). Implicitly this is sampling with replacement.	Parametric statistics (analytically). Parametric Monte Carlo simulations.
Resampling an empirical distribution (as represented by a sample) without replacement.	Randomization tests. Jackknife statistics.
Resampling an empirical distribution (as represented by a sample) with replacement.	Non-parametric Bootstrap. Non-parametric Monte Carlo.

4.8   COMPUTER PROGRAMMING

We will implement all of these computer intensive methods in Excel workbooks using surprisingly little macro coding. Excel macros, however, can often be too slow for serious analyses when they are doing a great deal of work (the whole point of computer intensive statistics). Ideally, in those cases, we might wish to write an executable program in some programming language such as Pascal, C++, or Fortran. Manly (1991) and Edgington (1987) provide program code for sub-routines in Fortran for randomization tests, while Efron and Tibshirani (1993), provide the necessary code in the *S* statistical package for carrying out bootstrapping routines. The Systat statistical package now includes bootstrapping as an option for many of its statistical routines. Many statistical packages can now be macro-programmed into conducting computer intensive statistics. For our purposes in this book, the Excel spreadsheet and its Visual Basic macro language will suffice.

5

Randomization Tests

5.1 INTRODUCTION

When used to test hypotheses, standard parametric statistics, such as analysis of variance (ANOVA), require the samples and data involved to adhere to at least one restrictive assumption. If data fail to meet the conditions laid down in such assumptions, any conclusions drawn from the analyses can be suspect. Randomization methods can also be used to test hypotheses but require fewer assumptions. Given this extra flexibility, it is surprising that there is not a greater awareness of randomization tests. In this chapter we will examine randomization or permutation testing and its potential value to ecological and fisheries research.

5.2 HYPOTHESIS TESTING

5.2.1 Introduction

Many hypothesis tests are determinations of the likelihood that the situation one has observed in nature could have arisen by chance events alone. What this means is that one is testing whether a pattern observed in nature is unusual. But something can only be unusual relative to something else, usually one group relative to one or more other samples, or relative to an hypothesis of how the sample ought to be. This is tested by determining whether there is a good chance that a perceived pattern between groups could have come about by random variations of whatever components make up the pattern of interest. For example, one could ask whether differences observed in the size frequencies of a fish species found on the open coast and in coastal bays arose due to chance mixing or if the observed pattern is highly unlikely to be due to random movements between sites.

5.2.2 Standard Significance Testing

When attempting to test an hypothesis, at some stage one invariably calculates a test statistic. Given a value for a particular test statistic (e.g., a t-statistic comparing the difference between two means), one then wishes to know if the value one has denotes a significant difference. This is such a common event that it is simple to forget what lies behind the steps leading to the test statistic and the meaning of its test of significance. Three things are needed when testing an hypothesis statistically:

1) An hypothesis, ideally should be stated formally and is often translated into a null-hypothesis that is the inverse or complement of the hypothesis we wish to test. This inversion is necessary because of the logical asymmetry between disproof of a general statement (possible) and its proof (not possible). Hypotheses are testable theoretical constructs e.g., average fork-length of fish from the open coast and coastal bays are the same (are not different).

2) A test statistic, any single valued (and smooth and continuous) function of the data. Commonly used test statistics include the F-ratio, the t-statistic, χ^2, and r the correlation coefficient.

3) Some means of generating the probability distribution of the test statistic under the assumption that the null-hypothesis is true is required to permit the determination of how likely the observed value of the test statistic is if the null hypothesis is true.

The first two requirements are straightforward and need little discussion. The third requirement may not be immediately familiar but is strongly related to the second requirement of using a particular test statistic. The many, very commonly used parametric statistics are used because their underlying assumptions are known, which means the expected probability density function (e.g., Fig. 5.1) of a test statistic under the conditions of a particular situation can be calculated analytically. The value of the test statistic obtained from nature can be compared with the expected values given the circumstances of the particular test (degrees of freedom, etc.).

In this way, assuming the sampled population adheres to the assumptions of the test statistic, it is possible to determine how likely the observed value would be assuming the null-hypothesis to be true. For common parametric test-statistics the comparison of calculated values with the theoretical probability density function is made simple by its translation into tables relating degrees of freedom to likelihood or significance (Fig. 5.1).

By convention, if the value of a test statistic would only be expected to occur less than once in 20 sets of replicated samples (i.e., < 5% of the time) then usually a "significant" difference is claimed. All this means is that unusual patterns (e.g., four heads in a row when tossing a coin) are not impossible, they are just unlikely chance events relative to the many other possible events that can occur. Because convention has established the levels of significance at 5%, 1% (1 chance in 100), and 0.1% (1 chance in 1000), values of test statistics at these levels tend to be printed in statistical tables (Example Box 5.1). By comparing the observed value of the test statistic with these published values one determines whether one has a significant difference, and at what level.

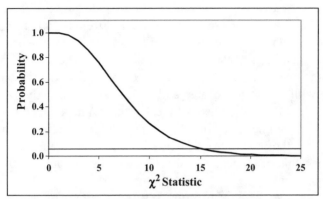

Figure 5.1 A one-tailed probability distribution curve (probability density dunction—PDF), for the χ^2 statistic with 8 degrees of freedom. The fine horizontal line indicates a probability of 0.05. One could utilize this curve to determine the likelihood of obtaining a χ^2 value as large as the one observed between two populations, if one assumes they are from the same population. An observed χ^2 statistic greater than 15.507, would be expected to occur less than 1 time in 20. The shape of the curve varies with the degrees of freedom.

Example Box 5.1 Three different test statistics and their respective cumulative density function indicating the probability that the value of the observed test statistic for the observed pattern being tested could have arisen through chance alone. Examine the Excel help files for descriptions of the functions used. Try varying the test statistic value and the degrees of freedom for each statistic and see the response. For the χ^2 statistic with 1 degree of freedom a value of approximately 3.84 should give a probability of 0.05. Compare the outputs from the Excel functions with published tables.

	A	B	C	D
1	Test Statistic	χ^2	F	t
2	Test Statistic value	2.9	2.9	1.98
3	Degrees of Freedom 1	1	1	120
4	Degrees of Freedom 2		25	
5	Probability	=chidist(B2,B3)	=fdist(C2,C3,C4)	=tdist(D2,D3,2)

5.2.3 Significance Testing by Randomization Test

Unfortunately, the theoretically derived tables of the various test statistics are only valid if the data adhere to the assumptions of the statistical test employed. The important problem is that if they do not apply, then the analytically derived probability density function will not usually be applicable validly and erroneous conclusions can be made. The effect of such failure to match assumptions is an increased chance of rejecting a real difference between groups or of accepting a difference where none exists. This is one reason it behoves an ecologist to know about the effects of data transformations. For a t-test, the assumptions would be that the two groups being compared are independently and normally distributed random variables with constant means and variances. The only part of this hypothesis that one really wishes to test is the "independently distributed" part (i.e., they come from independent populations), but all the rest is necessary else the probability density distribution of the test statistic (the t distribution) cannot be used validly (though it is, in fact, relatively robust to such departures).

We need to know the probability distribution of the selected test statistic to determine the likelihood of the value observed/calculated assuming the null-hypothesis to be true. A problem with using a theoretically derived probability density function (PDF), as in the t-test example above, is that if the observed test statistic value is not significant we cannot tell without further analyses whether the test failed because the samples are not independently distributed (the thing being tested) or because the samples were not from normally distributed populations (the data or sampled populations failed to conform to the assumptions necessary for the test to be valid). This is where randomization tests exhibit their strength. They are independent of any analytically determined (parametric) probability density function because, during the test, the randomization procedure generates an empirical PDF for the test-statistic from the available data (Manly, 1997).

If the hypothesis to be tested claims to explain a particular pattern in a set of data (e.g., in-shore fish sizes are smaller on average than those off-shore), then the null-hypothesis would claim that the observed pattern found in the data is typical of any random allocation of the available data (fish sizes) among the in-shore and off-shore groups. The actual pattern observed is represented in the test by a single value of a test statistic (perhaps a t-statistic). Obviously, the test statistic should be chosen to be sensitive to the pattern of interest.

Given a null-hypothesis, the expected probability density function for the test statistic chosen can be generated by repeatedly randomizing the data with respect to the sample group membership and recalculating the test statistic. In the inshore/offshore example the membership of inshore and offshore groups is randomized. Fisher (1936) would have suggested that all the fish lengths should be written on separate cards, the cards shuffled, and then the pack dealt out into the inshore/offshore groups in their original relative frequencies. When this process of randomization and calculation of test statistic is done very many times (generally a minimum of 1000 randomizations are used) the frequency of occurrence of different values of the test statistic can be tabulated and this may be compared with the original value observed from the original un-randomized data. If the original value of the test statistic is found to be an unusual event relative to the values generated by the permutations of the data then the null-hypothesis may be rejected. The null-hypothesis is, in effect, that the groups being compared are random samples from the same population. Speaking in terms of shuffling cards Fisher (1936, p. 59) wrote: "Actually, the statistician does not carry out this very simple and very tedious process, but his conclusions have no justification beyond the fact that they agree with those which could have been arrived at by this elementary method."

A test of significance is thus really an attempt to answer whether the observed samples could have been drawn at random from the same population. The answer is a probability and if it is large then it is likely that the pattern could have arisen due to chance and one could answer that the samples might have derived from the same population (we could never claim they definitely were from the same population). However, if the probability of the pattern arising through randomly sampling the same population is small we can be more definite and can claim they were not likely to be from the same population. This asymmetry is why one uses a falsifiable null-hypothesis.

5.2.4 Mechanics of Randomization Tests

An example will illustrate the ideas that have been discussed above and hopefully make more sense of the methods (Fig. 5.2; Example Box 5.2).

The group allocations (inshore/offshore) of the sample observations were randomized 1000 times and the mean difference was recalculated (algorithm shown in Fig. 5.3). The randomization test found that the original mean difference only occurred 25 times out of 1000 so evidence of a real difference exists (Fig. 5.4). If this analysis is repeated a number slightly different from 25 might arise but a significant difference should still be found. It is the weight of evidence that matters not whether we have a difference significant at 5%, 1%, or whatever. The probability value indicates how likely it was to obtain a value like the observed test statistic. When outliers in the data are ignored (effectively removed from the data set), the randomization test agrees very closely with the conventional *t*-test.

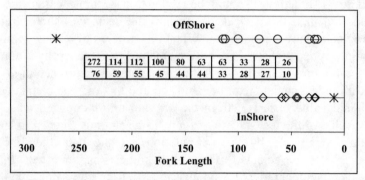

Figure 5.2 Two samples showing the data both numerically and graphically. Note the extreme values specially marked in both groups. Conventional t-tests of the two groups, both including and excluding the extreme values (and assuming unequal variances), indicate that the groups are not significantly different despite the mean difference being 47. With the extreme values included, t = 1.99 and $P = 0.0746$, and without the extreme values, t = 1.8 and $P = 0.0993$. The samples are clearly too small in this example.

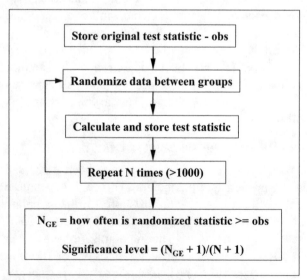

Figure 5.3   Algorithm for conducting a randomization test and for calculating the consequent significance test. One always includes the original test statistic value; this prevents a probability of zero occurring.

When the algorithm in Fig. 5.3 is followed, one obtains N randomized replicate values. When these are sorted in ascending order they may be plotted to give a visual representation of the distribution of values (Fig. 5.4; Example Box 5.2).

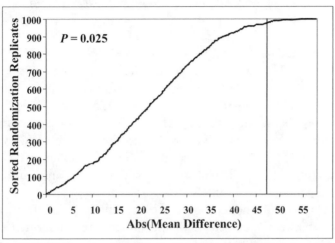

Figure 5.4.   One-thousand sorted randomization replicates of the absolute mean difference between the groups in Fig. 5.2, using the algorithm in Fig. 5.3. The outer vertical line is at 47, the original mean difference. Including the original 47, there were 25 replicates that >= 47 (i.e., lie on or outside 47), therefore $P = 0.025$. With the extreme values excluded the P value tends to be more like 0.094. More replicates would lead to a smoother curve.

5.2.5 Selection of a Test Statistic

With the release from the requirement of having one's test statistic adhere to a theoretical probability distribution there comes the freedom to select the most appropriate statistic. A test statistic should be chosen because its value is sensitive to the substantive theory being tested (i.e., the hypothetical property being checked or compared). However, some researchers see this freedom as a disadvantage because with the same sample, different test statistics can give different levels of significance. They are mistaken in thinking this invalidates the process. Rather it emphasizes that one should be careful when selecting and using a non-standard test statistic.

Basu (1980) was aiming to criticize Fisher and took as an example a simple data set of five values with which it was desired to determine whether the expected population value was significantly different from zero. The data were a set of five numbers the values of which could have been either positive or negative but were, in fact, (1, 2, 3, 4, 7) which have a mean of 3.4. Basu (1980) argued that the question of whether this is significantly different from zero can be tested by a randomization test. He states that the significance of this test (is the sample mean > 0) can be determined by comparing the mean with all the means possible from the 32 possible unique combinations of (± 1, ± 2, ± 3, ± 4, ± 7). It is quickly shown that a mean of 3.4 is the largest value out of all possible values with the data available and so is significant at the 1/32 level ($P = 0.0313$). However, the median is 3, and when used as the test statistic there are four combinations, namely, (± 1, ± 2, 3, 4, 7), with a median as high as 3 implying a significance of 4/32 = 1/8 = 0.125, which is not generally recognized as being significant. This disparity between the mean and median works for any set of five positive numbers tested in this way. In the end, this does not imply that the randomization procedure is not workable but, instead, that this particular test statistic, the median, can be unstable and should not be used (Example Box 5.3).

Examples of non-standard test statistics could include the differences between parameter estimates of two growth curves (K_1 - K_2, or $L_{\infty 1}$ - $L_{\infty 2}$, etc. although a randomization test to compare growth curves should not focus on single parameters and would have other complications – see later example in Chapter 8). Other examples could include the comparison of parameters from other models such as stock-recruitment relationships. Such direct comparisons would not be possible with traditional statistics. One can investigate questions which were previously un-testable. For example, Lento et al. (1997) examined haplotype diversity across the geographical range of southern fur seal species using a randomization test for testing the reality of apparent genetic structure between populations. This tests whether the H_{st} value (a measure of the diversity and evenness of genetic variation between populations) provides evidence of geographical structure.

Multivariate comparisons are also possible and are limited only by imagination. Having said that, one must be extremely careful to determine exactly which hypothesis is being tested by the test statistic developed; when in doubt be conservative.

5.2.6 Ideal Test Statistics

Ideally one should select the test statistic that has the greatest statistical power for the situation being studied. This relates to the two types of error (Table 5.1). Making decisions under circumstances of uncertainty is very much concerned with the different types of statistical inference error it is possible to make. The implications of making each type of error in a particular situation should be made explicit before deciding on a test statistic.

Example Box 5.2 The mechanics of a randomization procedure to compare two groups of data. The artificial data is from Fig 5.2. A copy of the original data is placed in column A (see Fig 5.2 for all data). Column B starts with a copy from column A, Column C is filled with random numbers from the function =rand(). Cells C1 and D1 contain the average of the first and second group, respectively, while E1 contains the test statistic, the absolute mean difference. The objective is to test whether the observed samples could have arisen by chance from a single population. To preserve a copy, the original values are pasted from column A into B and the observed test statistic pasted into D5. The randomization works by sorting columns B and C on the column of random numbers. This has the effect of randomly reordering the available data into the two groups (inshore and offshore). The test statistic alters accordingly and each new value is copied into column D.

Create a macro using Tools/Macro/Record New Macro menu item. Call it Do_Rand and start by using absolute references. Press <Shift><F9> to recalculate the sheet. Select and copy A5:A24 and paste their values into B5:B24. Copy E1 and paste it into D5. Select B5:C24 and sort them on Column C. Copy E1. Switch to relative references and paste the values into cell D6. Stop recording the macro. Press <Alt ><F11> and maximize the macro window. The modifications necessary to the macro are shown on the next page. Assign the Do_Rand macro to a button created from the Forms toolbar. If you change the number of iterations in the macro you will also need to change cell D3. Exclude the extreme values by altering C1 to B6:B14, D1 to B15:B23, and D3 to ">=23.1111", and alter the sort command in the macro.

	A	B	C	D	E
1	Do_Rand		=average(B5:B14)	=average(B15:B24)	=abs(C1-D1)
2					
3				P = =countif(D5:D1005,">=47")/1000	
4	Original	Values	Randomize Rows		
5	272	272	=rand()		
6	114	114	=rand()		
7	112	112	=rand()		
8	100	100	Copy Down		
9	80	80	To Row 24		
10	63	63	0.729468		
11	63	63	0.450439		
12	33	33	0.023943		
13	Continue down, data from Fig. 5.2.				

Example Box 5.2 [cont.] The macro contents as recorded following the directions in the first part of this Example Box. You will need to add the lines and changes shown in Italics. The only tricky bit is the replacement of the step of 5 rows down in the ActiveCell.Offset statement to become the dynamic 4 + i; thereby using the counter variable to identify the pasting location for the results from the randomizations. To conduct the randomizations without the extreme values alter the calculations for the group averages and alter the select command just prior to the sort to read Range("B6:C23").Select. This isolates the extreme values and the results should be rather different. Try modifying the macro by omitting commands such as the Application.Screenupdating=False (a tip - run fewer than 1000 replicates). How many replicates are needed to obtain consistent results?

```
Sub Do_Rand()
' Do_Rand Macro
  Range("A5:A24").Select
  Selection.Copy              ' copy the original data into the randomization block
  Range("B5").Select
  Selection.PasteSpecial Paste:=xlValues
  Range("E1").Select
  Selection.Copy                 ' store the original test statistic
  Range("D5").Select
  Selection.PasteSpecial Paste:=xlValues    ' unnecessary defaults removed
  Application.CutCopyMode = False
  Application.ScreenUpdating = False        ' necessary for speed and sanity
   For i = 1 To 999                         ' plus the original = 1000
     Range("B5:C24").Select          ' randomize by sorting random numbers
     Selection.Sort Key1:=Range("C5"), Order1:=xlAscending, Header:=xlGuess, _
       OrderCustom:=1, MatchCase:=False, Orientation:=xlTopToBottom
     Range("E1").Select
     Selection.Copy                          ' copy and store test replicates.
     ActiveCell.Offset(4 + i, -1).Range("A1").Select  ' use counter to identify cells.
     Selection.PasteSpecial Paste:=xlValues
   Next i
Range("A1").Select                           ' returns to top of sheet
Application.ScreenUpdating = True            ' not necessary, but good practice
End Sub
```

If the null hypothesis of negligible effects of fishing was incorrectly accepted, this would be a Type II error that could have dangerous implications. No mitigating management actions would be implemented in the false belief that the stock was healthy. Conversely if it were incorrectly concluded that stock damage was accruing when it was not (Type I error) then managers could unjustifiably reduce the potential earnings of fishers (null-hypothesis would be that no stock damage was detectable and this is mistakenly deemed false). After viewing even part of the history of commercial fishing it is hard to escape the conclusion that more Type II errors than Type I errors, as above, have been made in the past.

Example Box 5.3 An example of a randomization procedure to compare two test statistics. Example from Basu (1980). Column B is filled with the random number generator =rand(). Cells C1 and D1 contain the average and median of the top 5 cells (A5:A9). The objective it to test whether the collection shown is significantly different from zero when the values can be (±1, ±2, ±3, ±4, ±7). The observed values of the two test statistics are copied as values into C5 and D5. The randomization works by sorting columns A and B on the column of random numbers (keep a copy of the raw data and its order somewhere safe on the worksheet). This has the effect of randomly reordering the available data into the two groups (used and un-used). The test statistics alter accordingly and these new values are copied down below the starting observed values. After sufficient replicates we can count the number that are the same as or greater than the original observed values. Create a macro using Tools/Macro/Record New Macro menu item. Call it Do_Randz and start by using absolute references. Select C2:D2 and delete the contents. Press <Shift><F9>. Select A5:B14 and sort them on Column B. Copy C1:D1. Switch to relative references and paste the values into cell C6 (and D6). Switch to absolute references. Select C2 and type =countif(C5:C1004,">=3.4"), select D2 and type =countif(D5:D3203,">=3"). Stop recording the macro. Press <Alt ><F11> and maximize the macro window. The modifications necessary to the macro are shown on the next page. Assign the Do-Randz macro to a button created from the Forms toolbar. If you change the number of iterations you will also need to change cells C3 and D3.

	A	B	C	D	E
1			=average(A5:A9)	=median(A5:A9)	
2	Do_Randz		1	10	
3			=C2/1000	=D2/1000	
4	Values	Reorder			
5	7	=rand()	3.4	3	
6	4	=rand()			
7	3	=rand()			
8	2	Copy Down			
9	1	To Row 14			
10	-1	0.729468			
11	-2	0.450439			
12	-3	0.023943			
13	-4	0.97243			
14	-7	0.312233			

Example Box 5.3 [cont.] The macro contents as recorded following the directions in the first part of this Example Box. You will need to add the lines and changes shown in Italics. You can either delete the struck out text or leave it, whichever you prefer. If you decide on more than 1000 iterations then you will need to alter the COUNTIF statements to reflect your choice. The only tricky bit is the replacement of the step of 5 rows down in the ActiveCell.Offset statement to become the dynamic 4 + i; thereby using the counter variable to identify the pasting location for the results from the randomizations. Try running the macro and comparing the significance of the comparison with a mean of zero when using the average and when using the median. There is a clear difference. In the text, complete evaluations of all possible combinations were discussed. In small discrete cases like this, complete evaluation is a reasonable option but with larger numbers of observations a complete evaluation becomes onerous and a random sampling provides sufficient resolution (Manly, 1997).

```
Sub Do_Randz()
'
' Do_Randz Macro
' Macro recorded 30/07/2000 by Malcolm Haddon
'
Dim i As Integer                  ' Not strictly needed, but good programming practice
'
Application.ScreenUpdating = False                    ' vital for sanity and speed
    Range("C2:D2").Select
    Selection.ClearContents            ' saves recalculating these each iteration
For i = 1 To 1000                      ' number of iteration, usually >=1000
    ActiveSheet.Calculate          ' from the Shift F9; provides new random numbers
    Range("A5:B14").Select
    Selection.Sort Key1:=Range("B5"), Order1:=xlDescending, Header:=xlGuess, _
        OrderCustom:=1, MatchCase:=False, Orientation:=xlTopToBottom
    Range("C1:D1").Select
    Selection.Copy
    ActiveCell.Offset(5, 0).Range("A1").Select            'Replace 5 with 4 + i so it reads:
    ActiveCell.Offset(4 + i, 0).Range("A1").Select
    Selection.PasteSpecial Paste:=xlValues, Operation:=xlNone, SkipBlanks:= _
        False, Transpose:=False          ' defaults can be deleted for clarity and speed
Next i
    Range("C2").Select
    Application.CutCopyMode = False
    ActiveCell.FormulaR1C1 = "=COUNTIF(R[3]C:R[1002]C,"">=3.4"")"
    Range("D2").Select
    ActiveCell.FormulaR1C1 = "=COUNTIF(R[3]C:R[1002]C,"">=3"")".
Range("A1").Select                                ' returns to top of sheet
Application.ScreenUpdating = True   ' not necessary, but good practice
End Sub
```

The *significance* of a test is the probability of making a Type I error. Because we would expect a particular test statistic value to arise about 1 time in 20 at a probability of 0.05, accepting a significance level of 5% implies that 1 time in 20 we are likely to be claiming to have found a significant difference between groups where one does not exist (Type I error). This is why a difference that is significant at the 5% level is less convincing than one at the 0.1% level.

The *power* of a test is the complement of the probability of making a Type II error. To make a Type II error is to claim no differences when differences exist, therefore the complement of this probability is that of deciding when differences exist (the complement of δ, is $1 - \delta$). In short, the power of a test is the probability of making the correct decision.

In practice, one would fix the significance level as small or large as is acceptable and then choose a statistic that maximizes the power of the test. A test is said to be *unbiased* if in using the test one is more likely to reject a false hypothesis than a true one (Good, 1994).

Table 5.1 Type I and Type II error types (after Sokal and Rohlf, 1995). A null-hypothesis would normally state that no differences existed between groups being compared (for example, before and after fishing). The columns represent the real state of nature and the rows represent the conclusions of statistical tests. In short, if one rejects the null hypothesis by concluding that differences exist when they do not, this is a Type I error. If one incorrectly accepts the null hypothesis that there are no differences, one is committing a Type II error.

	No differences exist Null true	Difference exist Null false
Null accepted	Ok	Type II error
Null rejected	Type I error	Ok

5.3 RANDOMIZATION OF STRUCTURED DATA

5.3.1 Introduction

The fact that randomization tests are restricted to tests which are basically comparisons between groups may be considered a major limitation; for example, they cannot be used for parameter estimation. On the other hand, they are very good at comparison tests.

Not every comparison is a comparison of means; sometimes we would wish to compare variation between samples. It turns out that simple randomization tests of a mean difference between two samples can be influenced by differences in variation, particularly if this is related to sample size. For comparisons of variation, Manly (1997) recommends that one should randomize the residuals of the two samples instead of the original data values. That effectively standardizes the means to zero and concentrates the test upon the variation within each sample.

Many comparisons are of more than two groups and the conventional approach is to analyze these using ANOVA. Randomization tests have been used in conjunction with ANOVA for a few decades. They are best used when the assumptions of ANOVA fail badly. The basic rule is that with unbalanced and highly non-normal data one should use a randomization procedure to determine the significance of one's analyses.

With more complex ANOVA models such as three-way, perhaps with nested factors, or repeated measures, there is some debate in the literature over the correct procedure to use in a randomization test. The question at issue is whether one should just randomize one's data across all categories and treatments [Manly (1997) says yes, Edgington (1987) emphatically says no], or whether there should be some restrictions placed upon what is randomized (randomize the treatments and categories separately). Manly also claims that one can test for interactions as one would normally, while Edgington claims the contrary and says one cannot test interactions. In Edgington's (1995) latest edition he continues the discussion and provides many insights into the analysis of structured data. Investigations into the most efficient method of conducting randomization tests on structured data are an open field in need of further research (Anderson and Legendre, 1999). The question that tends to be addressed when considering structured data is whether one should randomize the raw data, sub-sets of the raw data, or residuals from the underlying model. There is no simple answer to this as the debate between Manly and Edgington demonstrates.

What this all means is that when one's data is structured or non-linear (often the case with fisheries models and relationships), care needs to be taken in deciding what components should be randomized during a test. Anderson and Legendre (1999, p. 302) capture the present situation when they conclude the following:

> Substantial computer power is now available, enabling researchers to investigate the behaviour of computationally intensive methods. Obtaining empirical measures of type I error or power allows direct practical comparisons of permutation methods. Current theoretical comparisons of the methods cannot provide us with complete

information on how the methods will compare in different situations in practice.

5.3.2. More Complex Examples

Consider the problem of comparing growth curves (this will be considered in detail in the chapter on growth). If one wanted to compare age-length data from two populations of fish then one might fit two von Bertalanffy curves and wish to determine whether any of the parameters of the two curves are significantly different. At present one would generally use a likelihood ratio test (Kimura, 1980; see Chapter 8) but to avoid some of the assumptions concerning underlying distributions one could utilize a randomization test. However, it is not immediately clear what one should randomize.

One could randomize the available data pairs (ages plus associated lengths) between the two populations keeping the number of observation pairs in each sample the same as in the original data. However, by chance many of the older animals may be collected in one population and many of the younger in the other; this would obviously distort any growth curve fitted to this randomized data. On the other hand, this randomization design could be used when testing whether the proportional age composition or length composition of the different data sets was significantly different.

If one wanted to test the difference between the average growth curves for the two samples, then one requires a slightly more complicated scheme of randomization. The original data sets need to be stratified, in this case into discreet ages, and then the data should be randomized between populations but within age strata, thereby maintaining the original numbers of individuals within each age class for each population. In this way the underlying age structures of the two populations could be maintained while the dynamics of growth compared. In this case, it would be important to make sure that every age class was represented in each data set, though the numbers in each age class would not need to be the same. An alternative approach, which should produce equivalent results, would be to randomize ages between populations but within length classes. This example will be considered explicitly when we consider the comparison of growth curves in the chapter on growth.

6

Statistical Bootstrap Methods

6.1 THE JACKKNIFE AND PSEUDO-VALUES

6.1.1 Introduction

In this chapter, we will describe the use of the bootstrap to generate estimates of bias, standard errors, and confidence intervals around parameter estimates. We will also include a short treatment on the jackknife, which pre-dated the bootstrap as a method for generating estimates of bias and standard errors around a parameter estimate. While the jackknife and bootstrap share some common uses, they are based on very different principles of resampling.

6.1.2 Parameter Estimation and Bias

A random sample from a population may not necessarily be representative of the whole population (in the sense of reflecting all of the properties of the sampled population). For example, a sample from a normal random deviate [N(0,1) i.e., mean = zero, variance = 1] could have values almost solely from one arm of the bell-shaped curve (Fig. 6.1) and in this way the sample would not represent the full range of possible values.

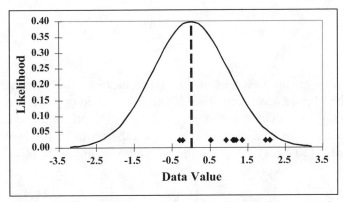

Figure 6.1 The solid line shows the expected distribution from a standard random normal deviate of mean = 0 and variance = 1. The expected mean of samples from this distribution is shown as the dashed line. However, the diamonds are a random sample of 10 having a sample mean of 0.976 and variance of 0.642.

Such a sample (Fig. 6.1) could well be random (and indeed, was) but

153

might also come about because the sampling is not strictly random. Truly random samples can be difficult to arrange, for example, the one in Fig 6.1 only arose after repeated trials. But without the samples being "truly" random, there is a possibility that the sample could produce biased estimates of such population parameters as the mean and variance or other parameters with less well-known behaviour. There can also be other sources of bias. With the example above, a visual inspection is enough to inform us that our sample is likely to give poor estimates of such statistics. But what can be done when we are unsure of the exact probability density function of the population from which we have a sample? One obvious answer is to take a larger sample, or replicate samples, but this is frequently not possible for reasons of time, funding, or circumstances.

An early solution to the investigation of bias, termed the half-sample approach (because it split the original data randomly into two equal sized groups), gave unreliable estimates of the statistic of interest and its variance (Hinkley, 1983). Quenouille (1956) produced a relatively sophisticated extension of this half-sample idea in an effort to estimate bias. Tukey (1958) went a step further by generalizing Quenouille's particular approach, calling it the jackknife, and recommending it be used to produce estimates of parameters along with approximate confidence limits.

The jackknife method is relatively straightforward. Assuming one has a random sample (iid - independent and identically distributed) of n values $x_1, x_2, ..., x_n$, then the sample mean is

$$\overline{x} = \frac{\sum x_i}{n} \tag{6.1}$$

This is also an un-biased estimate of the mean of the population from which the sample was taken. The jackknife methodology sub-sets the data sequentially by calculating the required statistic with one original value missing each time. In a data set of n values there will be n subsets of $n-1$ data points (Table 6.1). Thus, for the sample mean minus the jth value

$$\overline{x}_{-j} = \frac{\left[\left(\sum_1^n x_i\right) - x_j\right]}{(n-1)} \tag{6.2}$$

where x_j is the jth observation out of the n values. Equations 6.1 and 6.2 are related thus

$$(n-1)\overline{x}_{-j} = \left(\sum_1^n x_i\right) - x_j = n\frac{\left(\sum_1^n x_i\right)}{n} - x_j \tag{6.3}$$

which, when we convert the sum of x over n to the average of x, gives us

$$(n-1)\overline{x}_{-j} = n\overline{x} - x_j \tag{6.4}$$

and then, rearranging

$$x_j = n\overline{x} - (n-1)\overline{x}_{-j} \tag{6.5}$$

which is simply a way of showing how the x_j values relate to the sample mean and the mean with the x_j values removed. Equation 6.5 is important because it is the basis for generating the pseudo-values upon which the jackknife calculations are based; we thus go one step further than a simple resample.

The jackknife estimate of the mean is simply the mean of these x_j values, known as pseudo-values, which is clearly the same as the original un-biased mean estimate

$$\tilde{x} = \frac{\sum x_j}{n} \tag{6.6}$$

As Manly (1991, p. 25) said, "Obviously, this is not a useful result if the sample values are known in the first place. However, it is potentially useful in situations where the population parameter being estimated is something other than a sample mean." These other parameters might be a standard error or measure of kurtosis or skewness, or some other statistic or parameter.

Analogous to the jackknife estimate of the mean, the jackknife estimate of standard error of the parameter using the pseudo-values would be either of (i.e., the standard deviation of the pseudo-values divided by root n). Thus

$$\tilde{s}_{jack} = \sqrt{\frac{\sum(x_j - \tilde{x})^2}{(n-1)}} \Bigg/ \sqrt{n} \quad \text{or} \quad \sqrt{\frac{\sum(x_j - \tilde{x})^2}{(n-1)n}} \tag{6.7}$$

If other population parameters, such as the variance, were estimated, then the equivalents to the x_j pseudo-values would not necessarily be known. It is, however, possible to generalize the jackknife procedure to such population parameters. Given a population parameter, θ, which is estimated as a function of a sample of n values of x_i

$$\hat{\theta} = f(x_1, x_2, ..., x_n) \tag{6.8}$$

Example Box 6.1 A simple Jackknife example. The five jackknife samples are in columns C to G. The average and standard deviation of all columns is estimated in rows 8 and 9 with the count in row 10 (copy C8:C12 across to column G and into column B). These are used with Eq. 6.5 to create the required pseudovalues for the mean and standard deviation in rows 11 and 12. By taking putting =average(C11:G11) and =average(C12:G12) into D14 and D15 we generate the jackknife estimates of the two parameters. Try altering the data in column B to see the impact on the estimates.

	A	B	C	D	E	F	G
1		Original	JK1	JK2	JK3	JK4	JK5
2		1		1	1	1	1
3		2	2		2	2	2
4		3	3	3		3	3
5		4	4	4	4		4
6		5	5	5	5	5	
7							
8	Mean	3	=average(C2:C6)	3.25	3	2.75	2.5
9	StDev	1.5811	=stdev(C2:C6)	1.708	1.826	1.708	1.291
10	Count	5	=count(C2:C6)	4	4	4	4
11	PV Mean		=($B$10*$B8)-(C$10*C8)	2	3	4	5
12	PV StDev		=($B$10*$B9)-(C$10*C9)	1.074	0.603	1.074	2.742
13							
14			Mean of PseudoValue Mean Values	3			
15			Mean of Pseudo Value StDevs	1.647			

Table 6.1 An illustration of the generation and use of jackknife samples and replicate observations. Note that the pseudo-values when estimating the mean are simply the original data values while those when estimating the standard deviation are not directly related to the data values. The mean is an unbiased estimator of a sample mean hence the mean of the pseudo-values for the mean equals the original sample mean. The sample is not taken from a normal distribution and contains some consequent bias when estimating the standard deviation. This is reflected in the difference between the sample StDev (1.581) and the Mean of the StDev pseudo-values (1.647) (Example Box 6.1).

	Original	Data Series Systematically Minus One Observation				
	1		1	1	1	1
	2	2		2	2	2
	3	3	3		3	3
	4	4	4	4		4
	5	5	5	5	5	
Average	3	Mean$_{-j}$	3.5	3.25	3	2.75	2.5
St Dev	1.581	StDev$_{-j}$	1.291	1.708	1.826	1.708	1.291
n	5		4	4	4	4	4
PV Mean	3	PV for Mean	1	2	3	4	5
PV Mean	1.647	PV for St Dev	2.742	1.074	0.603	1.074	2.742

Just as before, there are n estimates of this θ with the jth observation removed, and just like Eq. 6.4 we can define the set of pseudo-values θ_j (Efron and Tibshirani, 1993) to be

$$\theta_j = n\hat{\theta} - (n-1)\hat{\theta}_{-j} \qquad (6.9)$$

these θ_j values act in the same manner as the x_j values when estimating the mean. To produce the jackknife estimate of the parameter θ

$$\tilde{\theta} = \frac{\sum \theta_j}{n} \qquad (6.10)$$

If θ is the sample mean then the 'pseudo-values' are exactly the same as the x_j values. If we extend this special case then we can calculate the jackknife estimate of the standard error of the jackknife replicates as

$$\tilde{s}_{jack} = \sqrt{\frac{\sum(\theta_j - \tilde{\theta})^2}{(n-1)n}}$$ (6.11)

This is treating the n 'pseudo-values' as independent data values. Efron and Tibshirani (1993, p. 145) state: "Although pseudo-values are intriguing, it is not clear whether they are a useful way of thinking about the jackknife." Partly, I believe they are saying this because they are advocates of using bootstrapping for producing better estimates of such parameters.

Some people suggest using such jackknife parameter estimates along with their standard errors to produce jackknife confidence intervals

$$\tilde{\theta} \pm t_{n-1}\tilde{s}_{jack}$$ (6.12)

where t_{n-1} is the percentile value of the t-distribution (eg. 95% value) with $n - 1$ degrees of freedom. Efron and Tibshirani (1993) suggest that confidence intervals produced in this way are not significantly better than cruder intervals based on asymptotic standard errors. It is thought that uncertainty over exactly how many degrees of freedom are involved in jackknife standard errors is part of the problem with this approach.

6.1.3 Jackknife Bias Estimation

The jackknife procedure was originally introduced in an effort to estimate and remove bias from parameter estimates. The sample parameter value estimate $\hat{\theta}$, is compared with the mean of the jackknife pseudo-values of the θ statistic; thus where the jackknife replicates are defined as a function of the jackknife sample data (note the missing x_j value)

$$\tilde{\theta}_{-j} = f\left(x_1, x_2, .., x_{j-1}, x_{j+1}, .., x_n\right)$$ (6.13)

In addition, each jackknife replicate has its complementary pseudo value as in Eq. 6.9 above, so that the jackknife estimate of bias is defined as

$$\tilde{b}_{jack} = (n-1)\left(\tilde{\theta}_m - \hat{\theta}\right)$$ (6.14)

where the mean of the jackknife pseudo values is

$$\tilde{\theta}_m = \frac{\sum \tilde{\theta}_j}{n} \qquad (6.15)$$

Of course, with these formulations one will see that when $\hat{\theta} = \bar{x}$, the estimator is unbiased (assuming a truly random sample).

Exactly what one does once one has an estimate of bias is not clear. The usual reason for its calculation is to correct $\hat{\theta}$ so that it becomes less biased:

$$\bar{\theta} = \hat{\theta} - \tilde{b}_{jack} \qquad (6.16)$$

where $\bar{\theta}$ is the bias corrected estimate of $\hat{\theta}$.   However, Efron and Tibshirani (1993, p. 138) finish a discussion of bias estimation thus:

> To summarize, bias estimation is usually interesting and worthwhile, but the exact use of the bias estimate is often problematic. ... The straightforward bias correction [Eq. 6.16] can be dangerous to use in practice, due to high variability in [the bias estimate]. Correcting the bias may cause a larger increase in the standard error ... If the bias is small compared to the estimated standard error [Eq. 6.11], then it is safer to use $\hat{\theta}$ than $\bar{\theta}$. If bias is large compared to standard error, then it may be an indication that the statistic $\hat{\theta}$ is not an appropriate estimate of the parameter θ.

This appears to suggest one should never apply Eq. 6.16 in practice! It is useful to see if there is bias but if detected, it appears it is best not to try to correct it (presumably one should try again to obtain an unbiased sample if the bias is too large).

6.2 THE BOOTSTRAP

6.2.1 The Value of Bootstrapping

There are a number of analytical strategies for producing estimates of parameters with confidence intervals from samples that have an unknown probability distribution. Efron and LePage (1992) state the general problem thus:

> We have a set of real-valued observations $x_1, \ldots x_n$ independently sampled from an unknown probability distribution F. We are interested in estimating some parameter θ by using the information

in the sample data with an estimator $\hat{\theta} = t(x)$. Some measure of the estimate's accuracy is as important as the estimate itself; we want a standard error of $\hat{\theta}$ and, even better, a confidence interval on the true value θ.

Since Efron (1979) first discussed the idea of bootstrapping it has become relatively popular, even trendy, at least among statisticians; three reasons have been suggested (Kent, 1988):

1)   Elegance: the principle behind the bootstrap, that of resampling from the empirical distribution function (as represented by a sample) instead of the actual probability density function, is simple and elegant, and yet very powerful.

2)   Packaging: the catchy name "bootstrap" makes it easy for people to recognize the product, though the potential for confusion exists now that parametric resampling, as in classical Monte Carlo simulations, is now sometimes included in the term bootstrapping.

3)   Ease of use (Kent, 1988): "for the practitioner there is the possibility of a fairly automatic and clear methodology that can be used without the need for any thought."

The last reason seems frightening and I would suggest that this latter idea be discouraged vigorously in any analyses. Bootstrap resampling is a general form of resampling in that it is resampling with replacement to produce bootstrap samples of size n. When one resamples with replacement there are many more possible arrangements (n^n in fact) of the available data (Fig. 6.2).

If one had a sample from a known normal distribution then there would be no advantage to using a bootstrap method for estimating the standard error of a parameter; "normal" theory would be best. However, in situations where the sampled population cannot be adequately represented by a normal distribution, and especially where the underlying population distribution is unknown, bootstrapping becomes most useful.

6.2.2 Empirical vs. Theoretical Probability Distributions

In fact, given a sample from a population, the non-parametric, maximum likelihood estimate of the population distribution is the sample itself. Expressed precisely, if the sample consists of n observations [x_1, x_2, x_3, ... x_n], the maximum likelihood, non-parametric estimator of the population distribution is the probability function that places probability mass $1/n$ on each of the n observations x_i. Take note that this is not saying that all values have equal likelihood, just that each observation has equal likelihood. One expects, some of the time, to obtain the same or similar values in different observations if the population distribution being sampled has a mode.

Original Sample		Bootstrap Samples				
3	8	5	3	8	3	
5	5	8	11	9	3	
7 $\Rightarrow$	11	5	5	3	11	....etc.
8	7	8	7	7	9	
9	8	7	7	3	8	
11	8	11	11	9	11	
Mean 7.17	7.83	7.33	7.33	6.5	7.5	

Figure 6.2 An original sample of six numbers with their average. From this are drawn five bootstrap samples each with a separate average. It is clear that with a sample of size n there are n^n possible bootstrap combinations. Replacement implies it is easily possible for a single observation to appear more than once in the bootstrap sample; it is also possible that some original observations will not occur in the bootstrap samples. The average of the five bootstrap replicates is 7.292 (Example Box 6.2).

The implication, first suggested by Efron (1979), is that when a sample contains or is all the available information about the population, why not proceed *as if* the sample really *is* the population for purposes of estimating the sampling distribution of the test statistic? That is, apply Monte Carlo procedures, sampling with replacement but from the original sample itself as if it were a theoretical distribution. Sampling with replacement is consistent with a population that is essentially infinite. Therefore, we are treating the sample as representing the total population.

In summary, bootstrap methods are used to estimate a parameter of an unknown population by summarizing many parameter estimates from replicate samples derived from replacing the true population by one estimated from the population (the original sample from the population).

Example Box 6.2 A simple bootstrap example (see Fig. 6.2). One can generate a bootstrap sample in Excel using the vlookup function (or the offset function). This entails generating random integers that match the numbers in an index list as in column A. The vlookup function uses these index numbers to return the data value next to each respective index. There is nothing to stop the same integer from arising, which will lead to sampling with replacement. In C7 put the function =vlookup(trunc(rand()*6)+1,$A$7:$B$12,2,false). Use the Excel help to understand this function. The trunc(rand()*6)+1 term will generate random integers between 1 and, in this case, 6. Try typing just this term elsewhere in the sheet and examining its performance each time you press F9 (i.e., recalculate the sheet). Copy C7 down to C12. Then copy C7:C12 across to IV7:IV12. Copy B1 across to IV1. The four averages in C2:C5 are bootstraps with different numbers of replicates (20 to 254). Keep pressing F9 to generate new random numbers and hence new bootstrap samples, and examine the differences between the observed mean and the bootstrap means. Which sample size provides the better estimates? The option of duplicating the parameter calculation either up or down the sheet (perhaps use hlookup instead) can be fast but will generate large worksheets. Plot C2:C5 against A2:A5.

	A	B	C	D	E	F	G
1	n	=average(B7:B12)	5.833	8.833	7.667	7.500	7.833
2	20	=average(C1:V1)	=B1-B2				
3	50	=average(C1:AZ1)	=B1-B3				
4	100	=average(C1:CX1)	=B1-B4				
5	254	=average(C1:IV1)	=B1-B5				
6	Index	Data	BS1	BS2	BS3	BS4	BS5
7	1	3	8	11	9	5	3
8	2	5	3	7	9	8	9
9	3	7	5	8	5	7	8
10	4	8	7	7	7	9	11
11	5	9	3	9	11	5	8
12	6	11	9	11	5	11	8

6.3 BOOTSTRAP STATISTICS

"Standard error" is a general term for the standard deviation of a summary statistic. So one may have the standard deviation of a sample and a standard error of a mean, but one could not have a standard error of a sample. With a sample from a normally distributed variable, one can estimate the standard error of, for example, the mean of a sample of n observations, analytically by using:

$$se_{\bar{x}} = \frac{StDev}{\sqrt{n}} = \sqrt{\frac{\sum(x_i - \bar{x})^2}{(n-1)n}} \tag{6.17}$$

Alternatively, one could take a large number of independent samples from the same population and find the standard deviation of the means of these samples. This latter process is exactly what one does for a bootstrap estimate of standard error for any parameter. A general approach for producing the bootstrap estimation of the standard error of any parameter θ is as follows:

1) Generate b independent bootstrap samples $x_1, x_2, x_3,........ x_b$, each consisting of n data values drawn randomly with replacement from the n values in the original sample (the empirical distribution). Efron and Tibshirani (1993) recommend b to be at least in the range of 25 to 200 for estimating a standard error.

2) Calculate the bootstrap replicate of the parameter or statistic $\hat{\theta}_b$ for each of the b bootstrap samples, x_b. The statistic must be a continuous function of the data (Eq. 6.18).

$$\hat{\theta}_b = f(x_b) \tag{6.18}$$

3) Estimate the standard error se_θ of the parameter θ by calculating the standard deviation of the b bootstrap replicates

$$se_\theta = \sqrt{\frac{\sum(\hat{\theta}_b - \bar{\theta}_b)^2}{b-1}} \tag{6.19}$$

where $\bar{\theta}_b$ is the mean of the bootstrap replicates of θ, which is the bootstrap estimate of the statistic θ

$$\bar{\theta}_b = \frac{\sum \hat{\theta}_b}{b} \tag{6.20}$$

Such estimates would be called *non-parametric bootstrap* estimates because they are based upon an empirical distribution instead of a theoretical probability distribution. It would be unusual today to use the bootstrap to estimate a standard error. More commonly, the bootstrap is used to generate percentile confidence intervals around parameter estimates for which other methods of obtaining confidence intervals would present difficulties.

6.3.1 Bootstrap Standard Errors

Efron and Tibshirani (1993) provide a nice example of using a bootstrap to estimate standard errors. They had a collection of data on relative student aptitudes as measured by two indices. A sub-sample from this data exhibited a correlation between the two indices of $r = 0.7764$ (Fig. 6.3). The question they asked was how to calculate the standard error of the correlation coefficient using the bootstrap (a notoriously difficult thing to do). As they had all the available data, they were able to estimate the standard error by making many independent random samples of 15 from the original 82 data points and calculating the standard deviation of the sample means.

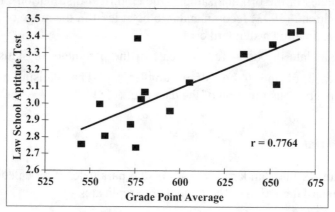

Figure 6.3 Scatterplot relating particular aptitude test scores of students from 15 law schools from the 82 that are in the U.S. (Efron and Tibshirani, 1993).

To determine if the bootstrap standard error estimate improved with increasing numbers of bootstrap replicates, more and more bootstrap replicate samples can be made for each of which the correlation coefficient is calculated. We have repeated their exercise a few times to determine the repeatability of their results (Table 6.2; Fig. 6.4; Example Box 6.3).

The standard error estimates differ between series and with b, the number of bootstrap replicates, but all estimates where b is 100 or greater, are reasonably close to the actual value of 0.131, and in some cases smaller numbers of bootstrap replicates also come close to the actual value.

Efron and Tibshirani (1993) suggest that a sample size of between 25 and 200 is sufficient when calculating standard errors. A sample size of 25 would appear to be rather low but 100 and more are clearly enough.

Hinkley (1988) suggests that the minimum number of bootstrap samples will depend on the parameter being estimated but that it will often be 100 or more. With the large increases in the relative power of computers in recent years, the number of replicate bootstraps to conduct should no longer be an issue for most problems.

Table 6.2 Standard error estimates for the correlation between two performance indices (Efron and Tibshirani, 1993) repeated ten times for each of the eight bootstrap replicate numbers. The standard error estimates become less variable after about 500 replicates (Fig. 6.4). The true standard error was approximately 0.131. This Table could be generated using Example Box 6.3; expand the macro to automate the process.

25	50	100	200	500	1000	1500	2000
0.1145	0.1293	0.1430	0.1204	0.1287	0.1343	0.1353	0.1293
0.1606	0.0976	0.1059	0.1251	0.1353	0.1341	0.1377	0.1344
0.2013	0.1436	0.1224	0.1227	0.1272	0.1353	0.1392	0.1313
0.1466	0.1196	0.1366	0.1300	0.1316	0.1319	0.1353	0.1353
0.1449	0.1257	0.1590	0.1434	0.1290	0.1351	0.1318	0.1327
0.1104	0.1725	0.1166	0.1382	0.1384	0.1346	0.1414	0.1315
0.1083	0.1307	0.1368	0.1373	0.1343	0.1310	0.1303	0.1321
0.1105	0.1242	0.1278	0.1510	0.1242	0.1310	0.1334	0.1371
0.1507	0.1483	0.1208	0.1303	0.1360	0.1260	0.1298	0.1339
0.1585	0.1545	0.1775	0.1356	0.1395	0.1336	0.1341	0.1358

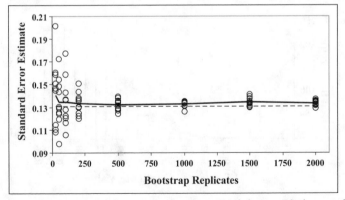

Figure 6.4 Ten estimates of the standard error around the correlation coefficient at each of eight different numbers of bootstrap replicates (25, 50, 100, 200, 500, 1000, 1500, 2000). The fine dashed line is the actual standard error of 0.131, the solid line is the mean of the estimates. Use Example Box 6.3 to generate an equivalent graph.

6.3.2 Bootstrap Replicates

Beyond using the bootstrap to estimate standard errors, it can also provide an estimate of an empirical frequency distribution (Fig. 6.5) of possible values for the statistic or parameter in question. The distribution of correlation coefficients derived from the bootstrap replicates is clearly non-normal with a significant skew.

So far, we have considered exactly what a bootstrap sample is and how it is produced. We have also seen how to determine a bootstrap estimate of a parameter (the average of the bootstrap replicates) and its standard error (standard deviation of bootstrap replicates). However, one of the major areas of research on bootstrapping has been the consideration of different approaches to calculating statistical confidence intervals for estimated parameters (DiCiccio and Romano, 1988). With non-normal populations, these can be difficult to fit in a valid or un-biased way. Using standard errors to generate confidence intervals always leads to symmetrical intervals. If the distribution of the parameter of interest is not symmetric, such confidence intervals would be invalid. The bootstrap provides us with a direct approach that can give rise to excellent approximate confidence intervals. All these direct methods rely on manipulating in some way the empirical frequency distribution of the bootstrap replicates (Fig. 6.5).

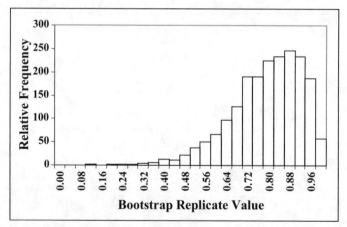

Figure 6.5 Frequency distribution of bootstrap replicate values of the correlation coefficient. The observed value was 0.7764. The median value of the illustrated frequency distribution was 0.7921. The bootstrap estimate of the correlation coefficient was 0.7739, a difference of 0.0025 from the sample estimate (this provides an estimate of any bias).

6.3.3 Parametric Confidence Intervals

Where the parameter being estimated is expected to exhibit a normal distribution of expected values (here the central limit theorem may play a part) then confidence intervals around the parameter may be obtained from the usual

$$CI = \theta \pm t_{n-1,\alpha/2} se_\theta \qquad (6.21)$$

where θ is the sample parameter estimate, $t_{n-1,\alpha/2}$ is the student's t-distribution value for n-1 degrees of freedom (where n is b, the number of bootstrap replicates), and $\alpha/2$ is the percentage confidence limits desired; with $\alpha = 0.05$, $100(1-\alpha)\%$ provides the 95% intervals where $\alpha/2 = 2.5$ and 97.5. Because the number of replicates in bootstrapping is likely to be high, then instead of using the t-distribution many statisticians simple replace it with the z value (e.g., 1.96 for the 95% confidence intervals). The bootstrap estimates of standard error can be used in this way to generate confidence intervals. In the example we have been considering the observed correlation coefficient was 0.7764 and the standard error was approximately 0.131. If we used a normal approximation, we would expect the confidence intervals to be: $0.7764 \pm 1.96 \times 0.131 = 0.5200$ and 1.0332. Clearly, with a correlation coefficient a value greater than one is nonsense and illustrates the problem of using normal theory to estimate confidence intervals for parameters with expectations from non-symmetrical distributions.

6.3.4 Bootstrap Estimate of Bias

If the sample estimate is biased it is possible to remove this bias before adding and subtracting the requisite z value to find the confidence intervals

$$\theta - \left(\overline{\theta}_b - \theta\right) \pm z_{\alpha/2} se_\theta \qquad (6.22)$$

or, equivalently

$$\left(2\theta - \overline{\theta}_b\right) \pm z_{\alpha/2} se_\theta \qquad (6.23)$$

which is the sample estimate minus the bootstrap bias, $\pm$ the standard normal distribution for $\alpha/2$, times the bootstrap standard error estimate (standard deviation of the bootstrap replicates).

Example Box 6.3 Bootstrap standard errors around a correlation coefficient [data from Efron and Tibshirani (1993); *cf.* Fig. 6.3]. We will use vlookup again, but this time because we need to return data pairs we will have a separate column of random index values (column D) and use them in both vlookup functions in columns E and F; e.g., in E5: =vlookup(D5,$A$5:$C$19,2,false) and in F5 put =vlookup(D5,$A$5:$C$19,3,false), note the only change is the column from which to return data. Copy these down to row 19. Each press of F9 will renew the random numbers in column D, which will lead to new data pairs in columns E and F. In C2 put =correl(B5:B19,C5:C19), which is the original correlation, and in F2 put =correl(E5:E19,F5:F19), which is the bootstrap sample correlation. Record a macro through Tools/Macro/Record New Macro, and call it Do_Boot. While in absolute references copy F2, change to relative references and paste values into G5. Stop recording. Place a button on the worksheet from the Forms toolbar and assign Do_Boot to it. Edit the macro (via Alt-F11) and make the changes listed in italics to make the macro functional. Change the number of bootstrap replicates to conduct by altering C1. Note the answers and determine a reasonable number of replicates to estimate the standard error of approximately 0.131. You will need to run each set a number of times (LSAT is law school aptitude test and GPA is grade point average).

	A	B	C	D	E	F	G
1	TRIALS		500				
2	Original r		0.7764	Do_Boot		0.9735	=average(g5:g2006)
3	Original Data pairs						=stdev(G5:G2006)
4		LSAT	GPA	Bootstrap Index	lsat	gpa	Bootstraps
5	1	576	3.39	=trunc(rand()*15)+1	558	2.81	0.930064
6	2	635	3.3	=trunc(rand()*15)+1	578	3.03	0.833836
7	3	558	2.81	8	661	3.43	0.90485
8	4	578	3.03	5	666	3.44	0.679865
9	5	666	3.44	2	635	3.3	0.708266
10	6	580	3.07	8	661	3.43	0.857945
11	7	555	3	12	575	2.74	0.341376
12	8	661	3.43	8	661	3.43	0.726493
13	9	651	3.36	12	575	2.74	0.876033
14	10	605	3.13	9	651	3.36	0.75519
15	11	653	3.12	8	661	3.43	0.948527
16	12	575	2.74	8	661	3.43	0.771965
17	13	545	2.76	9	651	3.36	0.790429
18	14	572	2.88	15	594	2.96	0.728108
19	15	594	2.96	14	572	2.88	0.628797

Example Box 6.3 [cont.] Some of the changes needed could be recorded as a separate macro and then copied into this one (e.g., The cell clearing lines just below the Dim statements). The MsgBox will soon lose its novelty and can be turned off by converting it to a comment with a '. Try running this a few times for 15 replicates without the screenupdating turned off (comment out the necessary statements). Plot the bootstrap replicate samples of column E against F and observe how they change.

```
Sub Do_Boot()
'
' Do_Boot Macro
' Macro recorded 12/08/2000 by Malcolm Haddon
'
Dim i As Integer, b As Integer
Dim start As Double, endtime As Double
'
Range("G5").Select
Range(Selection, Selection.End(xlDown)).Select
Selection.ClearContents
Range("G6").Select
Application.ScreenUpdating = False
start = Timer
b = Range("C1").Value        ' get the number of replicates from the sheet
For i = 1 To b
   ActiveSheet.Calculate
   Range("F2").Select
   Selection.Copy
   ActiveCell.Offset(2 + i, 1).Range("A1").Select
   Selection.PasteSpecial Paste:=xlValues
Next i
Range("A1").Select
endtime = Timer
Application.ScreenUpdating = True
MsgBox Format(endtime - start, "##0.000"), vbOKOnly, "Bootstrap Complete"
End Sub
```

Of course, if the statistic being considered were far from normal then Eq. 6.23 would produce erroneous results. The example of the correlation coefficients (Fig. 6.5) is one where a statistic has an expected distribution that is far from normal. If the estimates are corrected for bias prior to estimating the confidence intervals, the distortion on the confidence intervals becomes slightly greater. Using Eq. 6.23 we have: (2 x 0.7764 – 0.7739) ± 1.96 x 0.131 = 0.5221 and 1.0357. Thus, the upper limit becomes slightly further removed from one (the maximum possible value for a correlation coefficient).

6.4 BOOTSTRAP CONFIDENCE INTERVALS

6.4.1 Percentile Confidence Intervals

Given b bootstrap replicate samples we can generate b bootstrap estimates of the parameter of interest θ_b. An estimate of the 100(1-α)% confidence limits around the sample estimate of θ is obtained from the two bootstrap estimates that contain the central 100(1-α)% of all b bootstrap estimates (Efron, 1979). Thus, the 97.5 and 2.5 percentiles of the 2000 bootstrap estimates of the correlation coefficient, summarized in Fig. 6.5, estimate the confidence intervals around the correlation coefficient (Fig. 6.6).

With 2000 bootstrap replicates the 95% confidence intervals values would be found at the 50th and at the 1950th position in the sorted bootstrap estimates. In this case the intervals are between 0.4591 and 0.9629. The bootstrap estimate of the correlation coefficient was very slightly to the left of the sample estimate (0.7739 vs. 0.7764), which shows that the bias in the estimate was small. Note, in this case, the sample estimate is also close to the median value of the bootstrap estimates (Fig. 6.6).

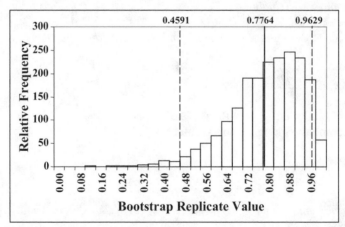

Figure 6.6 Two thousand separate bootstrap estimates of a correlation coefficient grouped into classes of 0.04 with the observed correlation indicated as a solid line and the bootstrap percentile confidence intervals indicated as dashed lines. The 95% bootstrap confidence intervals were at the 51st (2000 x 0.025) and the 1950th (2000 x 0.975) sorted bootstrap replicate. The bootstrap estimate of r was 0.7739, while the median was 0.7921. Unlike normal confidence intervals these confidence intervals are clearly asymmetric and do not suffer from suggesting an upper limit greater than one. In Excel these can be found using =percentile(g5:g2006,0.025) and =percentile(g5:g2005,0.975).

In general terms, one would not wish to attempt to fit percentile bootstrap confidence intervals with less than 1000 bootstrap replicates.

Even with 2000 replicates the histogram is not as smooth as one could desire for it to be used as a probability density function. However, there is a limit to the precision with which it is useful to estimate these simple bootstrap percentile confidence intervals because they are only ever approximate.

6.4.2 Bias-Corrected Percentile Confidence Intervals

Often the bootstrap estimate of the parameter of interest differs somewhat from the observed parameter estimate indicating that there may be evidence of bias. The validity of any confidence intervals should be improved if they take into account any bias present. Fitting bias-corrected percentile confidence intervals is slightly different to fitting usual bootstrap percentile intervals around a parameter. The difference lies in having first to determine exactly which percentiles should be used after removing any bias that arises because the observed parameter value is not the median of the distribution of bootstrap estimates. Thus, with a slightly biased sample, one might use percentiles such as 0.0618 and 0.992, or perhaps 0.0150 and 0.960 instead of 0.025 and 0.975.

The determination of bias-corrected percentiles is relatively simple (Efron, 1987; Efron and Tibshirani, 1993, provide a justification). After generating the bootstrap sample estimates, one determines F, the fraction of bootstrap replicates that are smaller than the original parameter estimate. Thus if 56% of the bootstrap replicates were smaller than the original parameter estimate, F would be equal to 0.56. From this fraction, a constant z_0 is calculated to be the probit transform of F

$$z_0 = \Phi^{-1}(F) \qquad (6.24)$$

where Φ^{-1} is the inverse, standard cumulative normal distribution [=norminv(0.56,0,1) in Excel]. From Eq. 6.24, the appropriate percentiles for the 95% confidence intervals are calculated by the following

$$P_{lower} = \Phi(2z_0 - 1.96)$$
$$P_{upper} = \Phi(2z_0 + 1.96) \qquad (6.25)$$

where Φ is the cumulative normal distribution function. The 1.96 is, of course, the critical value from the inverse normal curve for the 95% confidence intervals. This can be altered appropriately for alternative intervals (e.g., The value would be 1.6449 for 90% intervals). After using

Eqs. 6.25, one would then determine the bias corrected percentile confidence intervals by taking the values from the bootstrap distribution that align with the calculated upper and lower percentiles.

For an example where 56% of bootstrap replicates are smaller than the original parameter estimate this gives $F = 0.56$ so that $z_0 = 0.151$, leading to $P_{lower} = \Phi(-1.65806) = 4.87\%$, and $P_{upper} = \Phi(2.26194) = 98.8\%$ [using =normdist(0.988,0,1,true) in Excel]. Thus, from 1000 ordered values in a bootstrap distribution the 95% confidence intervals would be the 49th and 988th values instead of the 25th and 975th values. If there is no bias at all then $F = 0.5$ which would imply a $z_0 = 0$ which would finally lead to the lower and upper bounds being the default of 2.5% and 97.5% (Example Box 6.4).

Example Box 6.4 Simple percentile and first-order bias-corrected bootstrap percentile confidence intervals. This example extends the worksheet created in Example Box 6.3. Column G contains 1000 bootstrap replicates of a correlation coefficient with the mean bootstrap estimate in G2 and the standard error estimate in G3. The additions are in columns I, J, and K. The simple percentile intervals are the easiest to implement, in K10 and K11 put the functions =percentile(G5:G1004, 0.975) and =percentile(G5:G1004, 0.025). The values that arise will depend upon the particular bootstrap replicates that are on the sheet but they will be similar to that shown. J3 contains the estimate of F, the fraction of bootstrap replicates less than the observed value, J4 contains Eq. 6.24, J5 and J6 contain the $2z_0 - 1.96$ and $2z_0 + 1.96$, and J7 and J8 contain Eq. 6.25, the percentile points where the bias-corrected confidence intervals will be found. In the example sheet, the values were 0.9528 and 0.4324, which have been shifted to the left of the simple percentile intervals in K10 and K11. Run the macro a few times with 1000 bootstrap replicates and notice the impact on the different percentile confidence intervals. Increase the number of replicates to 2000 and modify the equations in J3, J4, and J10:K11. Does the increased number of replicates stabilize the estimated intervals?

	G	H	I	J	K
1					
2	0.77441		**Original r**	0.776374	
3	0.12621		**Fraction F**	=countif(G5:G1004,"<0.776374")	
4	Bootstraps		**norminv(F)**	=norminv(j3/1000,0,1)	
5	0.93006		**P_Lower**	=2*J4-1.96	
6	0.83384		**P_Upper**	=2*J4+1.96	
7	0.90485		**Percentile L**	=normdist(J5,0,1,true)	
8	0.67987		**Percentile U**	=normdist(J6,0,1,true)	
9	0.70827			**Bias-Corrected Intervals**	**Simple**
10	0.85795		**Upper95**	=percentile(G5:G1004,J8)	0.9621
11	0.34138		**Lower95**	=percentile(G5:G1004,J7)	0.4765

6.4.3 Other Bootstrap Confidence Intervals

We have only considered the simple bootstrap percentile and first order bias-corrected bootstrap percentile confidence intervals but many other algorithms exist for generating confidence intervals from bootstrap samples. When using confidence intervals (for example 90% intervals) one would want them to fail to cover the parameter θ exactly 5% of the time in each direction. There have been many comparative studies conducted on the variety of methods of constructing confidence intervals around a parameter (Efron, 1992; Manly, 1997). Because they use data sets whose properties are known, these studies are able to indicate the strength and weaknesses of the various methods available. There are two methods that usually perform rather better (generate confidence intervals that work – see Example Box in Chapter 7) than the simple and first-order bias-corrected algorithms considered here; these are named the bootstrap-t method and the accelerated bias-corrected percentile methods (Efron, 1987). Both of these extensions are more computer intensive than the two we have considered here in that they both require the use of a jackknife for best performance (Efron and Tibshirani, 1993; Manly, 1997).

Manly (1997) provided a listing of published biological work using the bootstrap. Generally, in fisheries, the simple percentile confidence interval is most commonly used (Kimura and Balsinger, 1985; Sigler and Fujioka, 1988) though as the first-order bias-corrected percentile intervals involve only a small amount of extra work and the confidence intervals are generally improved we would recommend that these be used instead. Further examples of the use of the bootstrap will be given when we consider other more complex models including the surplus-production models that are the simplest stock assessment models available.

6.4.4 Balanced Bootstraps

In the process of running a bootstrap procedure the resampling with replacement means it is possible that not all observations in the original sample will be resampled the same number of times across however many replicates are used. While this is not seen as a problem when there are many thousands of bootstrap replicates, if there are only, say, 1000, then a better algorithm might be to use a balanced bootstrap. In the balanced bootstrap one first copies the values to be resampled the same number of times as there will be bootstrap replicates, then one resamples at random, without replacement, from within the multiple copies. This process will automatically mean that all observations occur an equal number of times while bootstrapping.

The balanced bootstrap does not appear to have been used very often. It would certainly be easier to use with a single series of data than with more complex models or with structured data. If a model had more than one series of data to be bootstrapped a balanced design would still be possible but would require some sort of cross-tabulation between data series. For example, two-dimensional balance could be achieved using a classic orthogonal Latin square design.   By balancing the use of each observation the amount of resampling needed to estimate the expectation and other moments of a parameters distribution, with acceptable accuracy, is reduced by up to fivefold (Hinckley, 1988). However, Hinckley (1988) reports that balanced bootstraps are not so effective for estimating bootstrap percentiles, especially for $p < 0.05$ or $p > 0.95$. This is an area deserving of further research.

6.5 CONCLUDING REMARKS

Bootstrapping offers the ability to generate confidence intervals around parameters and model outputs in situations that were previously impossible to approach. The ability to test for bias around a parameter estimate is valuable but of most value is the ability to estimate the uncertainty around the parameter estimates by estimating standard errors or confidence intervals and the underlying frequency distribution of the parameter of interest.

Where it is valid to use parametric statistics, then probably the best strategy is to use them. However, if one is uncertain what the underlying distribution of a parameter is then bootstrapping is to be recommended. Fisheries data are generally so variable and uncertain that the debate over the optimum algorithm, whilst interesting, neglects the fact that the real limitation is most often in the quality of the available data.

7

MONTE CARLO MODELLING

7.1 MONTE CARLO MODELS

7.1.1 The Uses of Monte Carlo Modelling

Monte Carlo models and simulations are computer intensive methods that can be used to test specific hypotheses and to answer less specific "what-if" type questions, including projections into the future. Whichever task is at hand one requires a simulation model of the fishery that can be compared with nature. It can differ from an assessment model in that it may include hypothetical relationships, parameters, and other modifications. Any modelling task that requires a projection or extrapolation of a model's implications into the future is best carried out using Monte Carlo simulation techniques. Risk assessment, harvest strategy evaluation, and the more general management strategy evaluation, are all relatively recent developments that are becoming more commonly applied to fisheries around the world. All use simulation models to explore the implications of manipulating the management of the fishery concerned. The impact of uncertainty concerning any part of a model may also be investigated using simulations.

In this chapter we will consider aspects of modelling that are required when implementing simulations that include hypothetical or unknown components. Like other computer intensive methods, Monte Carlo simulations require resampling from statistical distributions. In particular they require resampling with replacement from theoretical probability density functions (PDFs). Detailed examples will be used to illustrate the principles discussed.

Strict Monte Carlo resampling differs from bootstrapping in that the distributions being resampled are parameterized PDFs and are not just empirically determined. Beware of confusion over terminology; some people refer to Monte Carlo resampling as parametric bootstrapping and to bootstrapping as non-parametric bootstrapping. As always, terminology is unimportant as long as one is clear about what is being resampled and how.

7.1.2 Types of Uncertainty

A major requirement for developing Monte Carlo simulations is to devise methods for including uncertainty about parameter estimates and other components of a model. No ecological process is known perfectly and this constitutes a problem when attempting to create a simulation that adequately captures a system's dynamic behaviour. This appears to be especially true when attempting to understand fishery dynamics. Rather than work entirely with deterministic models, simulations invariably include stochastic elements in which random variation around some model component(s) is included. In this way the implications of uncertainty may be included in a model analysis. There are a number of different types of uncertainty that can influence ecological modelling:

1) The appropriateness of a model. Fundamentally different, independent models can be used to describe the dynamics of a natural system; it is uncertain which model is the best representation.
2) Process error or uncertainty arises where a model has been defined in a deterministic manner but some of the components actually vary randomly through time (for example, virgin biomass is usually defined as representing a constant but this index of stock productivity is likely to change with environmental conditions).
3) A model of how a particular system operates in nature is accepted but values for some of the parameters involved are unknown or cannot be independently estimated; these tend to be given fixed values (e.g., natural mortality is often given an assumed constant value in population models).
4) Observation error or uncertainty relates to how precisely a given model's parameters can be estimated given a particular set of data (this is what is estimated when constructing confidence intervals around a set of parameters).
5) Measurement error or uncertainty, where data being put into the model is only measured with error (e.g., ring count estimates from otoliths). This can easily be confounded with observation error.

When conducting a fisheries' stock assessment, large parts of the model are often deterministic (process error – type 2 above, is ignored). Parameters are estimated along with some indication of the precision of the parameter estimates (type 4 above, observation error is recognized). However, very often there are parameters, such as natural mortality, that are only poorly known and are not easily estimated. In these cases values for such parameters are often set at an assumed value (type 3 uncertainty, above). The model can be said to be conditioned on such parameters. One of the advantages of Bayesian methodology, as it is used in fisheries (Punt and Hilborn, 1997), is that it is possible to include a prior-distribution that attempts to capture our uncertainty for such parameters. A common alternative is to test the sensitivity of the assessment to these constant conditioning parameters (the data is repeatedly re-fit to the model and each time the conditioning parameters are given different values from a predetermined range of values). If there were more than one conditioning parameter, then testing the sensitivity of all possible combinations of values would become more and more difficult as the number of parameters and their ranges increased. Rather than completing such a set of discrete combinations it would be more efficient to conduct a simulation in which values for the conditioning parameters were selected at random from defined ranges. Such simulations would be Monte Carlo simulations.

By comparing simulated outcomes against observations from nature it is possible to test hypotheses. Simulation models would tend to be stochastic models because at least some of the variables or parameters have uncertain values. In these cases, the values for the variables or parameters are determined anew during each *run* of the Monte Carlo model, by sampling "with replacement" from a known probability distribution.

7.2 PRACTICAL REQUIREMENTS

7.2.1 The Model Definition

An obvious requirement for Monte Carlo simulation modelling is a formal simulation model. This may closely resemble an assessment model but in addition it will have a number of unknown components (variables, parameters, or sequence events) that need to be included. The final requirement is to define the expected probability density functions that are to be used to generate values for each uncertain model component.

7.2.2 Random Numbers

Fundamental to the process of stochastic simulation modelling is a requirement to obtain random values from under a variety of probability density functions. In Chapter 3, a number of standard PDFs were introduced. Given the parameters of a PDF for a given variable, we saw how to calculate the likelihood for a particular value of that variable. What we now require is a method of inverting that process. We want to be able to randomly select values of the variable concerned once we have selected a parameterized PDF.

There are numerous algorithms for generating random values from given PDFs (Hastings and Peacock, 1975) and modellers who use Monte Carlo simulations tend to collect different techniques for doing this (Press *et al.*, 1989). Fundamental to many methods is the generation of random numbers from a standard uniform variate (i.e., values from 0 to 1 with equal probability). Press *et al.* (1989) discuss a number of algorithms for generating uniform random numbers.

The pseudo-random number generators used in computers are of variable quality. Such generators are so fundamental to computer intensive methods that it is a good idea to test any random number generator that you may use. A simple test is to count the number of times the random number generator gives particular values and compare those with the number of times such values would be expected; one could use either a G test or a χ^2 test (Example Box 7.1).

In tests of the efficiency of random number generators, if one chooses a significance level of 5% (i.e., $\chi^2 > 16.919$, for 9 degrees of freedom), the comparison of the frequency of classes of random numbers with their expected values would be expected to fail 5% of the time. The random number generator in Excel (Example Box 7.1) will generate failure rates that average 5%; in some particular tests it is greater than 5%, and in others it is less than 5% (e.g., the average of 10 trials of 1000 replicates each was 0.0504).

7.2.3 Non-Uniform Random Numbers

The fundamental idea of sampling under a PDF curve is relatively simple, although how it is implemented for each PDF is not necessarily straightforward. For continuous variates the area under a PDF curve must sum to one. If there is a method for generating random numbers from 0 to 1, then these can be used to generate target areas under a curve. For example, given a normal distribution, $N(0,1)$ i.e., mean zero and variance = 1, then we know that a random number (area under the curve) of 0.05 will give an X-variate value of -1.6448, a random area of 0.5 will give a value of 0, and one of 0.9 will give 1.2815 (Fig. 7.1). To obtain these inverse PDF values one requires a mechanism by which the PDF curve is integrated until the required random probability is attained (generated from a random number taken between 0 and 1) and then the respective target value of the X – variate is determined (Fig. 7.1).

It must be remembered that for many statistical distributions there are either direct equations for generating the required random numbers (Hastings and Peacock, 1975), or different, more efficient algorithms for generating the required numbers.

If someone wishes to write an independent computer program to conduct a simulation then a formal algorithm will be necessary once it has been decided which PDF to use for the problem being considered (Press *et al.*, 1989). Fortunately, in Excel many inverse PDFs are provided as built in functions (Example Box 7.2; Fig. 7.2), and one could write other user-defined functions.

Example Box 7.1 A Monte Carlo test of the random number generator function rand(). The =rand() function is copied down as far as A1001 (exactly how far is up to you). Enter the number of Monte Carlo trials in E15. In cells C2:C11 enter the array function =frequency(A2:A1001,B2:B11), then press <Ctrl><Shift><Enter>. This function counts the number of times a value <= each respective bin value occurs in the column of random numbers. The χ^2 test is of whether the observed frequencies match the expected and column E has the (Obs-Expected)^2/Expected. This test should fail 1 time in 20 at $P = 0.05$. Instead of repeatedly pressing the F9 button one should generate a macro to count the frequency of successes/failures. The only part of the macro that can be recorded is the activesheet.calculate (as <Shift><F9>). The rest of the algorithm must be added. The ability to write simple programs like this is required for Monte Carlo simulation modelling. Does using bins of 0.05 affect the result?

	A	B	C	D	E
1	=count(A2:A1001)	Bin	Observed	Expected	Chi2
2	=rand()	0.1	41	=$A$1/10	=(C2-D2)^2/D2
3	=rand()	0.2	38	=$A$1/10	=(C3-D3)^2/D3
4	=rand()	0.3	42	CopyDown	0.1
5	Copy Down	0.4	38	40	0.1
6	0.693034	0.5	39	40	0.025
7	0.502615	0.6	38	40	0.1
8	0.395583	0.7	46	40	0.9
9	0.230871	0.8	43	40	0.225
10	0.598551	0.9	45	40	0.625
11	0.579494	1	30	40	2.5
12	0.421189			Chi2	=sum(E2:E11)
13	0.705528	Do_Test		P	=chidist(E12,9)
14	0.197339			Significant	0.0430
15	0.275854			Replicate	1000

```
Sub Do_Test()                                    ' attached to the button on the sheet
  Dim i As Integer, n As Integer                 ' unnecessary, but good practice
  Randomize                                      ' resets the random number generator
  n = 0                                          ' Sets counter to zero
  For i = 1 To 1000
    ActiveSheet.Calculate                        ' only recalculates a single sheet
    If Cells(13, 5).Value < 0.05 Then n = n + 1  ' no  End If  as only one line
    Cells(14, 5).Value = n / i                   ' interact with the worksheet directly
    Cells(15, 5).Value = i                       ' enables awareness of which replicate
  Next i                                         ' is running
End Sub
```

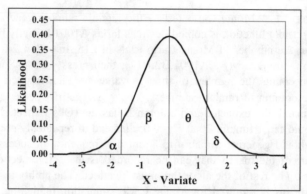

Figure 7.1 Three different values of a variate, X, taken from a standard normal distribution N(0, 1), using three different cumulative probabilities: $\alpha + \beta + \theta$. The area under the curve for α was 0.05, for $\alpha + \beta$ was 0.5, for $\alpha + \beta + \theta$ was 0.9, and $\alpha + \beta + \theta + \delta$ was 1.0. The X – value for each of the first three probabilities were - 1.6448, 0, and 1.2815, respectively.

How one conducts the integration under the curve will depend upon the probability density function being used (Press *et al.*, 1989). This approach to generating random numbers is, in fact, only one out of many possible. It has only been given here to assist in the development of intuitions about the process.

7.2.4 Other Practical Considerations

The selection of a suitable probability density function to represent a particular variable or parameter in a simulation is obviously an important step. There are many PDFs that were not considered in Chapter 3 but which are useful in simulations because they are so flexible and general. As usual, one can use whatever is most appropriate as long as one can defend the selection.

Commonly, Monte Carlo replicates are often more time-consuming than randomization or bootstrap replicates and, in the literature it is often found that the number of Monte Carlo replicates are limited. This is risky as few replicates rarely produce a smooth distribution of expected outcomes. As with the randomizations and bootstraps, the reduction of noise in the results should improve with increasing numbers of Monte Carlo replicates.

Example Box 7.2 The generation of normally distributed random numbers. Instead of using the variance in their normal equations Microsoft programmers elected to use the standard deviation. This unusual usage possibly derives from them using the standard normal distribution (N[0,1]) in which the variance equals the standard deviation. The frequency bins are extended down to B36 generating a range from 1 to 9 in steps of 0.25. Select C5:C37, type =frequency (A5:A1004,B5:B37), and press <Ctrl><Shift><Enter> to enter the array formula. Put the following function into D5 and copy down to D37: =($C$3/4)*normdist (B5,$B$1,$B$2,false). The first term accounts for the relative numbers and the bin sizes. Plot column C against B as a histogram and add column D to the same graph but make the graph type for that data series a line graph (to mimic Fig. 7.2). By repeatedly pressing F9 the random numbers will be recalculated, generating new normally distributed random numbers. Note the differences between the sample mean and standard deviation (C1:C2) and the parameters from which the sample was generated (B1:B2). Note how the graph alters each time and how closely or otherwise it resembles the expected. Try reducing the number of normally distributed random numbers (e.g., delete A105:A1004) and note the impact on the quality of the generated normal distribution. Try altering the value given in B2 and note its impact on the quality of the distribution generated. Clearly, large samples are best when there are many frequency categories. If you implement a χ^2 test on this sheet remember that the test will be more severe than usual as there will be many expected frequencies of less than 1, as in D5:D9 (Sokal and Rohlf, 1995). Note the use of rand() in column A. Compare C3 with D3.

	A	B	C	D
1	**Mean**	5	=average(A5:A1004)	
2	**StDev**	1	=stdev(A5:A1004)	
3	Count		=count(A5:A1004)	=sum(D5:D36)
4	Inverse Normal Values	Bins	Frequency	Expected
5	=norminv(rand(),$B$1,$B$2)	1	0	0.017
6	=norminv(rand(),$B$1,$B$2)	1.25	0	0.044
7	=norminv(rand(),$B$1,$B$2)	1.5	0	0.109
8	Copy down to row 504	1.75	0	0.254
9	4.829436	2	0	0.554
10	5.430025	2.25	0	1.137
11	3.227922	2.5	4	2.191
12	4.424067	2.75	4	3.967
13	6.961762	3	7	6.749

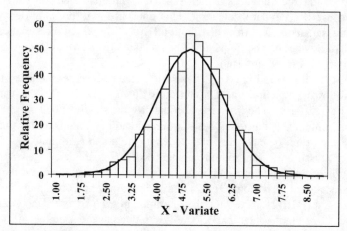

Figure 7.2 A normal distribution generated from 500 random numbers sampled from under a normal PDF with a mean of 5 and variance of 1. The solid curve denotes the expected relative frequency (Example Box 7.2). Note the imperfections despite there being 500 random numbers contributing to the frequency distribution. Each sample of 500 would generate a different relative frequency distribution. This should reinforce the idea that one should try to use as many Monte Carlo replicates as is practically possible.

7.3 A SIMPLE POPULATION MODEL

As an example we will consider a population growing without density dependence

$$N_{t+1} = rN_t \tag{7.1}$$

where N_t is the population size at time t, and r is the population growth rate. As was seen in Chapter 2, such a population has three possible behaviours depending on the value of the growth rate r. If r > 1, then the population grows exponentially to infinity, if r = 1 the population is stable, and if r < 0, then it declines exponentially to zero. This assumes that the growth rate is a constant. It seems much more likely that a density-independent species will be greatly affected by the environment and the growth rate is likely to vary above and below a value of one. The question to be answered is whether a population with a randomly varying growth rate (having a mean value of 1 and a given variation σ) would be able to survive. Put another way, we could ask how often would we expect such a species to go extinct over a given number of generations and how is that extinction rate influenced by the variability in growth rate? The model becomes

$$N_{t+1} = r_{N(1,\sigma)}N_t \tag{7.2}$$

where the $r_{N(1,\sigma)}$ is a random normal deviate having a mean of 1 and a standard deviation of σ; we also need an artificial lower limit to the r value, say 0.01.

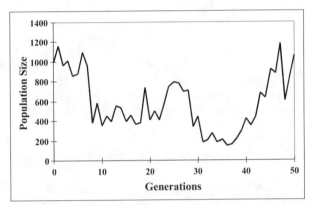

Figure 7.3 A typical population trajectory through 50 generations when the growth rate for Eq. 7.2 is set to a mean of 1 and a standard deviation of 0.3. A series of r-values greater than 1 would lead to massive population growth and similarly a series of values less than 1 would lead to a marked reduction. While variation is low most model runs do not go extinct and the population trajectory reflects the random walk of the r-values around the mean value.

We need to limit the value of r in the simple model to positive numbers because if unconstrained normal random deviates were used it would be possible to obtain a negative growth rate, which would mean instant extinction. We also need to define extinction; in this case if the population falls below a single individual (i.e., $N_t < 1$) it can be regarded as extinct.

With the simulation model definition in Eq. 7.2 (and the listed assumptions), it would be possible to generate the information necessary to illustrate the relationship between the proportion of simulation runs that go extinct in a given number of generations and the variability of the growth rate (Fig. 7.4; Example Box 7.3). In a real exercise, one would also monitor the range and variability of population sizes, along with other measures of the model's performance. Despite the extreme simplicity of this model investigation the power of the method should be clear (Fig. 7.4).

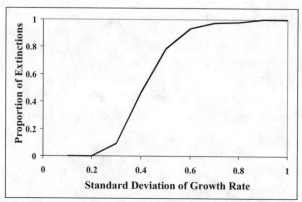

Figure 7.4 Outcome of 10,000 Monte Carlo replicates of a population growing in a density-independent fashion with a growth rate described by a normal distribution with a mean of 1 and a standard deviation of σ. The graph compares the proportion of model runs that went extinct over 50 generations against the standard deviation of the growth rate.

7.4 A NON-EQUILIBRIUM CATCH-CURVE

7.4.1 Ordinary Catch-Curve Analysis

A more complex example will illustrate many of the ideas already discussed. As we saw in Chapter 2, if a single cohort were exposed to a constant mortality rate then the numbers would be expected to decline exponentially following the relationship

$$N_t = N_0 e^{-(M+F)t} = N_{t-1}e^{-Z} \tag{7.3}$$

where N_t is the cohort size at time t, and Z is the instantaneous total mortality rate (natural and fishing combined).

   If the total population concerned received constant recruitment and a constant mortality rate each year, then all cohorts would be identical and the numbers in each age-class would be expected to decline exponentially. Equation 7.3 would refer irrespective of whether N_t referred to a cohort or the total population (Fig. 7.5). While it is obviously unlikely that any fished population adheres to any of these conditions (constant recruitment and mortality through time), the logarithmic transform of this relationship is sometimes used to make an estimate of the total mortality

$$Ln(N_t) = Ln(N_0) - Zt \tag{7.4}$$

Example Box 7.3 A density-independent population growth model in which the growth rate is described by a normal distribution: N(1, σ). Plot the numbers through time (column B) against time (column A) as a connected scattergram to obtain the equivalent to Fig. 7.3. By varying the value in B1 and pressing F9 to recalculate the sheet the population trajectory will alter and may go extinct. If the population goes extinct, then the text in C4 will appear (format it as bold and red). Cell B55 is duplicated in B4 for visual convenience. The use of the max function in column C is to prevent negative deviates being produced. Instead of manually replicating model runs, one could use the macro listed at the bottom of the box. This will run 1000 Monte Carlo replicates and paste the results onto the sheet when finished. You should try to extend this macro by adding an outer loop so that it automates the investigation of the relationship between σ and the proportion going extinct (Fig. 7.4). Investigate how this population model responds to lognormal random deviates, put =loginv(rand(),0,$B$1) into C7, copy down, and repeat your analysis. This is computer intensive work so try to automate as much as possible.

	A	B	C
1	sigma σ	0.3	
2	Prop	0.30	
3	Initial	1000	
4	N50	=B55	=IF(B4=0,"EXTINCT"," ")
5	Time	Nt	**Normal Random Deviate**
6	0	=B3	
7	1	=IF(B6*C7>1,B6*C7,0)	=max(norminv(rand(),1, $B$1),0.01)
8	2	=IF(B7*C8>1,B7*C8,0)	=max(norminv(rand(),1, $B$1),0.01)
9	3	=IF(B8*C9>1,B8*C9,0)	=max(norminv(rand(),1, $B$1),0.01)
10	4	Copy down to row 55	Copy down to row 55
11	5	1685.231	0.887778
12	6	1765.256	1.047486

```
Sub Do_Monte()
Dim i As Integer, n As Integer
Randomize
Application.ScreenUpdating = False
n = 0
For i = 1 To 1000
  ActiveSheet.Calculate          ' Only part recordable
  If Cells(4, 2).Value = 0 Then n = n + 1
  Cells(2, 2).Value = n / i
Next i
Application.ScreenUpdating = True
End Sub
```

Given Eq. 7.4 and the assumptions of constant recruitment and mortality, a linear regression of the log of numbers-at-age against age should produce a plot with a negative gradient of slope Z. Such an analysis is referred to as a catch-curve (Fig. 7.6), and by definition these relate to a fishery in equilibrium.

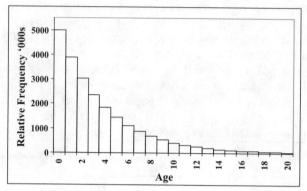

Figure 7.5 A population in equilibrium with a constant recruitment of 5 million animals and a constant rate of mortality, in this case $Z = 0.25$, being applied instantaneously leading to an exponential decline. As recruitment is constant the decline in the total population numbers reflects the decline in each cohort i.e., there is equilibrium.

7.4.2 The Influence of Sampling Error

As well as assuming constant, equilibrium levels of recruitment and mortality, standard catch-curve analyses also assume that samples precisely represent the population. Sadly, it is difficult to obtain a sample from a population (in which the fish are measured and aged) that provides a perfect representation of the relative abundance of each age class (the catch-at-age). It is natural to ask what difference sampling error would make to the catch curve analysis. The simplest way of determining this would be to take the perfect sample of size N, which gives rise to the true relative frequencies f_t for each age class t, and then add some random variation to each estimate of the relative frequency. This variability would need to be smaller for the older ages classes in absolute terms else unrealistically large relative frequencies of older animals would be possible. Also, no frequencies less than one should be permitted

$$\hat{f}_t = f_t + \varepsilon \times \frac{f_t}{\bar{f}} \qquad (7.5)$$

where f_t is the true frequency, f-bar is the overall mean frequency, f-hat is the expected frequency, ε is a normal random variate with mean zero and standard deviation σ (i.e., $N[0, \sigma]$). The last term weights each normal deviate so that the variation around larger true frequencies is greater than around the smaller true frequencies (i.e., we have a constant coefficient of variation and not a constant variance).

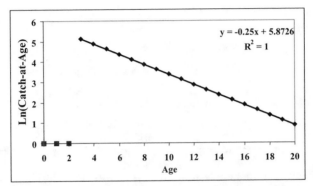

Figure 7.6 A population in equilibrium with a constant recruitment of 5 million animals and a constant rate of mortality of $Z = 0.25$. The first three year-classes are not selected by the fishery and hence are not sampled. The log of the relative frequency of each age-class (the relative catch-at-age) plotted against age, generates a straight line whose gradient is an estimate of Z. This is the classical catch curve analysis (Example Box 7.4).

When this is done, the first set of random errors added to the "true" relative frequencies will give rise to a non-perfect straight line with a slightly different gradient from the "true" mortality rate (Fig. 7.7; Example Box 7.4).

The actual gradient observed in a single trial, has little meaning on its own. However, one can easily imagine taking the same age-structure and adding a new set of random errors and repeating the analysis to derive further mortality estimates. If this process were to be repeated many times we would end up with, say, 1000 Monte Carlo estimates of the Z value. Using these values to form a frequency distribution of outcomes we could then determine the effect of different degrees of sampling error upon the estimation of Z (Fig. 7.8). In addition, as with bootstrapping, we could determine approximate percentile confidence intervals on the estimate of Z (Example Box 7.5).

The catch curve analysis conducted in Example Boxes 7.4 and 7.5 assumed both a constant annual recruitment and total mortality applied to each recruited age-class. Sampling error was added in the form of normal random variation with a constant coefficient of variation. By implementing a Monte Carlo simulation of the fishery and the sampling, we were able to investigate the importance of obtaining a representative sample of the relative catch-at-age.

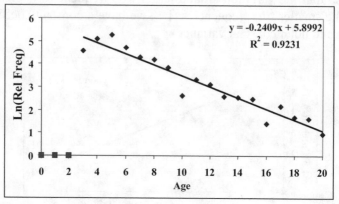

Figure 7.7 The same population data as in Figs. 7.5 and 7.6 except that the true relative frequencies in a sample of 750 (*cf.* Fig. 7.6) have had random variation added as described in Eq. 7.5, with a σ of 15. Note that, in the Monte Carlo replicate illustrated, the linear regression now has a gradient of only 0.241 instead of the true value of 0.25.

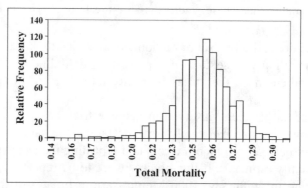

Figure 7.8 Frequency distribution of a simple Monte Carlo simulation of a catch curve analysis. The hypothetical population had constant recruitment and total mortality but was sampled with random error in the estimates of numbers-at-age (Example Box 7.4). The errors had a constant coefficient of variation instead of a constant variance, which allowed for greater variation at high relative frequencies and lower variation at low frequencies. The distribution of Z estimates is slightly skewed but approximately normal. While the median was 0.24997 (the expected was 0.25), the 2.5 and 97.5% percentiles were 0.205 and 0.286, respectively (55.4% below 0.25 and 44.6% above).

Example Box 7.4 A classical catch curve analysis with sampling error included. N0 is the constant annual recruitment. Age_c is the age at which the fish is first vulnerable to capture. Z is the total mortality (name cell B6 as Z), n is the sample size, and σ_s is the standard deviation of the sampling error. Column D takes the age at first capture into account. Sample size (B8) times the proportion of each age class in the total fished population =IF(D11>0,D11*$B$8/$D$9,"") is the "Perfect Sample" in E11:E31. To add sampling error, put =if(isnumber(E11),E11+ (norminv(rand(),0,$B$7)*E11/$E$9),0) into F11:F31. The total sample may now be smaller or greater than n. The E11/$E$9 generates the constant coefficient of variation that prevents excessively large numbers occurring in the older age classes (the lower frequencies). In column H, repeat the formula without the first E11 to see the range of random variation decreasing with age. In G11 enter =if(F11>1,Ln(F11),0), which prevents the sampled frequency dropping below 1 and produces the required log of catch-at-age. Plot G14:G31 against column A as a scattergram; add a linear trend line to the data to mimic Fig. 7.7. Enter =linest(G14:G31,A14:A31) into E1:F1 as an array function (<CTRL><SHIFT> <ENTER>) to produce the catch curve regression. By pressing F9, the rand() functions are recalculated and a new sample will be generated with its associated gradient. You can observe the impact of varying the sampling error variance (in B7) on the quality of the observed catch curve. If you alter the age at first capture you will need to alter E1:E2 and the plot.

	A	B	C	D	E	F	G
1				**Gradient**	-0.269	5.8317	
2			Z	=abs(E1)			
3							
4	N0	5000000					
5	Age$_c$	3					
6	Z	0.25					
7	σ$_s$	15					
8	n	750		**Tot Vuln**	**Average f**		
9				=sum(d11:d31)	=average(e11:e31)		
10	Age	Nt		**Vuln N**	**Perfect Sample**	**+Error**	**Ln(n)**
11	0	=B4		=if(a11>=b5,b11,0)		0.0	0
12	1	=b11*exp(-Z)		=if(a12>=b5,b12,0)		0.0	0
13	2	=b12*exp(-Z)		=if(a13>=b5,b13,0)		0.0	0
14	3	=b13*exp(-Z)		Copy Down	167.8	184.1	5.2155
15	4	Copy Down		To Row 31	130.7	168.5	5.1267
16	5	To Row 31		1432524	101.8	93.6	4.5386

The Monte Carlo simulation in Example Box 7.4 could also be used to investigate the impact of obtaining a biased catch-at-age sample, or of bias in the ageing methodology (ageing error), or other sources of error. By investigating which sources of error exert the greatest impact on the final estimates, researchers can identify weak links in their chain of reasoning and focus their research on reducing uncertainty in those areas.

Example Box 7.5 A macro that can be added to Example Box 7.4 to automate the execution of multiple repeats of the Monte Carlo simulation. This will generate a column of estimates of Z that can be summarized using the =frequency function. Implement this and attempt to obtain a version of Fig. 7.8.

```
Sub montecarlo()
  Dim i As Integer
  ActiveSheet.Calculate
  Application.ScreenUpdating = False
  For i = 1 To 1000
    Range("D2").Select
    Selection.Copy
    ActiveCell.Offset(i, 9).Range("A1").Select 'pasting recalculates the sheet
    Selection.PasteSpecial Paste:=xlValues     ' unless automatic calculation is
  Next i                                        '   turned off
  Application.ScreenUpdating = True
  Range("a1").Select
End Sub
```

7.4.3 The Influence of Recruitment Variability

The assumption of constant recruitment and equilibrium in the catch curve analysis grates against our intuitions about reality. To investigate the significance of this assumption we could arrange a Monte Carlo simulation where stochastic recruitment variation was included. In this case, the population model would need to be able to step its cohorts through time with each annual iteration having a randomly selected number of recruits. An efficient method of arranging this would be to use some deterministic stock recruitment relationship, such as the Ricker equation (see Chapter 9), and then adding random error (Eq. 7.6)

$$\hat{R}_{t+1} = aS_t e^{-bS_t} e^{N(0,\sigma)} \tag{7.6}$$

where R_t is the recruitment at time t, S_t is an index of the spawning stock size at time t, a and b are constants of the Ricker equation (a is the recruits-per-spawner at low stock levels, and b relates to how quickly the level of recruits-per-spawner drops as S increases). As with most stock-recruitment relationships, we should use multiplicative lognormal errors (hence the e^ε). The spawning stock is the sum of all mature aged animals. In practice, this would entail calculating the deterministic recruitment expected from a given stock size and including the random variation. The impact of total mortality on each age class would be calculated and then the whole copied and stored ready for the next iteration (Example Box 7.6).

Example Box 7.6 A non-equilibrium, variable recruitment catch curve analysis. This worksheet is identical to that in Example Box 7.4 except in cells A1:B3, B9, B11:B31, and C12:C31. B11:B31 are now just numbers, C12:C31 now represent the population suffering mortality. Name B1 as a and B2 as b. Column D now refers to column C, which represents the latest population structure. Otherwise the worksheet remains the same. Cells A1:B3 represent the parameters of the Ricker stock recruitment relationship and the standard deviation of the recruitment error. Put =a*D9*exp(-b*D9)*exp(norminv(rand(),0,B3)) into B9 to calculate the expected recruitment. One could have used loginv(rand(),0,B3) to obtain the same result. D9 represents the spawning stock size that is used in the stock recruitment relationship. Try varying both σ_r and σ_s, and determine the effect on the catch curve analysis. Implement a macro to repeat the analysis 1000 times and construct a relative frequency of Z estimates (Example Box 7.7; Fig. 7.11).

	A	B	C	D
1	Ricker a	6		Gradient
2	Ricker b	1.826E-07	Z	0.2365
3	σ_r	0.3		
4	N0	5000000		
5	Age_c	3	Step_Model	
6	Z	0.25		
7	σ_s	15		
8	n	750		Mature Pop
9	Recruits	6673370		=sum(D11:D31)
10	Age	Nt	Nt+1	Mature N
11	0	8774442		
12	1	5760803	=B11*exp(-$B$6)	=IF(A12>=$B$5,C12,0)
13	2	2726971	=B12*exp(-$B$6)	=IF(A13>=$B$5,C13,0)
14	3	3374618	=B13*exp(-$B$6)	=IF(A14>=$B$5,C14,0)
15	4	2540555	Copy Down	Copy Down
16	5	1520554	to row 31	to row 31
17	6	1293433	1184209	1184209
18	7	878412	1007327	1007327

Given such a model, even with no sampling error we would expect to see variation in the relative frequencies of ages in the population. This variation would reflect the non-constant recruitment to the population. An implication of variable recruitment is that a population is likely to exhibit evidence of strong and weak year classes (Fig. 7.9). If this data were used in a catch curve analysis, we would expect to see noise about the hypothetical straight line, even in the absence of sampling error (Fig. 7.10; Example Box 7.6).

Example Box 7.7 A macro and Monte Carlo procedure to add to Example Box 7.7 to automate the simulations and enable percentile analyses of the results. Set the sampling error to 0.001 and the recruitment error to 0.3, and run the Monte Carlo analysis, generate a histogram of the results using the =frequency function and store the actual numbers away from column K. Then set the sampling error to 15 and rerun the Monte Carlo analysis. Add the Z estimates from this to the histogram already generated to produce something akin to Fig. 7.11.

```
Sub Step_Model()                          ' best attached to a button on the sheet
' Conducts a single population growth cycle ' watch for strong year classes
    Range("C12:C31").Copy                 ' copy the latest population and
    Range("B12").Select                   ' save it as values ready for the
    Selection.PasteSpecial Paste:=xlValues ' next iteration
    Range("b9").Copy                      ' copy the recruitment
    Range("B11").Select                   ' ready for the next iteration
    Selection.PasteSpecial Paste:=xlValues
    Application.CutCopyMode = False       ' removes the copy highlights
    Range("A1").Select
End Sub
```

```
Sub Do_Monte_Carlo()
' automates the simulations required for Monte Carlo analyses
Dim i As Integer
For i = 1 To 1000
    Application.ScreenUpdating = False    ' for peace of mind and speed
    Step_Model                            ' call the other macro
    Range("D2").Copy
    ActiveCell.Offset(i, 10).Range("A1").Select ' paste Z estimates into
    Selection.PasteSpecial Paste:=xlValues       ' column K
Next i
    Range("a1").Select                    ' go back to top of sheet.
    Application.ScreenUpdating = True
End Sub
```

A Monte Carlo analysis of catch curves with only recruitment variation leads to a symmetric distribution of total mortality estimates, while recruitment and sampling variation leads to a skewed distribution with much wider percentile confidence intervals (although the median remains centrally located about 0.25). The earlier analysis of sampling variation alone (Fig. 7.8) indicated that the distribution of the Z-estimates only increased its range in a small way after the addition of recruitment variation. With only sampling error the 95% percentile confidence intervals were 0.205 and 0.286 (Fig. 7.8). With both sampling error and recruitment variation the confidence intervals around the Z-estimate were 0.1995 and 0.292 (Fig. 7.11). The importance of recruitment variability does not appear as important as the presence of sampling error on the number-at-age.

It would not be valid to draw conclusions after so few investigations. The ranges of variation assumed for the sampling and recruitment error would need to be altered and their interaction more thoroughly determined before any conclusions should be drawn.

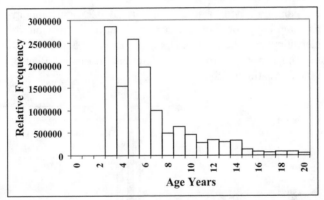

Figure 7.9 A non-equilibrium age-structured population model with a constant rate of mortality of $Z = 0.25$ and variable recruitment as in Eq. 7.6, with $\sigma = 0.3$. The first three year-classes are not selected by the fishery and hence are not sampled. Strong year classes appear evident from 3, 5, 6, and 9 years before present (*cf.* Fig. 7.5). If this population were to be sampled, especially if there were sampling error, we would not expect a particularly good fit of a straight line from the log of the relative catch-at-age against age, as required by the standard catch curve analysis (Fig. 7.10; Example Box 7.6).

7.5 CONCLUDING REMARKS

Monte Carlo simulation modelling allows one to investigate many parts of nature that are currently not amenable to direct observation of experiment. Its value in risk assessment and management strategy evaluation is its most important use in fisheries today (Francis, 1992). Risk assessments involve projecting the stock dynamics forward under different simulated management regimes and of course, this is beyond experience.

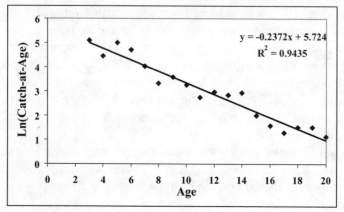

Figure 7.10 Catch curve analysis of the data, given in Fig. 7.9, from a non-equilibrium population having a total mortality of 0.25 and random recruitment. There was no sampling error. Note the scatter about the regression line and, in this case, the underestimate of the Z value (Example Boxes 7.5 and 7.6). Repeating this analysis many times would permit a characterization of the effect of recruitment variation on the estimation of Z.

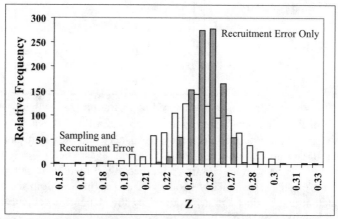

Figure 7.11 The impact of sampling error relative to recruitment variation on the estimation of total mortality using a catch curve analysis (Example Boxes 7.6 and 7.7). The model settings are recruitment variability = 0.3 and sampling variability = 15. The darker columns reflect the catch curve analysis when only recruitment variability is present. The empty bars are where both types of variation are present. The sampling errors dominate the analysis (*cf.* Fig. 7.8 on p.186).

Many skills are needed to conduct such simulation modelling, including a knowledge of probability density functions and which to use in what circumstances, as well as a facility with computer programming, in whatever language suites you best. Most importantly, however, is a need to have an understanding of the present state of knowledge about the system being modelled. Not all simulation models need be realistic and for those situations a simple mathematical view of the system would be all that was required. However, for a simulation model to be realistic any hypothetical additions would need careful construction to ensure biological possibility. Even if a model has been inappropriately constructed it will still generate the implications of particular actions. Generally, it is better to investigate a range of model structures than rely on a single formulation.

Whenever a simulation model is used and it contains parts that are strictly invention it must not be mistaken for reality; it is only a model. When conducting what-if trials, we would be testing what would be the case if the system operated in a particular way. It is important not to forget that even if the results reflected observable nature closely this would not imply that the particular model being used is a true representation of the operation of nature. As always, while a good fit certainly describes it does not necessarily explain.

8

Growth of Individuals

8.1 GROWTH IN SIZE

8.1.1 Uses of Growth Information

Ignoring immigration and emigration, stock production is a mixture of recruitment of new animals to the population and the growth of the individuals already in the population. This is one reason there is a huge literature on the growth of individuals in fisheries ecology. In addition, many aspects of a species' life history characteristics tend to be reflected in how it grows. Thus, long-lived species tend to be relatively slow growing, while short-lived species tend to grow much more quickly. In this chapter, we will ignore the biology of growth and focus on the mathematical description of growth.

In ecology and fisheries, it is very common to gather information about how organisms grow in size (especially as length or weight at age). Besides recruitment of juveniles, increases in the stock biomass vulnerable to fishing come about by the growth of individuals already recruited to the stock. This aspect of production is used in many stock assessment analyses. For example, yield-per-recruit analyses, as discussed in Chapter 2, ask: what average or minimum size or age at first capture leads to the optimal yield? This is a trade-off between the loss of biomass through the natural mortality of individuals and the gain to biomass through the growth of individuals. Without this sort of information and analysis there is a risk of growth over-fishing (taking too many fish when they are still too small for optimum yield). Ecologically, growth information can provide insights into how a species interacts with its environment. Mathematical relationships between age, length, and weight are often used to translate the outputs of models based upon animal numbers into outputs based upon biomass.

The literature on individual growth is too extensive (Summerfelt and Hall, 1987) to permit anything other than a review of the most important aspects from a stock assessment point of view. In this chapter, we will introduce various models of individual growth and the methods used to estimate their parameters for particular populations. These will include growth transition matrices as used in size-structured models (most common with species that are difficult to age). Very commonly, it is useful to be able to compare growth curves (indeed, any set of non-linear relationships). Strategies and methods for making such comparisons are also addressed in this chapter, including a randomization test that may improve matters when the ageing data is sub-optimal.

8.1.2 The Data

Generally, the data in studies of growth consist of estimates and measurements of age, size (often length or width), and weight. However, tagging data can also be collected and that, generally, consists of the dates of tagging and recapture, the initial length at tagging, and the length at recapture. It is very unusual to have the age of the individuals in tagging studies. Different methods are required for fitting growth models to these two distinct types of data.

We are interested in the relationships formed between all three possible combinations of direct data: age vs. length, age vs. weight, and length vs. weight. Of these, the latter usually has the simplest model and is usually the most straightforward to fit to raw data. Obtaining data relating to length or weight is relatively simple, although weight may be affected by many things other than length (maturity and gonad development stage, gut contents, freshness, etc.). Obtaining data relating to age tends to be far more problematical.

Many methods have been suggested for ageing aquatic organisms and the particular approach that best suites a particular species depends upon many factors. Many species do not grow at a uniform rate throughout the year. Instead, there tends to be a slowing of growth in the winter or less productive months. This can be relatively unmarked in tropical areas but tends to be very clear in temperate or freshwater environments. Many body parts can be affected by this differential growth and these effects can be used to age the fish. Literal growth rings (analogous to annual tree rings) can be found in such body parts as the scales, the vertebrae, fin-ray spines, but especially in the various ear bones known as otoliths (Summerfelt and Hall, 1987). As well as annual rings there have been more recent developments using daily growth rings. Analogous rings can be found in some invertebrates and although these may not necessarily be annual, they can also be used to age the animals concerned (Moltschaniwskyj, 1995).

There is an enormous literature concerning the development and structure of otoliths. However, here all we need to know is that when a fish to be aged is captured it is first measured for length and/or weight and then commonly its otoliths are removed (how they are removed can vary from species to species; this can be a highly skilled art when done properly).

There are a number of ways in which the otoliths (or other structures to be aged) can be treated to make the yearly rings visible but, whichever is used, the aim of the ageing is to determine the age of the fish in years. More precise work can provide estimates of fractions of years. Clearly, using such methods relies upon many assumptions, the most important being that the rings can be clearly identified and that they are, in fact, annual; this latter can be validated using tagging experiments or a variety of other techniques (Summerfelt and Hall, 1987).

8.1.3 Historical Usage

A variety of mathematical descriptions have been applied to growth. At the start of the century various people used either a constant proportional increase or a linear increase in size with age. In the 1930s the observed size at age was used, and, later still, Ricker is reported as assuming an exponential increase in weight (Smith, 1988). In the 1950s, there was a search for a general mathematical model of growth with a biological basis, that is, an explanatory model. Beverton and Holt (1957) introduced the idea of von Bertalanffy growth curves to fisheries. This was an approach in which growth was defined as the balance between positive and negative processes within organisms. Von Bertalanffy derived an equation that could be used to predict the length of an organism as a function of its age (von Bertalanffy, 1938). The cube (or close approximation to the cube) of this equation could then be interpreted as the weight of an organism in relation to age (changes in one dimension being reflected in three dimensions). The validity of applying this model to the average growth of collections of individuals, when it was designed to describe the growth of single individuals, was not attended to at the time (Sainsbury, 1980).

A common alternative to the von Bertalanffy equation is simply to have a look up table of mean lengths (or weights) at a given age or the proportional distribution of numbers at different sizes for given ages. If sufficient information is available, such age-length keys permit estimates of the uncertainty around each mean length-at-age value. However, an empirical table does not readily permit interpolation of missing or under-represented ages so most models use some mathematical representation of average individual growth. Not all growth equations currently used have interpretable parameters; there are now examples of the use of polynomial equations to describe growth in fish (Roff, 1980).

8.2 VON BERTALANFFY GROWTH MODEL

8.2.1 Growth in Length

Despite a wide array of criticisms the model of growth in length most commonly used in fisheries is the three-parameter equation developed by von Bertalanffy (1938).

$$L_t = L_\infty \left(1 - e^{-K[t-t_0]} \right) + \varepsilon \tag{8.1}$$

where L_t is the length at age t, L_∞ (pronounced L-infinity) is the asymptotic average maximum body size, K is a growth rate coefficient that determines how quickly the maximum is attained, and t_0 is the hypothetical age at which the species has zero length (t_0 fixes the position of the curve along the x axis, and can affect the steepness of the curve; see Fig. 8.1). It should be remembered that t_0 is an extrapolation from available data and can be difficult to interpret. The epsilon (ε) denotes the belief that residuals would be distributed normally about the expected growth line. When fitted using least-squared residuals this curve represents the average growth of the population members.

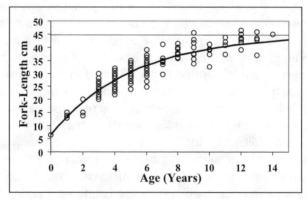

Figure 8.1 Von Bertalanffy growth curve for fish length (cm) against age in years, representing a sample of snapper, *Pagrus auratus*, from the Bay of Plenty, New Zealand in 1985. The curve parameters are $L_\infty = 44.79$, $K = 0.196$, and $t_0 = -0.81$. The L_∞ asymptote is illustrated by the fine horizontal line. Notice that some points lie above this line. This is because, with Eq. 8.1, L_∞ is interpreted as the average length at the maximum age. Notice also that most data occur where the curve is turning over (ages 3 to 6), that there are few data points for the youngest ages, and the number of observations is reducing for the older fish; this is typical of many data sets. This lack of data points for the younger and older animals can distort or bias all of the estimated parameters (L_∞ and K should be bigger, while the t_0 value should be smaller).

Continuous and serial spawning species add variation into the length at age relationship. To some extent the t_0 parameter glosses over uncertainties relating to the date of spawning and metamorphosis from larval form to juvenile.

The parameter values derived from a single sample may not provide an adequate representation of the growth properties of the sampled population (Fig. 8.1). The L_∞ and t_0 parameters are at the extremes of the curve and this is where the data tend to be least adequate. As with all things, care must be taken to obtain a representative sample. The curve may be fitted to raw data using non-linear least squares methods (Example Box 8.1). Whether the assumption of a constant variance is valid will depend on the quality of the data used.

8.2.2 Growth in Weight

The relation between length (L_t) and weight (w_t) for many animals is best described by a power function

$$w_t = aL_t{}^b \tag{8.2}$$

where b is the allometric growth parameter (often close to the value 3) and a is a scaling constant. By combining Eqs. 8.1 and 8.2, and defining the asymptotic maximum expected weight to be w_∞ we can produce the von Bertalanffy growth equation for body weight

$$w_t = w_\infty \left[1 - e^{-K[t-t_0]}\right]^b \tag{8.3}$$

which is identical in form to the equation for length, Eq. 8.1, but replacing L_∞ with the equivalent of Eq. 8.2 at L_∞ (i.e., $w_\infty = a.L_\infty{}^b$) and the addition of the allometric growth parameter b (the two equations are identical when a and b are set to one). Often the constant b is set to a value of 3. The shape of the two curves is clearly different (cf. Figs. 8.1 and 8.2). The curve describing growth in length has a rapid increase that slows down to reach the asymptote at the L_∞. The curve describing growth in weight, however, can have two inflections producing a sigmoidal curve (Fig. 8.2).

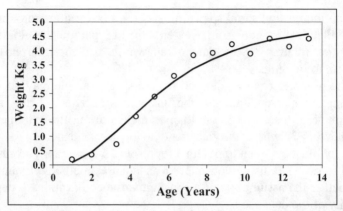

Figure 8.2 Von Bertalanffy growth curve of invented mean body weights against age in years for female Pacific hake (length and age data from Kimura, 1980). The sigmoidal shape of the curve, which contrasts with that for growth in length, is typical. ($w_\infty = 4.85$ and $b = 3.078$).

As with the von Bertalanffy curve of length at age, one can use non-linear least squares to fit this curve to a data set (Fig. 8.2; Example Box 8.1). Of course, if the residuals were deemed to have a distribution that is anything other than a normal distribution (this would be uncommon), then one would need to use maximum likelihood methods. Where there may be a change from the standard least squares strategy is with the variance of the residuals. It may well be the case the variance of the residuals is not constant but increases with age. If this were so, it might be necessary to use residuals having a constant coefficient of variation rather than a constant variance (Example Box 8.1).

8.2.3 Seasonal Growth

The growth of many organisms in highly seasonal waters does not necessarily proceed at the same rate throughout the year. Growth rings in otoliths and other hard parts come about through differences in the metabolism and growth rate of the species concerned. In the tropics, with reduced seasonal variation in the environment, annual rings are difficult to detect because growth is more continuous. This is one reason why age-related fisheries techniques can be far less useful in tropical regions than in temperate or boreal systems (although there has been some progress using daily growth rings; Choat and Axe, 1996).

Example Box 8.1 Fitting length- and weight-at-age, von Bertalanffy growth curves to length data for female Pacific hake (Kimura, 1980). The weight-at-age data were invented for this example. To calculate the expected lengths in column D, put =$B$1*(1-exp(-$B$2*(A5-$B$3))) into D5 and copy down to D17. Similarly, in F5 put =$F$1*(1-exp(-$B$2*(A5-$B$3)))^$F$2 and copy down to F17. Finally, put =average(D5:D17) into cell F3. Plot column C against A as a scattergram and add column F as a line (*cf.* Fig. 8.2) to observe the relation between age and weight. In addition, plot column B against A as a scattergram and add column D to the graph to relate length and age (*cf.* Fig. 8.3). First use the solver to minimize D1 by altering B1:B3 (the values given here are close to but not the optimum). Then, minimize D2 by varying cells F1:F2. Copy and store the values from A1:F3 to the right of the workings below. Then minimize the total sum of squared residuals in D3 by varying B1:B3 and F1:F2. Are the answers different? If you believe the variance of the residuals around the curve increases with age we could use a constant coefficient of variation and not a constant variance. This means we need to increase the weight of the residuals from the younger fish. To do this, put =($F$3/D5)*(B5-D5)^2 into E5, copy down, and re-solve. The curve fits the younger ages more closely than the older ages (the residuals are allowed to spread more widely with age).

	A	B	C	D	E	F	G
1	L_∞	61.2	SSq_L	=sum(E5:E17)	W_∞	4.85	
2	K	0.3	SSq_W	=sum(G5:G17)	B	3.078	
3	t_0	-0.06	Tot_SSq	=D1+D2	Avg Ex(L)	48.853	
4	Age	Obs(L)	Obs(W)	Ex(L)	SSqL	Ex(W)	SSqW
5	1	15.40	0.20	16.671	=(B5-D5)^2	0.0883	=(C5-F5)^2
6	2	28.03	0.35	28.212	=(B6-D6)^2	0.4466	=(C6-F6)^2
7	3.3	41.18	0.72	38.865	=(B6-D6)^2	1.1978	=(C7-F7)^2
8	4.3	46.20	1.70	44.654	Copy Down	1.8370	Copy Down
9	5.3	48.23	2.40	48.942	To Row 17	2.4366	To Row 17
10	6.3	50.26	3.12	52.119	3.4571	2.9574	0.0265
11	7.3	51.82	3.82	54.473	7.0378	3.3883	0.1864
12	8.3	54.27	3.93	56.216	3.7886	3.7336	0.0607
13	9.3	56.98	4.22	57.508	0.3342	4.0042	0.0466
14	10.3	58.93	3.88	58.465	0.2163	4.2129	0.1109
15	11.3	59.00	4.42	59.174	0.0302	4.3723	0.0023
16	12.3	60.91	4.13	59.699	1.4666	4.4929	0.1317
17	13.3	61.83	4.42	60.088	3.0345	4.5837	0.0268

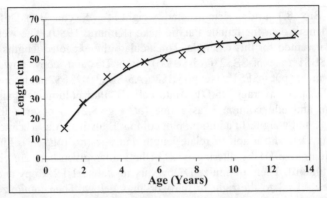

Figure 8.3 The optimal von Bertalanffy curve fitted to data for female Pacific hake (Kimura, 1980; weight data invented). The curve may be fitted using simple least squares with a constant variance for the residuals for all ages, or using weighted least squares to give rise to a constant coefficient of variation (Example Box 8.1).

In non-tropical regions, and especially in freshwater systems, differences in growth rate within a year are sometimes so marked that modifications are required to growth models so that seasonal variation can be described adequately. Once again, many different models have been proposed. One possible modification to the von Bertalanffy curve is (Pitcher and MacDonald, 1973)

$$L_t = L_\infty \left(1 - e^{-\left[C\sin\left(\frac{2\pi(t-s)}{52}\right) + K(t-t_0) \right]} \right) \tag{8.4}$$

where C is related to the magnitude of the oscillations above and below the non-seasonal growth curve, s is the starting point in time for the sine wave (relates to phase), and the other constants, K, L_∞, and t_0, are defined as before (Fig. 8.4). The variable t is the age at length L_t. The value of 52 in Eq. 8.4 indicates that the time scale of events is that of weeks (thus, s, t, and t_0 will be measured in weeks). In effect, this equation is the von Bertalanffy curve with a sine wave added.

By changing the period units (52, plus s, t, and t_0) to 12, one could just as easily work with months (though obviously with less precision). Such seasonal adjustments to the growth model are less likely to be necessary in tropical areas, and are more likely to be required in freshwater environments where temperatures and the productivity of the habitat are highly seasonal.

Table 8.1 A subset of length at age data for minnows measured from a figure in Pitcher and Macdonald (1973). The ages are in weeks and the lengths are in mm. This data is illustrated in Fig. 8.4 and used in Example Box 8.2.

Weeks	Obs L	Weeks	Obs L	Weeks	Obs L
1	3	52	30	114	65
4	9	60	36	122	67
8	15	64	38	132	68
10	12	70	48	134	64
14	19	72	45	138	65
24	24	82	49	146	67
28	24	90	49	152	68
30	21	94	52	158	71
38	21	104	59	172	75
48	24	112	61	176	73

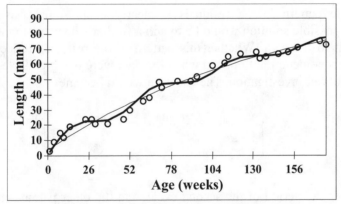

Figure 8.4 Data extracted from Pitcher and Macdonald (1973) for minnows. Note the x-axis is in weeks and covers just over three years. The thick curve is from Eq 8.4 with $L_\infty = 106.89$, $t_0 = -5.2$, $K = 0.00768$, $C = 0.06$, and $s = 4.007$. The fine non-oscillatory curve is the same von Bertalanffy curve without the imposed oscillation (Example Box 8.2). Both t_0 and s are in units of weeks.

As with all curve fitting, it is a good idea to examine the residuals after the optimum fit has been determined (Fig. 8.5). When this is done with the minnow data, it is clear that the optimum fit is missing an obvious cycle in the data.

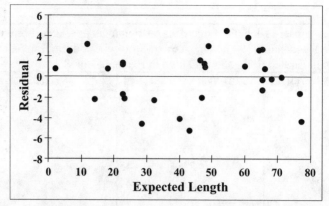

Figure 8.5   The residuals remaining after an optimum fitting seasonal growth curve has been fitted to the data in Table 8.1 (Fig. 8.4). There remains a clear sine wave in the data. Inspection of Fig. 8.4 suggests that it has a period of about two years so it may be worthwhile adding a further sine wave to the model (Example Box 8.2).

The pattern in the residuals is clearly a sinusoidal wave so an obvious possible solution would be to add a further sine wave to the model and re-fit to the data. Whether this pattern in the residuals reflects some natural phenomenon not noticed when the data were being collected would require further investigation. The equation would become

$$L_t = L_\infty \left(1 - e^{-\left[C_1 \sin\left(\frac{2\pi(t-s_1)}{52} \right) + C_2 \sin\left(\frac{2\pi(t-s_2)}{p} \right) + K(t-t_0) \right]} \right) \tag{8.5}$$

where p is the period of the second cycle, and the other parameters are as before, only duplicated for each cycle (Example Box 8.2).

The addition of a further cycle to the model may appear arbitrary but illustrates a vital issue when attempting to model a natural process. The objective is generally to obtain the optimal balance between generating a realistic model of a process while at the same time obtaining an optimal fit to the data. A regular pattern in the residuals indicates a trend remains undescribed by the model. Therefore, decisions need to be made about changing the model (usually to make it more complicated). The second sine wave presumably represents a cycle of longer period than the annual cycle of seasons. In this way, the model has indicated a valuable research direction to be followed in future work (Example Box 8.2).

Example Box 8.2 A seasonal varying growth curve fitted to data extracted from Pitcher and MacDonald (1973) for minnows (Fig. 8.4; Table 8.1). The data need to be copied into columns A and B, down to row 36. The model is relatively complicated being Eq. 8.5; put the following into C7 and copy down to C36: =$B$1*(1-exp(-($B$4*sin((2*pi()*(A7-$B$5))/52)+$E$1*sin((2*pi()*(A7-E$2)) /$E$3)+$B$2*(A7-$B$3)))). When the amplitude parameters c_1 and c_2 are set to zero, then the sine wave terms collapse to zero leaving the basic von Bertalanffy growth curve. Remember that each of the time-related parameters are in weeks. Plot column B against A as a scattergram and add column C to it as a thick line. Add column F to this as a thin line to mimic Fig. 8.4. Minimize F5 by altering cells B1:B5. Then plot the residuals as column D against C to mimic Fig. 8.5. Once the pattern in the residuals is apparent, re-fit the model but this time alter cells B1:B5, E1:E3. Observe the impact of this on the total sum of squares and on the residuals. Is there still a pattern in the residuals? Is there anything suggestive about the estimated period for the second curve? Is it worth adding three more parameters for the improved quality of fit?

	A	B	C	D	E	F
1	L_∞	106.9		c_2	0	
2	K	0.0068		S_2	-20	
3	t_0	-5.2		Period2	70	
4	c_1	0.06				
5	S_1	4.0			SSQ	=sum(E7:E36)
6	T	Obs_L_t	Model	Resid	Resid^2	Non-Seasonal
7	1	3	2.19	=B7-C7	=D7^2	=b1*(1-exp(-b2*(a7-b3)))
8	4	9	6.46	=B8-C8	=D8^2	=b1*(1-exp(-b2*(a8-b3)))
9	8	15	11.84	=B9-C9	=D9^2	=b1*(1-exp(-b2*(a9-b3)))
10	10	12	14.24	Copy	Down	Copy Down to Row 36
11	14	19	18.18	to	Row 36	13.056
12	24	24	22.64	1.363	1.857	19.210
13	28	24	22.78	1.221	1.491	21.557

8.2.4 Fitting the Curve to Tagging Data

So far, we have only considered fitting the von Bertalanffy curve to data where one has each fish's length at particular ages. Obviously, for this, one would need to be able to age the fish accurately. But there are other forms of data available that can be used to fit a von Bertalanffy curve. When one conducts a tagging experiment, it is common to obtain lengths when animals are first tagged and to re-measure them on recapture with the time interval between recaptures known. If the von Bertalanffy curve could be re-formulated in terms of size increments after a given time from a given initial size, it would be possible to use such data to fit the growth curve.

Fabens (1965) formalized the translation of the von Bertalanffy curve into a form where it could be used with the sort of information obtained from tagging programmes (see Appendix 8.1 for the full derivation). By manipulating the usual von Bertalanffy curve, Eq. 8.1, Fabens produced

$$\Delta L = \left(L_\infty - L_t \right)\left(1 - e^{-K\Delta t} \right)$$
$$\Delta L = L_{t+\Delta t} - L_t$$

(8.6)

where, for an animal with a initial length of L_t ΔL is the change in length through the period of Δt. By minimizing the squared differences between the observed ΔL and the expected ΔL for each point, using Eq. 8.6, estimates can be derived for the K and L_∞ parameters. The average length at a known age would be required to include an estimate of t_0 so, often, no estimate can be generated and the exact location of the growth curve along an age axis is not determined. In these cases, the t_0 parameter is often set to zero (Fig. 8.6; Example Box 8.3).

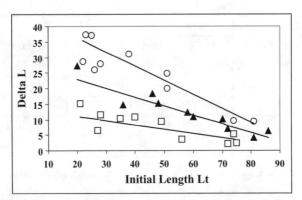

Figure 8.6 Plot of artificial data for tag returns from three different time intervals (squares = 170 days, filled triangles = 385 days, and circles = 690 days), each time interval has its own expected ΔL for a given starting length L_t illustrated by the three solid lines. If the lines were projected, they would meet the x-axis at L_∞ (Table 8.2; Exercise Box 8.3).

8.2.5 Extensions to Fabens Method

The use of the Fabens method appears to be straightforward (Example Box 8.3), but there are difficulties that are easily overlooked. Sainsbury

(1980) pointed out that, as originally developed, both Eqs. 8.1 and 8.6 relate to the growth of individuals and thus do not predict the average length at age t or the average growth increment for a given initial length and time passed, Δt. Instead, it is just assumed that these curves can be applied to collections of individuals. This ignores the fact that there will be variation in the growth of individuals.

Table 8.2 Artificial example tagging data simulated from an $L_\infty = 100$ and $K = 0.3$. Each set of three relate to different days at liberty, Δt, L_t relates to the initial size at tagging, and ΔL is the change is length during Δt. This data is illustrated in Fig. 8.6 and is used in Example Box 8.3.

Δt	L_t	ΔL	Δt	L_t	ΔL	Δt	L_t	ΔL
170	21	15.1	385	20	27.2	690	22	28.6
170	27	6.4	385	36	14.8	690	23	37.3
170	28	11.5	385	46	18.3	690	25	37.2
170	35	10.3	385	48	15.2	690	26	26.1
170	40	10.8	385	58	12.2	690	28	27.9
170	49	9.4	385	60	10.8	690	38	31
170	56	3.6	385	70	10.2	690	51	24.7
170	72	2.1	385	72	7.1	690	51	19.9
170	74	5.2	385	81	4.1	690	74	9.7
170	75	2.3	385	86	6.2	690	81	9.3

An inspection of the distribution of the length increments against the initial size at tagging (Fig. 8.6) indicates that the variation around the expected values appears to be greater at the larger ΔL values. Thus, as the expected ΔL declines with initial size the variability of the residuals appears to decline. One could use either a weighted least squares approach to fitting the model (having a constant coefficient of variation instead of a constant variance), or one could use a maximum likelihood method and directly estimate the variance. Francis (1988a) described just such a maximum likelihood approach that fitted the model to the data assuming the residuals were distributed normally. However, a number of different functional forms were suggested for the relationship between residual variance and expected ΔL. Thus, normal errors are used but the variance of the residuals is determined separately. Using ordinary normal random errors (i.e., constant variance) would provide identical answers to a least squares approach. The negative log-likelihood in this case would be

$$-veLL = -\sum Ln\left(\frac{1}{\sqrt{2\pi}\sigma}e^{-\frac{\left(\Delta L-\Delta\hat{L}\right)^2}{2\sigma^2}}\right) \qquad (8.7)$$

where σ^2 is the constant variance of the residuals between the observed and expected ΔL values. Francis (1988a) provided a number of different formulations for describing this variance, including an inverse linear relationship between the standard deviation and the expected ΔL

$$\sigma = \upsilon\left(\Delta\hat{L}\right) \qquad (8.8)$$

where υ (nu) is a constant multiplier on the expected ΔL, and would need to be estimated separately. The likelihood becomes

$$L\left(\Delta L \mid Data\right) = \sum_i\left(\frac{1}{\sqrt{2\pi}\upsilon\Delta\hat{L}}e^{-\frac{\left(\Delta L-\Delta\hat{L}\right)^2}{2\left(\upsilon\Delta\hat{L}\right)^2}}\right) \qquad (8.9)$$

Francis (1988a) also suggested lognormal residual standard deviation

$$\sigma = \tau\left(1-e^{-\upsilon\Delta\hat{L}}\right) \qquad (8.10)$$

where τ is a new estimable constant. Finally, Francis (1988a) suggested that the residual standard deviation might follow a power law

$$\sigma = \upsilon\Delta\hat{L}^\tau \qquad (8.11)$$

Francis (1988a, b) took his extension of the Fabens methods further with suggestions for how to account for consistent bias in the measurement of length, with additions to the model to estimate seasonal variation in growth rates and to account for outlier contamination. All of these extensions constitute valuable improvements over the simple Fabens method.

With each different formulation of the relationship between the standard deviation of the residuals and the expected ΔL, the constants τ and υ would change in their interpretation. As seen before, the parameter estimates obtainable from the same model can vary if different error structures are assumed (Example Box 8.3).

Example Box 8.3 The Fabens method of fitting a growth curve to tagging data extended using ideas from Francis (1988a). The t_0 is redundant (but could be used to plot a growth curve from a length-at-age perspective). Enter the data from Table 8.2 into columns A, B, and C, down to row 35. In E4 put =sum(H6:H35). To obtain the number of years at liberty put =A6/365.25 into D6. In E6 put =($B$1-B6)*(1-EXP(-$B$2*D6)). Copy D6:E6 down to row 35. In column H put =-Ln(normdist(C6,E6,G6,false)) and copy down to row 35 to obtain the negative log-likelihoods. The first model fit can be generated using least squares by getting the solver to minimize B4 through changing B1:B2. Record the answer somewhere else on the sheet. If you were to put =$E$3 into cell G6 and copy down you could obtain the same result via maximum likelihood as from least squares, by minimizing E4, by changing (B1:B2, E3). This is not particularly stable and you may need to try a number of starting points to obtain an answer. If it does fail and generate a #NUM error, look for the reason why it happened. Alternatively, you could put =$E$1*E6 into G6 and copy down to obtain Eq. 8.8, the linear relation between residual standard deviation and the expected ΔL (minimize E4 by modifying B1:B2, E1). Finally, try the two other residual structures =$E$1*E6^$E$2 and =$E$2*(1-exp(-$E$1*E6)). Compare the parameter estimates and the total log-likelihood. Which is the best fitting model? Examine the residuals in a graphical plot. Compare the various fits using the likelihood ratio test or using Akaike's information criterion (AIC = 2LL + 2p), where LL is the negative log-likelihood and p is the number of parameters. The smallest AIC wins.

	A	B	C	D	E	F	G	H
1	L_∞	100.39	Nu		0.5			
2	K	0.31	tau		0.5			
3	t_0	0	Sigma		2			
4	SSq	=sum(F6:F35)	-veLL		75.5175			
5	ΔT	Lt	ΔL	ΔT Yrs	E(ΔL)	Resids^2	StDev	LL
6	170	21	15.1	0.465	10.62	=(C6-E6)^2	3.00	3.133
7	170	27	6.4	0.465	9.82	=(C7-E7)^2	3.00	2.666
8	170	28	11.5	0.465	9.68	=(C8-E8)^2	3.00	2.201
9	170	35	10.3	0.465	8.75	Copy down	3.00	2.151
10	170	40	10.8	0.465	8.08	To F35	3.00	2.429
11	170	49	9.4	0.465	6.87	6.379	3.00	2.372

Unfortunately, how we select which error structure is most appropriate is not a question that is simple to answer. One could use a likelihood ratio test (Eq. 3.45) to compare the quality of fit obtained with the different numbers of parameters. Thus, if there were an improvement to the log-likelihood of the order of 1.92 (a χ^2 value of 3.84 divided by 2) from fitting the model using Eq. 8.11 instead of Eq. 8.8 (adding a parameter), this would constitute a statistically significant improvement. However, ideally, one would have other reasons for preferring one error structure over another for the residual variance.

8.2.6 Comparability of Growth Curves

The Fabens version of the von Bertalanffy equation derives directly from the classical equation (Appendix 8.1) and yet the parameters generated from size-at-age data have been given different interpretations to those generated from tagging data (Sainsbury, 1980; Francis, 1988b). This may appear paradoxical until it is realized that the curves are being fitted using very different residual error structures.

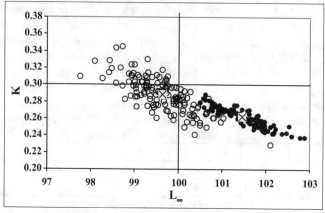

Figure 8.7 Growth curve parameters estimated from 200 Monte Carlo simulations of sampling an hypothetical population of individuals, each with their own von Bertalanffy growth parameters. Growth curves were fitted as with size-at-age data (open circles) and tagging data (small, solid circles). The two crosses indicate the mean of each set of parameter estimates. The mean values for L_∞ and K were 100 and 0.3, respectively, as indicated by the crossed lines. Generally, the estimates from tagging data were biased high for the L_∞ and low for the K parameter. From size-at-age data, both parameter estimates were biased slightly low.

With size-at-age data, the residuals are between observed size-at-age and expected size-at-age in a snapshot sample from the population. With tagging data the residuals are between the observed size increment and the expected for possibly different time intervals, for data collected at least some time after the initial observations. The net result is that the estimation of L_∞ tends to be biased upwards from tagging data, while the reverse seems true for the size-at-age data. That the estimates of the parameters obtained from length at age data and from tagging data are different (as claimed by Sainsbury, 1980; Francis, 1995) can be seen directly by simulating a population where the individuals each grow with their own particular von Bertalanffy growth parameters. If a population is set up with individuals each having their own L_∞ and K values (assume a t_0 of zero) varying normally about overall mean parameter values, then this hypothetical population could be sampled for both size-at-age data and time increment growth data. Note that the same population is being sampled. If there were no difference between the two methods, then, on average, we would expect their parameter estimates to coincide (Fig. 8.7). In fact, the parameter estimates do not coincide and both show biases with the tagging approach appearing to be more biased that the size-at-age approach. Clearly, care must be taken if comparisons are to be made between growth curves estimated differently.

8.3 ALTERNATIVES TO VON BERTALANFFY

8.3.1 A Generalized Model

Virtually every fishery paper concerned with growth uses the von Bertalanffy growth equation but this does not mean that it is the only possible growth function or even necessarily the best in a given situation. Many representations of growth have been used in the fisheries literature.

A very general model has been proposed by Schnute and Richards (1990). This starts as a four-parameter model of growth, which generalizes the classical logistic model as well as those by Gompertz (1825), von Bertalanffy (1938), Richards (1959), Chapman (1961), and Schnute (1981). It also generalises the Weibull (1951) distribution, which has been used with applied problems. All of these papers contain equations, which could easily be used as alternatives to the von Bertalanffy growth function (VBGF). They all share the property of permitting a description of increasing size leading to an asymptotic maximum. Schnute and Richards' (1990) model has the form

$$y^{-b} = 1 + \alpha \, e^{-ax^c} \tag{8.12}$$

By altering the parameters (a, b, c, and α), and the formal structure of the model, Eq. 8.12 can take the form of many of the popular growth models (Fig. 8.8). For example, if $b = c = 1$, then Eq. 8.12 becomes the classical logistic relationship

$$y = \frac{1}{1 + \alpha\, e^{-ax}} \tag{8.13}$$

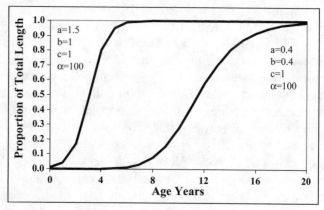

Figure 8.8 Two curves generated by substituting different sets of values for α, a, b, and c, in Eq. 8.12. The general equation generates a curve of relative size against age. Construct a worksheet to illustrate this growth curve.

Schnute and Richards (1990) implicitly assumed that that the y variable was scaled in units of y_∞. If this asymptote is included, the model finishes with five parameters and becomes extremely flexible

$$y_t = y_\infty \left(1 + \alpha e^{-at^c}\right)^{-\frac{1}{b}} \tag{8.14}$$

where the parameters are as before, t is age, and y_∞ is the average maximum value of the variable y. When parameter $c = 1$, then the model becomes equivalent to Schnute's (1981) generalized growth model. The Schnute and Richards (1990) model generalizes the Schnute (1981) model when the age t is replaced with t^c

$$Y_t = \left[y_1^b + \left(y_2^b - y_1^b\right) \frac{1 - e^{-a(t - \tau_1)}}{1 - e^{-a(\tau_2 - \tau_1)}} \right]^{1/b} \tag{8.15}$$

where t is age, y_i is the size at age τ_i, and the other parameters are as before. Once again, by varying the various parameters a wide range of curves can be generated, some of which are not necessarily asymptotic. Schnute's (1981) model appears to have been used more in the literature than the more general Schnute and Richards (1990) model (Gillanders *et al.*, 1999, although strictly, they used Francis', 1995, mark-recapture analogue to

Schnute's Eq. 8.15). Different model designs are still being developed (Francis, 1995; Wang and Thomas, 1995). This area of research remains open for further development.

8.3.2 Model Selection

With four parameters, Eq. 8.12 has one more parameter than most other growth models, which is one reason why it can embody so many models with fewer parameters as special cases. Such an equation is mathematically general and is useful for providing a unified analytical framework for a diverse literature. Knowing that all these models of growth constitute special cases of a single general model should make it clear that these growth models may be regarded solely as descriptions of growth with no absolute explanatory power. This is a very valuable lesson. What it means is that even if we manage to fit a particular growth model to a data set the interpretation of its parameters is not necessarily meaningful. For example, just because a model implies an asymptotic maximum mean length (because it calculates one) does not force the species concerned to actually have such a maximum. We will discuss this point further when considering the many criticisms directed against von Bertalanffy curves.

Whether one would use Schnute and Richards' (1990) general model when selecting an equation to describe growth, instead of one of the special cases, is a difficult question to answer. The special cases might be preferred to the general equation because having fewer parameters, fitting them to real data might be more straightforward, as would using them in a model. However, if these models truly are just descriptive black boxes, having an input to provide a particular output, then, with available computing power, it would make very little difference which equation was used. Presumably, one should use the equation that provides the best description of the growth process being described. Unfortunately, deciding what constitutes "best" is not as simple as one might hope. If one simply wants the closest description of one's growth data (according to some criterion such as likelihood ratio or AIC), then one could try fitting a wide range of models and error structures and proceed with the one producing the best fit. However, if one really wants to obtain biologically sensible interpretations concerning how a species grows from a growth model, then model selection cannot be solely dependent upon quality of statistical fit. Instead, it should reflect the theoretical viewpoint of growth that is being considered for reasons independent of its statistical fit to data.

8.3.3 Polynomial Equations

Polynomial functions, of three or four parameters, have been suggested as alternatives to the von Bertalanffy growth function. These polynomials are explicitly empirical as there is little to be gained from trying to interpret the parameters. Comparisons between curves would be reliant on them each having the same number of parameters. However, Chen *et al.* (1992) made an explicit comparison of a variety of polynomials against the VBGF and concluded that the VBGF performed better than any of the polynomial equations considered. In the process of doing this, they also suggested a neat way of comparing growth curves of any type (see later) to determine whether they differed statistically.

If all that is required is a description of how growth has occurred there is no reason not to use polynomial equations. However, while the interpolation of growth within the range observed is possible, attempts to extrapolate beyond the observed data would most likely lead to errors.

8.3.4 Problems with the von Bertalanffy Growth Function

With the wide range of alternative growth equations available, the very use of the VBGF to represent fish growth has been questioned by a number of people. Knight (1968) criticized the VBGF as often being more like fiction than a useful model of growth. His major complaint was that L_∞ was generally estimated through extrapolation, and sometimes extreme extrapolation (Fig. 8.9).

The estimation of L_∞ is especially a problem with fish species that do not exhibit an asymptotic maximum length (Fig. 8.9). The best fit von Bertalanffy curve to the minimal data provided by Knight (1968) suggests an L_∞ of 453 cm; that is, a cod 4.5 metres in length, which is biological nonsense. The validity of such extrapolation depends almost completely on the appropriateness of the assumptions of the von Bertalanffy model of growth for the species concerned. In the same year, Roff (1980) was critical of a number of people, including Beverton and Holt (1957), who applied the curve in an apparently unthinking manner. He also emphasized its inapplicability to fish species that do not appear to have an asymptotic maximum length. The same argument against extrapolation could have been expressed about the estimation of t_0.

Roff (1980) reviewed the problems associated with using the VBGF and suggested that people stop using it and turn to different functions instead. To his credit, Roff (1980) did not advocate any particular equation but stated that the choice should be dictated by circumstances. He also pointed out that the equation was very hard to fit to data in a statistically satisfactory manner, meaning, presumably, either that there was no simple or deterministic way of fitting the curve or that the comparison of growth curves was difficult. It is common to fit growth curves even with little data available for younger and older animals, and when the sampled population is fished and has a legal minimum size. In these cases, the data are not really representative and care must be taken with the generality of any resulting curves and comparisons with other populations.

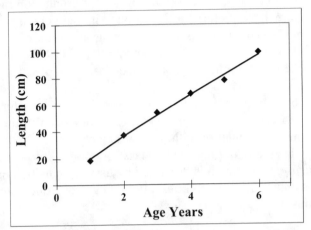

Figure 8.9 Plot of the growth of Atlantic cod through the first 6 years of life (after Knight, 1968). Shows the best von Bertalanffy curve fit with $K = 0.0391$, $t_0 = -0.126$, and $L_\infty = 453$ cm, which is clearly an exaggeration of possible reality.

Very obviously, there are many problems with using the VBGF; nevertheless, fisheries scientists continue to use it in their models. Given a model, although fitting the curve is no longer the problem it was, there is rarely any discussion about how the residuals are expected to be distributed about the expected curve. Generally, they are assumed to be normal-random and additive but there is no consideration of whether they really are symmetrical or otherwise, or whether the variance of the residuals is constant or varies with size. Analytical and statistical computer programs routinely have procedures for generating non-linear fits to particular models (e.g., SAS, Systat, even Excel has its non-linear solver). Putting aside Sainsbury's (1980) and Francis' (1988a, b) problem of different concepts having the same name, the major difficulty is that of statistically comparing growth curves, for example, from different stocks or sexes of the same species. Numerous methods have been suggested for doing this and in Section 8.4, we will consider some of the methods for conducting such analyses.

8.3.5 Growth in Size-Based Population Models

Commercial fisheries for abalone and rock lobster (and other invertebrates) often suffer from the fact that the species concerned are difficult or impossible to age using readily available technology. Many of these species are valuable products and thus require an assessment model to assist with the adequate management of the resource. The use of age-structured models for assessing these species is compromised so alternatives must be considered. It would be possible to use a surplus production model, which does not require age-structured information. Alternatively, a size- or stage-structured model could be used with the usual form of these models described by Sullivan *et al.* (1990) and Sullivan (1992). These models follow the fate of the numbers in each size-class as opposed to numbers in each age-class.

The population being modelled is described by a vector, say N_t, of the numbers in each size class at time t. Each time-step in the model, the individuals in the population either grow or not, which implies that they either stay in their original size class or move into a larger one. Mathematically, this growth is described by using a transition or projection matrix containing the probabilities of shifting from one size class into another. Thus if

$$
\mathbf{N_t} = \begin{bmatrix} N_{1,t} \\ N_{2,t} \\ . \\ N_{n,t} \end{bmatrix}
\tag{8.16}
$$

where N_t is the vector of n length classes indicating the numbers in length-class l at time t. The stage-structured model provides a description of the calculation of N_{t+1}, and through repeated application of the projection matrix, it describes the dynamics of the population.

The basis of these models is the transition matrix with which the population vector is repeatedly multiplied to describe the changing population size-structure through time. The transition matrix contains the probability that the individuals in length class l at time t have grown into length class $l*$ by time $t*$, and is often combined with survivorship probabilities and recruitment relationships. There are often variations in the exact contents of the transition matrix to reflect the particular circumstances of the fishery being modelled. In matrix notation the effect of growth alone would be represented thus

$$G = \begin{bmatrix} G_{1,1} & 0 & . & 0 \\ G_{1,2} & G_{2,2} & & \\ . & & . & 0 \\ G_{1,n} & . & G_{n-1,n} & G_{n,n} \end{bmatrix} \tag{8.17}$$

where G is the transition or projection matrix in which the $G_{i,j}$ are the probabilities that an animal in size class i will grow into size class j. The generation of the individual $G_{i,j}$ is described later. Multiplying the numbers-at-size vector, N_t by the transition matrix leads to

$$N_{t+1} = GN_t \tag{8.18}$$

which describes growth in the population without recruitment or mortality.

Natural mortality can be represented by the annual survivorship in year t in each length class l, $S_{l,t}$, and is easily included in the model as a square matrix with only the diagonal elements filled

$$S_t = \begin{bmatrix} S_{1,t} & 0 & 0 \\ 0 & S_{2,t} & \\ & & . & 0 \\ 0 & & 0 & S_{n,t} \end{bmatrix} \tag{8.19}$$

Annual survivorship, accounting for natural mortality would be

$$S_{l,t} = e^{-M_{l,t}} \tag{8.20}$$

where $M_{l,t}$ refers to the instantaneous rate of natural mortality for size class l during period t. The model now becomes

$$N_{t+1} = GSN_t \tag{8.21}$$

Care must be taken concerning the order of multiplication of the growth and survivorship matrices with the vector. It is the case that $G(SN_t)$ (survive first, then grow) is only the same as $S(GN_t)$ (growth first, then survive) if the survivorship is constant across all sizes. If the mortality schedule is not constant, then the order of multiplication with respect to when the growth occurs becomes important (Example Box 8.4).

Example Box 8.4 The importance of sequence of multiplication to the outcome of the size-structure model described in Eq. 8.21. Name the cells B3:F7 as **G**, B10:F14 as **S**, and B16:F20 as **T** (put =mmult(G,S) into B16:F20). In I3:I7 put =mmult(S,mmult(G,H3:H7) and copy across for a few columns. In I10:I14 put =mmult(T,H10:H14) and copy across the same number of columns. Remember that all array formulae must be entered using <Ctrl><Shift><Enter>. If the diagonal of **S** is not identical, then the numbers in each size class will differ depending on the multiplication sequence. LC is length class, the elements of $G_{i,j}$ are the probabilities that an animal in size class i grows to size class j, the elements of S_t are the annual survivorship values for each size class, and N_t is the vector of numbers-at-size. TN_t is equivalent to $G(SN)$. Note there is no recruitment operating, only growth and survivorship. The differences stem from the assumed time in which the survivorship is imposed. If it is at the end of a growth period this is equivalent to $S(GN)$, if at the beginning or before growth this relates to $G(SN)$. If animals grow into a size class possessing a different rate of mortality, then it will matter whether it grows first or dies first. Try varying the values in **S** and **G** and observing the result. Plot the changing N_t.

	A	B	C	D	E	F	G	H	I	J
1				G				N_0	N_1	N_2
2	LC	1	2	3	4	5			S(GN)	S(GN)
3	1	0.15	0	0	0	0		300	13.5	0.61
4	2	0.45	0.15	0	0	0		200	115.5	16.38
5	3	0.35	0.55	0.35	0	0		150	187.3	93.65
6	4	0.05	0.25	0.4	0.7	0		100	136.5	140
7	5	0	0.05	0.25	0.3	1		100	124.3	152.45
8								N_0	N_1	N_2
9	LC			S					G(SN)	G(SN)
10	1	0.3	0	0	0	0		300	13.5	0.6
11	2	0	0.7	0	0	0		200	61.5	8.3
12	3	0	0	0.7	0	0		150	145.3	60.7
13	4	0	0	0	0.7	0		100	130.5	115.6
14	5	0	0	0	0	0.7		100	124.3	142
15	LC			T						
16	1	0.05	0	0	0	0		Year		
17	2	0.14	0.11	0	0	0		Nt_S(GN)	=sum(I3:I7)	403.09
18	3	0.11	0.39	0.25	0	0		Nt_G(SN)	=sum(I10:I14)	327.10
19	4	0.02	0.18	0.28	0.49	0				
20	5	0	0.04	0.18	0.21	0.7				

One way to reduce the impact of when mortality is applied, relative to when growth is applied, is to break a year up into a number of seasons and generate a transition matrix for each season. Punt *et al.* (1997) used this strategy in generating transition matrices to describe the growth of the southern rock lobster, *Jasus edwardsii*, around Tasmania, Australia.

To complete the size-structured model one would also have to include recruitment (see Chapter 9), either as a separate term or included in the transition matrix. Finally, fishing mortality would need to be included and, if there were selectivity in the fishery, then the rates of survivorship by size class would not be the same. Thus, as with a non-constant natural mortality, where in the model this aspect of mortality is implemented with respect to growth would become important.

Sullivan *et al.* (1990) and Sullivan (1992) influenced many subsequent size-based population models. Most size-based stock assessment models have tried to use a minimal number of parameters to define the transition matrix for the growth of the members of each size class over each time interval. Sullivan *et al.* (1990) suggested using the gamma function to generate the transition matrix elements (Example Box 8.5; Fig. 8.10). However, one could use the normal distribution, lognormal distributions (Haddon, 1999), the beta distribution, and the empirical multi-nomial (Punt *et al.*, 1997); in fact, any left-skewed to symmetric curve could be used. Except when using an empirical multi-nomial distribution, the objective would generally be to use the Fabens version of the von Bertalanffy to estimate the average length increment for a given size-class, and then use the statistical distribution chosen to describe how individuals would be distributed around the average increment.

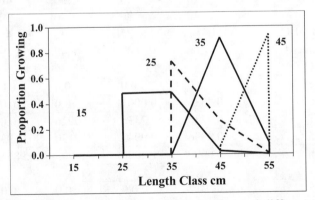

Figure 8.10 The expected proportions of the members of different size classes growing into the larger size classes in one time-step (Example Box 8.4). Thus, from the 15 cm size class approximately equal numbers grow into the 25 and 35 cm size classes with just very few growing into the 45 cm class. Note that no animals from size classes 15 or 25 stay in their initial size class, whereas a small proportion stay in size classes 35 and 45. Of course, 100% stay in size class 55, as this is the maximum.

Example Box 8.5 The generation of a growth transition matrix using the gamma function. Alpha and Beta are parameters of the gamma function. Copy B2 and B5 across to column F. To calculate the expected average increment put =max(($B$1-B$7)*(1-exp(-$D$1)),0) into B6 and copy across to F6. The max is to prevent any negative increments in case the length class is larger than the L_∞. The matrix elements are generated from the cumulative gamma function, thus, into cell B8 put =if(B$7<$B$1, if($A8>=B$7, gammadist($A8-$7+1,B$2,$F$1,true),0), if($A8>= B$7,1,0)) and copy down to B12 and then copy B8:B12 across to F6:F12. The if statements avoid the animals shrinking, and not growing beyond the L_∞. Note that the cumulative distribution does not necessarily attain a total of one. Hence the matrix below that (B14:F18) gives the final growth transition matrix. Copy B14 across to F14. Copy B15 down to B18, and then B15:B18 across to F15:F18. Plot the distributions for size classes 15 to 45 to mimic Fig. 8.10. Try altering the L_∞ and K values, as well as the Beta value (which controls the spread of the distributions). You could replace the Gammadist function with the Normdist function, replacing the alpha calculations with one of the equations governing the standard deviation Eqs. 8.9 to 8.11; you may need to add an extra constant. With so few size classes it makes little difference but would be important if one had sufficient size classes to discern details of growth. Combine with Example Box 8.4 and grow a population through a number of generations using B15:F18.

	A	B	C	D	E	F
1	**Linf=**	60	**K=**	0.3	**Beta=**	1.5
2	**Alpha**	=B6/$F$1	=C6/$F$1	4.3197	2.5918	0.8639
3	**Lower**	10	20	30	40	50
4	**Upper**	20	30	40	50	60
5	**AverageLC**	=(B4+B3)/2	25	35	45	55
6	**Increment**	11.66	9.07	6.48	3.89	1.30
7	**LC**	15	25	35	45	55
8	15	0.0	0.0	0.0	0.0	0.0
9	25	0.4836	0.0001	0.0	0.0	0.0
10	35	0.9733	0.7336	0.0026	0.0	0.0
11	45	0.9996	0.9942	0.9130	0.0593	0.0
12	55	1.0000	1.0000	0.9993	0.9865	0.5550
13				G		
14	15	=B8/B12	0.0	0.0	0.0	0.0
15	25	=(B9-B8)/B$12	0.0001	0.0	0.0	0.0
16	35	Copy Down	0.7335	0.0026	0.0	0.0
17	45	0.0263	0.2606	0.9111	0.0601	· 0.0
18	55	0.0004	0.0058	0.0863	0.9399	1.0000

8.4 COMPARING GROWTH CURVES

8.4.1 Non-Linear Comparisons

Because of the non-linear nature of the von Bertalanffy growth function (VBGF), we cannot use the standard analysis of covariance with which we might usually compare regression lines. Many early methods of comparing growth curves relied upon linearizing the VBGF and comparing the linear regressions produced. These methods often only produced approximate fits of the VBGF to the data and will not be discussed further. Misra's (1980) approach was rather more complicated but used a linearized re-parameratization of the VBGF, fitted it to data using multiple regression, and tested for differences between growth curves using ANOVA on the separate parameters. Alternatively, Bernard (1981) used a multivariate test, Hotelling's T, to compare all three VBGF parameters at once.

A summary and consideration of methods was provided by Cerrato (1990) who compared the ability of the t-, univariate χ^2, likelihood ratio, and Hotelling's T- tests to provide valid comparisons of VBGF curves. The last two methods are generally considered better than the others, not least because they consider all parameters at once. Because there are strong correlations between the parameters of the von Bertalanffy equation one should never compare individual parameters in isolation.

Cerrato (1990) provided a clear statement of the problems faced when attempting to compare non-linear growth curves using the likelihood ratio or Hotelling's T tests:

> Both approaches are approximate ones. They are taken from linear statistical theory, and their validity when applied to the von Bertalanffy equation depends on the degree of bias and non-normality in the parameter estimates caused by the non-linearity of the model. In addition, both are characterized in terms of asymptotic properties for which no exact small sample theory exists. Finally, at least as commonly practised, both approaches handle unequal and unknown error variances in an approximate way.

Despite these problems, Cerrato (1990), in an empirical test, found that the likelihood ratio comparison most often, and most accurately, reflected the true state of affairs and recommended that this should be the approach of choice. Moulton *et al.* (1992) went slightly further in that they recommended re-parameratizing the VBGF as recommended by Francis (1988a, b) and then using the likelihood ratio for comparative tests between the re-parameterized curves.

8.4.2 An Overall Test of Coincident Curves

A method of comparing a number of curves at once was proposed by Chen *et al.* (1992) and is called the analysis of the residual sum of squares (ARSS). The method is different from the likelihood ratio approach but it is useful in itself and, because there are analogies between the two approaches, the ARSS is a useful way to introduce the principle behind likelihood ratios. Both approaches can only be used to compare curves of the same type (i.e., VBGF or some other, but not VBGF with some other).

The analysis of residual sum of squares is a total comparison, meaning that it does not compare the parameters separately but simply tests whether two or more curves are statistically different (are coincident curves). A linear version of this test was described by Zar (1984) as an overall test for coincident regressions. Equation 8.22, below, is simply an extension to the non-linear case. There are four steps to the ARSS:

1) For each data set *i*, fit a curve and calculate the sum of squared residuals, RSS_i, and an associated degree of freedom, DF_i.
2) The resultant RSS_i and DF_i for each curve are added to give the ΣRSS_i and ΣDF_i.
3) Data for all curves are pooled, a new curve is fitted to the combined data and the total or pooled RSS_p and DF_p are calculated.
4) Using these statistics, an F-statistic is calculated as in Eq. 8.22.

$$F = \frac{\dfrac{RSS_p - \sum RSS_i}{DF_p - \sum DF_i}}{\dfrac{\sum RSS_i}{\sum DF_i}} = \frac{\dfrac{RSS_p - \sum RSS_i}{3.(K-1)}}{\dfrac{\sum RSS_i}{N-3.K}} \tag{8.22}$$

where F is the F-statistic with $3.(K - 1)$ and $(N - 3.K)$ degrees of freedom, K is the number of curves being compared, and N is the total or pooled sample size. This test can be applied to all classes of curve, not just the VBGF. This is a test of the hypothesis that the curves being compared are all equivalent descriptions of the data available; an example may serve to make this clear.

Kimura (1980) provides a table of age-length data relating to Pacific hake with separate data for both males and females (Fig. 8.11; Table 8.3). The question to be answered is whether the male and female Pacific hake exhibit different growth throughout their lives. In this case K = 2 and N = 24.

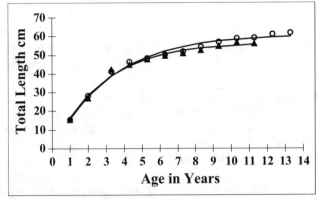

Figure 8.11 Average length-at-age for the Pacific hake for females (circles; L_∞ = 61.23, K = 0.296, and t_0 = -0.057) and males (triangles; L_∞ = 55.98, K = 0.386, and t_0 = 0.171). Data from Kimura (1980). The lines illustrated are the maximum likelihood best fitting curves. (Table 8.3; Example Box 8.6).

Table 8.3 Average length-at-age for male and female Pacific hake from the U.S. west coast. Data from Table 1 in Kimura (1980). Note the change to the years of age after age 2.

Age Years	Female Sample Size	Female Mean Length cm	Male Sample Size	Male Mean Length cm
1.0	385	15.40	385	15.40
2.0	36	28.03	28	26.93
3.3	17	41.18	13	42.23
4.3	135	46.20	83	44.59
5.3	750	48.23	628	47.63
6.3	1073	50.26	1134	49.67
7.3	1459	51.82	1761	50.87
8.3	626	54.27	432	52.30
9.3	199	56.98	93	54.77
10.3	97	58.93	21	56.43
11.3	44	59.00	8	55.88
12.3	11	60.91		
13.3	6	61.83		

As listed above one first has to find the best fitting curves for the males and females separately. For the von Bertalanffy curve this leads to: females L_∞ = 61.23, K = 0.2963, and t_0 = -0.0573, with 10 df, and a residual sum of squares RSS = 48.224; males L_∞ = 55.98, K = 0.3856, and t_0 = 0.1713, with 8 df, and RSS = 19.423. Then the male and female data are treated as one population and the pooled growth curve fitted which led to $L_\infty = 59.29$, $K = 0.3205$, and $t_0 = 0.0104$, with 21 df, and RSS = 79.764. When these figures are substituted into Eq. 8.22 the resulting $F = 3.924$ which is just significant ($P = 0.0256$). This analysis certainly indicates that the two curves (Fig. 8.11) are different but does not indicate in which way. A different test would be necessary if more details were required.

Example Box 8.6 The implementation of the Analysis of Residual Sum of Squares. The data are in Table 8.3. Put = sum(H14:H24) into C5. In cell B6 put =count(B14:B26)-3 or the female df, in C6 put =count(F14:F26)-3, and in D6 put =count(B14:B26, F14:F26)-3. In B10 put =abs((C8-B8)/(C9-B9))/(B8/B9)), and in C10 put =abs((C8-B8)/(3*(2-1)))/((B8/(24-(3*2))))); they should give the same F value. In B11 put =fdist(B10,3,B9). To calculate the expected lengths put =$B$2*(1-exp(-$B$3*(A14-$B$4))) into C14 and =(B14-C14)^2 into D14. Copy C14:D14 down to row 26. Similarly, put =(F14-G14)^2 into cell H14 and put =$C$2*(1-exp(-$C$3*(E14-$C$4))) into G14, and copy G14:H14 down to row 24. Using the solver, in turn, minimize B5 by altering B2:B4, and C5 by altering C2:C4. Copy B5:C6 into F2:G3 as values. Put =D2 into B2 and C2, and copy down to row 4. Then minimize cell D5 by altering D2:D4; copy D5:D6 into H2:H3, and the answer should be determined. Plot column B against A as points and add column C as a line, add columns E and F as points and column G as a further line to mimic Fig. 8.11.

	A	B	C	D	E	F	G	H
1		Female	Male	Total		Female	Male	Total
2	Linf	61.233	55.978	59.294	SSq	28.845	19.423	79.696
3	K	0.296	0.386	0.320	df	10	8	21
4	t0	-0.057	0.171	0.010				
5	SSQ	=sum(D14:D26)	19.423	=B5+C5				
6	df	10	8	21				
7		Individual	Pooled					
8	SSQ	=F2+G2	=H2					
9	df	=F3+G3	=H3					
10	F	3.907	3.907					
11	P	0.026						
12	Fem	Fem	Fem	Fem	Male	Male	Male	Male
13	Age	Length	E(L)	resid2	Age	Length	E(L)	resid2
14	1	15.4	16.47	1.14	1	15.4	15.31	0.01
15	2	28.03	27.94	0.01	2	26.93	28.32	1.94
16	3.3	41.18	38.58	6.73	3.3	42.23	39.22	9.03
17	4.3	46.2	44.39	3.27	4.3	44.59	44.58	0.00
18	5.3	48.23	48.71	0.23	5.3	47.63	48.23	0.36

8.4.3 Likelihood Ratio Tests

The general principle when using the ARSS involves comparing the outcome (RSS$_p$) of the hypothesis that all the curves are coincident (each data set is effectively a sample from the same population) with the outcome (ΣRSS$_i$) of the hypothesis that all data sets are from independent populations. The method of likelihood ratios does something very similar in that it sets up the outcome of the hypothesis of two or more independent curves as a base case against which to compare all alternatives hypotheses. For the von Bertalanffy curve the alternative hypotheses would be: 1) that all curves were coincident, as with the ARSS test, 2) that the L_∞ values were equal, 3) that the K values were equal, 4) that the t_0 values were equal, 5) that L_∞ and K values are equal, 6) that L_∞ and t_0 values are equal, and finally 7) that the K and t_0 values are equal. Strictly, if the test of coincident curves (hypothesis 1) were not rejected, one would not need to test hypotheses 2 to 7. However, if one were testing multiple curves for equivalence and the test of coincident curves proved false, then one would have to proceed by making subsequent pair wise comparisons between individual growth curves to discover the details of which aspects of the curves in question differed.

The standard description of the likelihood ratio methodology was given by Kimura (1980), who dealt exclusively with the von Bertalanffy growth curve. The method is more general than this, however, and can be applied to comparing any non-linear equations fitted to data as long as the residuals used are additive and normally distributed (N(0, σ^2)). The necessity for this restriction is illustrated in the derivation of the method (see later and Appendix 8.2). The important thing for our discussion is to determine how to calculate the likelihoods for each hypothesis to be compared. The test calculates a statistic that is compared with the χ^2 distribution with degrees of freedom equal to the number of constraints (i.e., the number of parameters being assumed equal). The method turns out to be extremely simple and, to start with the conclusion, the test is based on the following equation

$$\chi_k^2 = -N \times Ln\left(\frac{RRS_\Omega}{RRS_\omega}\right) \qquad (8.23)$$

where k is the degrees of freedom (equals the number of constraints placed upon the fit), N is the total number of observations from both curves combined, RRS$_\Omega$ is the total sum of squared residuals derived from fitting both curves separately (i.e., the minimum sum of squares from each curve added together), and RRS$_\omega$ is the total sum of squared residuals derived from fitting the curves with one of the hypothesized constraints (e.g., the K's are equal, or the L$_\infty$ are equal, the hypotheses 1 to 7 above). The base case (RRS$_\Omega$) will always provide the smallest residual sum of squares because it has the largest number of independent parameters (all parameters are assumed to be independent). When any of the possible constraints are tested the number of independent parameters will obviously decrease so the fit will be somewhat less precise. The question to be answered in each test is whether the decline in fit (i.e., the difference between the two residual sum of squares) is greater than would be expected by chance. If the χ^2 value calculated from Eq. 8.23 is not significant then the decline in the quality of fit is assumed to be no more than if the parameters being compared were random selections from the same origin (i.e., those parameters are not significantly different).

Equation 8.23 is relatively simple to calculate because, as with the ARSS method, it only requires the calculation of various sums of squares. Kimura's (1980) paper provides a brief description of why we can successfully use the specified residual sum of squares as likelihood estimates in this likelihood ratio test. Here (and in Appendix 8.2) we will try to make that demonstration clear.

Kimura (1980) starts by writing out the maximum likelihood estimator for a von Bertalanffy curve. With the usual notation, the length of the uth individual at age t_i is defined as

$$L_u = L_\infty \left(1 - e^{\left(-K(t_i - t_0)\right)}\right) + \varepsilon_u$$

$$(8.24)$$

$$= f\left(L_\infty, K, t_0, t_i\right) + \varepsilon_u$$

where L_∞ is the, possibly hypothetical, asymptotic length, K is the growth rate constant, t_0 is the hypothetical age at length zero, and the ε_u's are the independent, additive, normal random errors (N(0, σ^2)). The lower equation in Eq. 8.24 is simply a shorthand for writing that L_u "is a function of L$_\infty$, K, t$_0$, and time t$_i$". The residual sum of squares used in the likelihood calculations (Eq. 8.23), is therefore from

$$\left[L_u - f\left(L_\infty, K, t_0, t_i\right)\right]^2 = \varepsilon_u^2 \qquad (8.25)$$

In the following description, the f(L$_\infty$, K, t$_0$, t$_i$) referred to is the von

Bertalanffy equation but it should be noted that the argument holds irrespective of which growth or non-linear equation is used and compared. It should be remembered that there is more than one way in which the von Bertalanffy curve can be estimated (e.g., to individual fish data, to mean lengths at age, with or without constant variance) and this will influence the form of the residuals used (see later). Of course, one can only compare like with like. Although we use mean lengths in some of the examples discussed here, it is not recommended that these be used if the raw data is available when constructing growth curves.

As described in the chapter on parameter estimation, a general property of the maximum likelihood estimator for linear or non-linear equations that are fitted using additive, normal random errors ($N(0, \sigma^2)$) is that it is equivalent to the least squares estimator. Kimura (1980) proves this in a series of relatively complex algebraic manipulations, which are expanded in detail for increased clarity in Appendix 8.2. He concluded that, given an appropriate sum of squared residual term (Eq. 8.25) and assuming normal random errors (see Appendix 8.2), a maximum likelihood estimate of the variance of a particular hypothesis' fit is provided by

$$\hat{\sigma}_{\Omega}^{2} = \frac{\sum_{i=1}^{N}\left[L_{i} - f\left(L_{\infty}, K, t_{0}, t_{i}\right)\right]^{2}}{N} \quad (8.26)$$

where N is the total number of observations involved across the curves being compared. As noted before, the numerator of the right hand side of this relation is simply the sum of squared residuals which means that the problem of maximum likelihood estimation for the von Bertalanffy curve has been reduced to finding the least square estimates of L_{∞}, K, and t_{0} (Kimura, 1980). This variance estimate can be used in the likelihood ratio tests. The form of Eq. 8.26 is that required for the determination of the likelihoods of different hypotheses to be compared. The exact details of the structure of the numerator (the structure of the residuals) will be related to the form of the data available.

When fitting growth curves to length at age data, sometimes one would be fitting a curve to raw data and other times to mean lengths at age.

Kimura (1980) provides a variety of alternative sum of squared residual terms each appropriate to different types of data. We will follow Kimura's (1980) nomenclature and let L_{ij} be the length of the jth individual of age t_i, and $\overline{L}_i$ and s_i^2 be the sample mean and variance of the lengths of individuals of age t_i, based on a sample size of n_i. The appropriate sum of squares to be used when fitting the curve to data will differ depending on whether we are using raw data or mean lengths, and whether the variance is constant across ages. Kimura (1980) provides the following which would be the numerator in Eq. 8.26

$$\sum \left[L_{ij} - f\left(L_{\infty}, K, t_0, t_i\right) \right]^2 \tag{8.27}$$

to be used where all L_{ij} have a constant variance (a common assumption);

$$\sum \left[\bar{L}_i - f\left(L_{\infty}, K, t_0, t_i\right) \right]^2 \tag{8.28}$$

to be used where the mean lengths, $\bar{L}_i$, have a constant variance; and

$$\sum \left[\left(\frac{n_i}{s_i^2}\right) \left[\bar{L}_i - f\left(L_{\infty}, K, t_0, t_i\right) \right]^2 \right] \tag{8.29}$$

to be used where the variance of the mean lengths, $\bar{L}_i$, varies with age t_i, and at age t_i is equal to s_i^2.

The last equation is not strictly of the form described in the maximum likelihood equation but is acting, in effect, as a weighted least squares. Kimura (1980) also provided a further equation that was primarily useful as a short cut method of fitting the von Bertalanffy curve (it added an n_i to Eq. 8.28, thus weighting the mean lengths-at-age according to how many points contributed to each mean). With the easy availability of relatively powerful computers such computational devices are no longer necessary and the raw data should be used wherever possible.

Which residual error structure is most appropriate in Eq. 8.26 in a particular situation should primarily be determined by the validity of the particular error distribution, however, the purpose of fitting the curve can also play a part. Kimura (1980) suggests that curves fitted using Eq. 8.27 would do well in predicting the length of a randomly sampled individual, which might prove useful in modelling applications. He goes on to suggest that Eq. 8.28 might best describe the growth of a species over its entire lifespan, which might be best if one's purpose was to compare an array of growth curves. Finally, Eq. 8.29 would be best if the variance of the lengths at age changed markedly with age.

The likelihood equation for all available observations, $\tilde{L}$, would be

$$L\left(\tilde{L} \mid L_{\infty}, K, t_0, \sigma^2\right) = \prod_{i=1}^{i=N} \frac{1}{\sigma\sqrt{2\pi}} e^{\left(\frac{-\left[L_i - f\left(L_{\infty}, K, t_0, t_i\right)\right]^2}{2\sigma^2}\right)} \tag{8.30}$$

The σ^2_{Ω} and σ^2_{ω} from above are the σ^2 in Eq. 8.30. Appendix 8.2 illustrates how we can validly move from using Eq. 8.30 to the far simpler likelihood equation

$$L(\tilde{L} \mid L_{\infty}, K, t_0, \sigma^2) = \left(2\pi\hat{\sigma}^2\right)^{-\frac{N}{2}} e^{\left(\frac{-N}{2}\right)} \tag{8.31}$$

When comparing two or more curves using likelihood ratios one first sets up the general model against which all alternative simplified hypothesis are to be tested. This statistic is determined by comparing the ratio of the two likelihoods (Kimura, 1980)

$$\Lambda = \frac{\left(2\pi\hat{\sigma}^2_{\omega}\right)^{-\frac{N}{2}} e^{\left(\frac{-N}{2}\right)}}{\left(2\pi\hat{\sigma}^2_{\Omega}\right)^{-\frac{N}{2}} e^{\left(\frac{-N}{2}\right)}} = \left(\frac{\hat{\sigma}^2_{\Omega}}{\hat{\sigma}^2_{\omega}}\right)^{\frac{N}{2}} \tag{8.32}$$

Kimura (1980) states that under the hypothesis that the linear constraint ω is true then the test statistic:

$$-2Ln(\Lambda) = -2Ln\left(\left(\frac{\hat{\sigma}^2_{\Omega}}{\hat{\sigma}^2_{\omega}}\right)^{\frac{N}{2}}\right) = -2\frac{N}{2}Ln\left(\frac{\hat{\sigma}^2_{\Omega}}{\hat{\sigma}^2_{\omega}}\right) = -NLn\left(\frac{\hat{\sigma}^2_{\Omega}}{\hat{\sigma}^2_{\omega}}\right) \tag{8.33}$$

will have, asymptotically, a χ^2 distribution. This means that the larger the sample size the more likely the conclusions are to be valid. However, it should be clear that Eq. 8.33 is equivalent to Eq. 8.23 (the N's inside the brackets cancel to leave the RSS$_x$).

This section may have appeared relatively complex. In practice, what the likelihood ratio amounts to is calculating, separately, the total residual sum of squares for the base case and for the hypothesis of interest (e.g., the K values are equal) and substituting the values into Eq. 8.23 (or Eq. 8.33). The likelihood ratio is then compared with a table of the χ^2 distribution with k degrees of freedom (k is the number of constraint placed upon the fit; Example Box 8.7).

8.4.4 Kimura's Likelihood Ratio Test

Kimura (1980) provided a set of test data and the analytical outcomes so that anyone attempting to implement a likelihood ratio test would be able to ensure their computer program worked. He provided data relating to the average length-at-age for Pacific hake for each sex with the objective of determining whether the two sexes differed significantly in their overall growth pattern (Table 8.3, Fig. 8.11). The data were used in the ARSS-statistic example above and the curves were not found to be coincident. We will use the likelihood ratio test to determine how the curves differ (Example Box 8.7).

The method is simple. Firstly, find the best fitting separate curves for each data set separately. Then, in sequence compare the total residual sum of squares from this base case with that obtained by adding various constraints.

Perhaps the best order in which to impose these tests is to first assume the hypothesis that both data sets can best be described by a single line (coincident curves). If this indicates that a significant difference exists, then one should sequentially assume that single parameters are the same between the two lines and determine whether significant differences arise. If differences are found, then one would conclude that evidence exists that the curves do differ and be able to identify which parameters are different between the curves for each data set (Table 8.3).

When the likelihood ratios are calculated a difference between the curves was indicated by the comparison of the base case with the assumption of the same curve fitting both data sets. With the more detailed analyses, a difference between the L_∞ was strongly indicated, while there was some slight evidence that the K parameter might be different, however, there was no indication that the t_0 parameters differed significantly (Table 8.4).

8.4.5 Less than Perfect Data

Unfortunately, it is quite common with real fisheries to only obtain ageing data from a sample of fish obtained from commercial operations. Such a sample may be influenced by the existence of a legal minimum length, and there may be low numbers of animals taken in the younger or older age classes. Finally, with size selective fishing, the larger, faster growing fish of all ages may have been differentially removed from the population through fishing pressure (Fig. 8.12).

Table 8.4 Likelihood ratio test for the Pacific hake data from Table 8.3 (data from Kimura, 1980). The top three parameters relate to the female fish while the middle set of parameters refer to the male fish. The base case is where two separate curves are fitted independently, the coincident column is where the lines are assumed identical, and the remaining three columns are where the listed parameter is assumed equal between the two lines. RSS_ω refers to the total residual sum of squares for both curves together given the constraint ω. In the BaseCase column, the 48.224 refers to the unconstrained RSS_Ω, against which all the RSS_ω are compared. N in all these cases is 24, 13 females and 11 males (Example Box 8.7; Fig. 8.11).

	BaseCase	Coincident	=Linf	=K	=t0
$L_\infty f$	61.233	59.294	59.404	60.140	60.767
Kf	0.296	0.320	0.337	0.330	0.313
$t_0 f$	-0.057	0.010	0.087	0.095	0.057
$L_\infty m$	55.978	59.294	59.404	57.435	56.454
Km	0.386	0.320	0.297	0.330	0.361
$t_0 m$	0.171	0.010	-0.111	-0.021	0.057
RSS_ω	48.224	79.765	71.602	56.337	50.758
χ^2	-	12.077	9.487	3.732	1.229
df	-	3	1	1	1
P	-	0.0071	0.0021	0.0534	0.2676

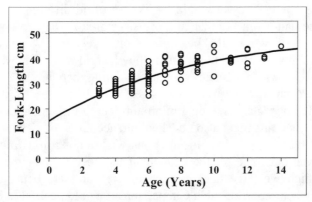

Figure 8.12 The same data as in Fig. 8.1, for snapper, *Pagrus auratus* from the Bay of Plenty, New Zealand, but with all animals less than 25 cm excluded as being below the legal size, and fewer older animals, as taken by the commercial fleet. Instead of $L_\infty = 44.79$, $K = 0.196$, and $t_0 = -0.81$, the best fitting von Bertalanffy growth curve was for $L_\infty = 50.45$, $K = 0.115$, and $t_0 = -3.06$; the L_∞ line is now above the data and the t_0 is obviously unrealistic. Clearly, there are many potential impacts of limiting the data in this way. It also makes drawing conclusions from curve comparisons dangerous.

Growth curves fitted to samples from a commercial fishery can have many biases and the parameters obtained may be very wide of the values obtained from a more representative sample (Fig. 8.12). However, if such data is all that is available it would be useful to know whether it was valid to compare growth curves derived from such imperfect data. For example, in Example Box 8.7 two curves were compared where one data set covered the ages 1 to 13.3 while the other only covered ages from 1 to 11.3 years. In that example, a significant difference was found between the two curves, driven primarily by differences in the L_∞ parameter. However, if the comparison if restricted to the first 11 data points in both data sets the hypothesis that the two curves are coincident is not proved incorrect (this can be tried in Example Box 8.7).

The von Bertalanffy curve is strongly determined by the values for L_∞ and t_0, which are at the extremes of the curve and usually have the least data. This is why so many people have spent time trying to reparameratize the curve (Sainsbury, 1980; Frances, 1988a, b) to have parameters closer to the data from which they are estimated. If the data we use to construct growth curves is biased or not representative (e.g., samples from commercially caught fish) then extra care needs to be taken when comparing growth curves. When comparing growth curves derived from non-representative data, it seems intuitively reasonable to conduct tests only over equivalent ranges of data. Thus, if we were to compare the data from Figs. 8.1 and 8.12 we would be likely to find differences. But if we restricted the data from Fig. 8.1 to those sizes greater than 25 cm and ages greater than two, we would be much more likely to find no differences (because, along with removing some of the older animals, that was how the data in Fig. 8.11 was generated). We are assuming there are no errors in the ageing so we should only compare ages with like ages. If there is an obvious difference such as a legal minimum length influencing only one of the samples, then this must also be taken into account.

What these suggestions mean is that we are not comparing complete growth curves but, instead, the growth of the species over the ranges of ages for which we have data in each of the data sets being compared. Obviously, when reporting such comparisons it would be vital to be explicit about exactly what was compared with what.

Example Box 8.7   Kimura's (1980) Likelihood Ratio test. Copy the data from Table 8.3 into columns B and C, female data followed by the male. In D14 put =$B$1*(1-exp(-$B$2*(B14-$B$3))) and copy down to D26. In D27 put =$B$4*(1-exp(-$B$5*(B27-$B$6))) and copy down to D37. In B8 put =sum(E14:E26) and in C8 put =sum(E27:E37). Put =count(B14:B26) into B12 and =count(B27:B37) into C12. In D12 put =B12+C12 for the total N. In D9 calculate the likelihood ratio =-D12*Ln(F8/D8). In D10 manually put the degrees of freedom = number of constraints being considered. In D11 put =chidist(D9,D10). In D9, note the reference to F8, which is merely the stored result of minimizing D8 by altering B1:B6 and storing B1:B6 into F2:F7 and B8 into F8 as values. This sets up the sheet ready for hypothesis testing. To test for coincident curves, put =B1 into B4 and copy down to B6. This sets the same parameters for both data sets. Then minimize D8 by altering B1:B3. Copy the answers as values from B1:B6 and from D8:D11 into G2:G11 for later reference. To test if the L_∞ are different put =B1 into B4 and minimize D8 by altering (B1:B3,B5:B6). Remember to alter the degrees of freedom in D10 to one. Copy the results to the right of column G. Try the other hypotheses by similar manipulations of which cells are set equal to which and by altering the cells to be altered by the solver. Try changing the analysis by restricting the sum of squares residuals for females to the first 11, like the males, by putting =sum(E14:E24) into B8 and =count(C14:C26) into B13. Are the results of the analysis the same?

	A	B	C	D	E	F	G
1	Linf-f	61.23			**Results**	**BaseCase**	**Coincident**
2	K-f	0.296			Linf-f	61.2331	59.2938
3	t0-f	-0.057			K-f	0.2963	0.3205
4	Linf-m	55.98			t0-f	-0.0573	0.0104
5	K-m	0.386			Linf-m	55.9779	59.2938
6	t0-m	0.171			K-m	0.3856	0.3205
7		Female	Male	Total	t0-m	0.1713	0.0104
8	SSQ	28.800	19.423	=B8+C8	SSQ	48.2238	79.7645
9			χ^2	0	χ^2		12.077
10			df	3	Df		3
11			P	1	P		0.0071
12	Count	13	11	24			
13		Age	Length	E (L)	Resid2		
14	Fem	1	15.4	15.54	=(C14-D14)^2		
15	Fem	2	28.03	27.70	=(C15-D15)^2		
16	Fem	3.3	41.18	38.76	Copy Down		
17	Fem	4.3	46.2	44.68	To row 37		

8.4.6 A Randomization Version of the Likelihood Ratio Test

When less than good data are used in a comparison, the assumption that the likelihood ratio adheres to the χ^2 distribution becomes invalid and alters the risk of Type I errors (saying there is a difference where one does not exist) and Type II errors (claiming no difference when one exists). For example, if, in Example Box 8.7, one restricts the comparison of curves to the ages 1 to 11.3 the coincident curve comparison indicates no statistical difference. However, if one then continued and compared the two curves with the L_∞ constrained to be identical, a slightly significant difference is found ($\chi^2 = 4.359$, $P = 0.0368$). Thus, one test concludes no difference while the other concludes a difference exists. An assumption in the analysis, that the likelihood ratio will approximate the χ^2 distribution, appears to be failing. One would never use mean length at age data in the way illustrated. At the very least on would weight each squared residual by the number of observations for each data point (Eq. 8.28), but usually one would use raw length at age data in all comparisons.

   With inadequate data, instead of using the likelihood ratio test as it stands, it is possible to arrange the data so that we can generate a randomization version of the likelihood ratio test that will generate its own empirical sampling distribution for the test statistic. What is being suggested is that the length at age data need to be randomized in some way between the two groups to test the hypothesis that the two data sets are equivalent to two random samples from the same statistical population. This has previously been suggested by Weinberg and Helser (1996) where they made the comparisons by randomizing age and length data pairs between different populations of surfclams. Unfortunately, with this design of randomization, if there are low numbers of the extreme age classes it would be quite possible to include tests where all the younger or older individuals were selected by chance into one of the randomized data sets. This would have the effect of overestimating the number of times large likelihood ratios would be expected to occur. This test appears to be comparing the proportional catch-at-age and not the growth curves from the two populations. If the data sets were large and the numbers at age were relatively evenly distributed this may not be a great problem. However, with Kimura's (1980) data (for ages 1 to 11.3), when a randomization test is conducted in this way the test of coincident curves suggests that likelihood rations greater than that observed (6.3986) occurred 223 times out of 1,000 (i.e., a $P = 0.223$), which is overly conservative. If we were to plot the proportion of the different likelihood ratio χ^2 values along with the hypothetical proportions from the χ^2 distribution, we would be able to see how closely the empirically determined statistic mimics the hypothetical (Fig. 8.13).

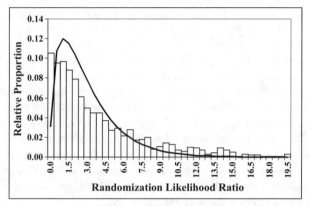

Figure 8.13 A comparison of the relative proportion of different χ^2 values obtained from a randomized likelihood ratio test (open columns) with the expected proportions derived from the χ^2 distribution (solid curve). The test was of whether the growth curves for Pacific hake were coincident for males and females when the ages being compared were restricted to ages 1 to 11.3 (Table 8.2, Fig. 8.12, Example Box 8.8). The randomization test, in this case, involved the randomization of age and length data pairs between the two sexes. If the critical value chosen is 0.05, then the empirical statistic would require a larger likelihood ratio than the true χ^2. In other words, the empirical statistic would claim not to have found a difference more often than it should (Example Box 8.8).

   Clearly, the randomization of data pairs between the sexes would lead to a greater number of Type II errors than expected. The test is too conservative and would have trouble detecting a difference even when one existed. If we are to use a randomization test, some other randomization schema would need to be adopted.

   The ages are assumed to have been measured without error and each age has an associated range of lengths observed in the population; in the Kimura example this is reduced to just a single length per age per sex, but with real data there would be a number of observations per age class. This structure to the data should be recognized and instead of randomizing data pairs the randomizations should be of lengths within age classes between sexes. Thus, for each randomization test one would need to conduct the same number of randomizations of data as one had age classes. With the Kimura data this would entail conducting 11 randomizations, one for each age class. In that case it would be merely deciding which data point was associated with which age. Clearly, this routine would be more complicated to implement (Example Box 8.9) than randomizing data pairs.

Example Box 8.8 The incorrect randomization algorithm can be introduced into Example Box 8.7 by adding some columns to the right of the calculations. Start in Row 1. In one column put a series of =rand(), next to that put the female data pairs of age and length for containing ages 1 to 11.3, under the female data put the male data for ages 1 to 11.3. Conduct the randomization as illustrated in Chapter 5. Sort the three extra columns on the column of random number to randomize the data pairs between the two sexes. Then copy, as values, the randomized ages and associated lengths into columns B and C of Example Box 8.7. Repeat that at least 1000 times, conducting the likelihood ratio test each time, and recording the χ^2 value each time. Use the frequency function to count the relative frequency of randomization χ^2 values into column N. Convert the cumulative χ^2 distribution into the standard distribution as shown in columns O and P. Plot column P and column Q against column L to mimic Fig. 8.12. If you do more than 1000 replicates be sure to alter column Q appropriately.

L	M	N	O	P	Q
Average Class	0	Obs Freq	χ^2 cumulative	χ^2	Prop Freq
=average(M1:M2)	1	197	=chidist(L2,3)	=1-O2	=N2/1000
=average(M2:M3)	2	173	=chidist(L3,3)	=O2-O3	=N3/1000
2.5	3	144	0.4753	=O3-O4	=N4/1000
3.5	4	73	0.3208	0.1545	0.073
4.5	5	51	0.2123	0.1085	0.051

By randomizing within age-classes but between sexes, the randomization test can avoid problems with the relative number of observations present in each age-class. These now remain the same as in the original data set and the proportional age-structure of each data set remains the same as the original. The only thing that changes is the distributions of lengths at age in each of the two data sets. In fact, with the Kimura data there are only 10 pairs of data points that differ as the data for the 1-year-old fish is the same for both sexes. By having so few data points that differ, the randomization test is relatively limited in the number of likelihood ratio values it can produce randomly. The upper and lower limits and the relative distribution did not alter after the first 4000 replicates (Fig. 8.14).

There are clearly some distortions away from the ideal χ^2 distribution for the test statistic. It illustrates that the reliance of the likelihood ratio test on the χ^2 statistic is only approximate, especially when the data are not as representative as one would wish it to be. When the data used are of better quality (Fig. 8.1), then the match between the empirical distribution and the hypothetical ideal can become closer.

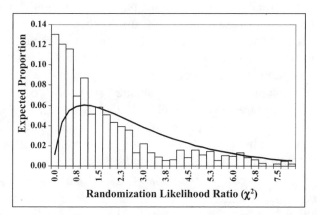

Figure 8.14 A comparison of the relative proportion of different χ^2 values obtained from a randomized likelihood ratio test (open columns) with the expected proportions derived from the χ^2 distribution (solid curve). The test was of whether the growth curves for Pacific hake were coincident for males and females when the ages being compared were restricted to ages 1 to 11.3 (Table 8.2, Fig. 8.12, Example Box 8.9). The randomization test, in this case, involved the randomization of length data within ages but between sexes. Despite the distortions below a ratio of 5.0, the cumulative distribution leads to similar decisions when comparing the two curves. The empirical distribution suggests that there is some evidence that the curves are not coincident (Example Box 8.9).

8.5 CONCLUDING REMARKS

The study of individual growth is fascinating for many people and this can lead them to draw strong conclusions from the form of the equations used to describe the growth of their favourite organism. Hopefully, it has been shown that growth equations are principally just descriptions of how growth proceeds, they are not explanations for the process of growth. Just because the equation has an asymptote does not imply that the fish species has such a thing. The move towards reparameratized growth curves should assist in avoiding this error.

Comparisons between growth curves are difficult to conduct in a valid manner and care must be exercised in any conclusions drawn from such tests. In very many publications, data are presented that only constitutes a sub-sample of the true population. While this is better than no data, it should be recognized for what it is, which is a non-representative sample from the population. The difficulties of sampling marine populations are such that this will be a persistent problem for fisheries biologists. It is quite possible to obtain representative samples of what is caught by commercial fishers. This must be recognized and its implications for comparisons of growth between different populations should be considered before such comparisons are made. The randomization test suggested in this chapter is complex to implement (but is possible even in Excel), but it permits one to conduct the test required while testing the validity of the assumption of the test statistic being equal to the χ^2 statistic. Where the representativeness of the data is at all suspect, it is recommended that some version of the randomization test be implemented. With a great deal of data this be relatively slow but it is a case of computer intensive statistics answering the questions for which answers are needed.

Example Box 8.9   A randomization of Kimura's (1980) Pacific hake data. The age data are set out into 11 columns with columns of random numbers (=rand()) beside them. Row 41 relates to female lengths and row 42 to males. The test first randomizes the row in which each data point sits. The age structured data is re-constituted in columns A and B. How the age-length data is reconstructed can vary but in the end one must be able to copy and paste the values into the required cells (C14:C35) prior to conducting the likelihood ratio test (assumes the females aged 12.3 and 13.3 have been removed). Age 1 is unchanged. For the rest one first has to randomize the lengths between sexes within ages. Of course, one would create a macro to do all of the separate sorts, first D41:E42 on column D, then F41:G42 on column F, and so on. Once completed the data in columns A and B would have been randomized. These data are copied and pasted as values into the correct cells for the likelihood ratio test to proceed. All this is repeated at least 1000 times and the results compared with the original analysis (Fig. 8.13).

	A	B	C	D	E	F	G
40	Age	Length	Age 1	Age 2	Age 2	Age 3.3	Age 3.3
41	1	=C41	15.4	=rand()	28.03	=rand()	41.18
42	2	=E41	15.4	=rand()	26.93	=rand()	42.23
43	3.3	=G41					
44	4.3	=I41					
45	5.3	=K41					
46	6.3	=M41					

APPENDIX 8.1

Derivation of the Fabens version of the Von Bertalanffy growth equation. This is the version that is used when attempting to estimate the growth curve parameters from tagging data. With such data, one tends to have the dates of tagging and recapture, the initial length at tagging, and the length at recapture. It would be very unusual to have the age of the tagged organisms. Hence, the standard length-at-age formulation cannot be used. Instead, we need an equation that generates an expected length increment in terms of the von Bertalanffy parameters and the length at time t (not age t). The standard formulation of the von Bertalanffy growth equation is given by Eq. A8.1

$$L_t = L_\infty \left(1 - e^{-K(t-t_0)}\right) \tag{A8.1}$$

where L_t is the length at time t, L_∞ is average maximum length of individuals in the population, K is a growth rate parameter, t is the age of the animals, and t_0 is the hypothetical age at length zero. The brackets can be expanded thus

$$L_t = L_\infty - L_\infty e^{-K(t-t_0)} \tag{A8.2}$$

The expected length of an animal of age t after the passage of time, Δt, is given by Eq. A8.3

$$L_{t+\Delta t} = L_\infty - L_\infty e^{-K(t+\Delta t - t_0)} \tag{A8.3}$$

Extract the Δt term from the exponential term

$$L_{t+\Delta t} = L_\infty - L_\infty e^{-K(t-t_0)} e^{-K\Delta t} \tag{A8.4}$$

The change in length over the time Δt is simply the difference between Eqs. A8.4 and A8.2

$$\Delta L = L_{t+\Delta t} - L_t = L_\infty - L_\infty e^{-K(t-t_0)} e^{-K\Delta t} - L_\infty + L_\infty e^{-K(t-t_0)} \tag{A8.5}$$

The stand-alone L_∞ terms cancel out and the order of the remaining two terms can be reversed

$$\Delta L = L_\infty e^{-K(t-t_0)} - L_\infty e^{-K(t-t_0)} e^{-K\Delta t} \tag{A8.6}$$

The $L_\infty\, e^{-K(t-t0)}$ term can be extracted

$$\Delta L = L_\infty e^{-K(t-t_0)}\left(1-e^{-K\Delta t}\right) \tag{A8.7}$$

To complete the transformation, the two extra L_∞ terms, that cancel each other out, are put back into to the primary term

$$\Delta L = \left(L_\infty - L_\infty + L_\infty e^{-K(t-t_0)}\right)\left(1-e^{-K\Delta t}\right) \tag{A8.8}$$

The final change is to recognize that, in the first bracket, the second and third terms combined are equivalent to the standard von Bertalanffy equation (Eq. A8.1) so we can substitute an L_t to leave the standard Fabens version of the von Bertalanffy growth equation.

$$\Delta L = \left(L_\infty - L_t\right)\left(1-e^{-K\Delta t}\right) \tag{A8.9}$$

APPENDIX 8.2

Derivation of the maximum likelihood estimator for the von Bertalanffy curve. Kimura (1980) starts by writing out the maximum likelihood estimator for a von Bertalanffy curve. We will do the same here but with a few more details to ensure clarity of exposition. With the usual notation the length of the uth individual at age t_u is defined as

$$L_u = L_\infty \left(1 - e^{(-K(t_u - t_0))}\right) + \varepsilon_u \qquad (A8.10)$$

Let L_∞ be the, possibly hypothetical, asymptotic length, K the constant relating to the rate of growth, t_0 the hypothetical age at length zero, and the ε_u's be the independent, additive, normal random errors ($N(0, \sigma^2)$). For brevity in the following equations, the expected values of L_u can be regarded simply as a function of L_∞, K, and t_0

$$\hat{L}_u = f(L_\infty, K, t_0, t_u) + \varepsilon_u \qquad (A8.11)$$

As Kimura states, the optimum way to fit a von Bertalanffy curve to a set of data is to use a maximum likelihood estimator and because we would be using normal random errors this means the likelihood function would derive from the normal distribution

$$L\left(L_u \mid L_\infty, K, t_0, \sigma^2\right) = \frac{1}{\sigma\sqrt{2\pi}} e^{\left(\frac{-\left[L_u - f(L_\infty, K, t_0, t_i)\right]^2}{2\sigma^2}\right)} \qquad (A8.12)$$

which is the likelihood of the single observation L_u. This is simply the usual normal distribution equation with the observed and expected lengths at t_u as the numerator in the major term. Note that the term involving the observed and expected values is identical to that which would be used in a least squares estimation. Given N different observations of age and length the overall likelihood $\tilde{L}$ is obtained by multiplying N the separate likelihoods together

$$L\left(\tilde{L} \mid L_\infty, K, t_0, \sigma^2\right) = \prod_{i=1}^{i=N} \frac{1}{\sigma\sqrt{2\pi}} e^{\left(\frac{\left(-\left[L_i - f(L_\infty, K, t_0, t_i)\right]^2\right)}{2\sigma^2}\right)} \qquad (A8.13)$$

This can be simplified by noticing that if we raise the $1/[\sigma\sqrt{(2\pi)}]$ term to the power N we can take it outside of the product and that the remaining product can be converted to a summation term as follows

$$L\left(\tilde{L}\mid L_{\infty},K,t_0,\sigma^2\right)=\left(2\pi\sigma^2\right)^{-\frac{N}{2}}\prod_{i=1}^{N}e^{\left(\frac{-\left[L_i-f\left(L_{\infty},K,t_0,t_i\right)\right]^2}{2\sigma^2}\right)} \tag{A8.14}$$

Equation A8.14 is a form that can be used for making maximum likelihood comparisons. However, Eq. A8.13 can be simplified by using log-likelihoods that serve to replace the product with a summation. This acts to reduce the chance of rounding errors occurring due to the very small likelihoods that can arise. The logarithmic transformation also has the effect of reducing the e term to its exponent. To make the impact of the transformation explicit note that

$$Ln\left(\prod_{1}^{N}\frac{1}{\sqrt{2\pi\sigma^2}}\right)=Ln\left(\left(\frac{1}{\sqrt{2\pi\sigma^2}}\right)^N\right)$$
$$=Ln\left(\left(2\pi\sigma^2\right)^{-N/2}\right)=-\frac{N}{2}Ln\left(2\pi\sigma^2\right) \tag{A8.15}$$

and that

$$Ln\left(\prod_{i=1}^{N}e^{\frac{-\left[L_i-f\left(L_{\infty},K,t_0,t_i\right)\right]^2}{2\sigma^2}}\right)=\sum_{i=1}^{N}\frac{-\left[L_i-f\left(L_{\infty},K,t_0,t_i\right)\right]^2}{2\sigma^2}$$
$$=\left(-\frac{1}{2\sigma^2}\sum_{i=1}^{N}\left[L_i-f\left(L_{\infty},K,t_0,t_i\right)\right]^2\right) \tag{A8.16}$$

We can then see that the log-likelihood estimator is the combination of Eqs. A8.15 and A8.16

$$LL\left(\tilde{L}\mid L_{\infty},K,t_0,\sigma^2\right)$$
$$=-\frac{N}{2}Ln\left(2\pi\sigma^2\right)-\frac{\sum_{1}^{N}\left[L_i-f\left(L_{\infty},K,t_0,t_i\right)\right]^2}{2\sigma^2} \tag{A8.17}$$

The maximum likelihood estimate of σ^2, denotes σ_φ^2, and is obtained from the log-likelihood estimator by calculating the partial derivative with respect to σ^2, and equating that to zero. Using Eq. A8.17

$$\frac{\partial \, \mathrm{LL}\left(\tilde{L} \mid L_\infty, K, t_0, \sigma_\varphi^2\right)}{\partial \sigma_\varphi^2}$$

$$= -\frac{N}{2\sigma_\varphi^2} + \frac{\sum_{i=1}^{N}\left[L_i - f\left(L_\infty, K, t_0, t_i\right)\right]^2}{2\left(\sigma_\varphi^2\right)^2} = 0 \tag{A8.18}$$

which leads by simple algebra to

$$\sum_{i=1}^{N}\left[L_i - f\left(L_\infty, K, t_0, t_i\right)\right]^2 = \frac{2\left(\sigma_\varphi^2\right)^2 N}{2\sigma_\varphi^2} \tag{A8.19}$$

and thereby

$$\sigma_\varphi^2 = \frac{\sum_{i=1}^{N}\left[L_i - f\left(L_\infty, K, t_0, t_i\right)\right]^2}{N} \tag{A8.20}$$

The numerator in this relation is the sum of squared residuals, thus the problem of maximum likelihood estimation for the von Bertalanffy curve is reduced to finding the least square estimates of L_∞, K, and t_0 (Kimura, 1980). As described in the chapter on parameter estimation, this is a general property of the maximum likelihood estimator for any linear or non-linear equations that are fitted using additive, normal random errors. The form of Eq. A8.20 is that required for the determination of the likelihoods of different hypotheses to be compared. This is done by substituting the maximum likelihood estimate of the variance. i.e., Eq. A8.20, into Eq. A8.16; which is equivalent to Eq. 8.30 in the main text:

$$L\left(\tilde{L}\mid L_\infty, K, t_0, \sigma^2\right) = \left(2\pi\hat{\sigma}^2\right)^{-\frac{N}{2}} e^{\left(\frac{-\sum(L_i - f(L_\infty, K, t_0, t_i))^2}{2\frac{\sum(L_i - f(L_\infty, K, t_0, t_i))^2}{N}}\right)} \tag{A8.21}$$

$$= \left(2\pi\hat{\sigma}^2\right)^{-\frac{N}{2}} e^{\left(\frac{-N}{2}\right)}$$

This last equation is extended in the main text to obtain the final form of the likelihood ratio test.

9

Stock-Recruitment Relationships

9.1 RECRUITMENT AND FISHERIES

9.1.1 Introduction

Ignoring immigration, recruitment, and individual growth are the major contributors to the production of biomass within a stock. Akin to the study of growth, some people have dedicated huge efforts towards the investigation of fisheries recruitment, especially the relationship between mature or spawning stock size and subsequent recruitment (Cushing, 1988). Recruitment to fish populations is naturally highly variable and the main problem for fisheries scientists is whether recruitment is determined by the spawning stock size or environmental variation or some combination of both. To conduct stock assessments that include a risk assessment involves projecting the population forward in time. To be able to do this a minimum requirement is to have information about a stock's productivity. It is best to have a stock recruitment relationship.

In this chapter, we will consider the mathematical description of stock-recruitment relationships but, as with growth, we will essentially ignore the biology behind the relationships. The biology will only be considered where it has a direct bearing on how the stock-recruitment relationships are described. We will review the most commonly used mathematical models of stock recruitment and will discuss their use in stock assessment models of varying complexity.

9.1.2 Recruitment Over-Fishing

Two types of over-fishing are commonly discussed in the fisheries literature. The first is termed **growth over-fishing**, and is where a stock is fished so hard that most individuals are caught at a relatively small size. This is the classic yield-per-recruit problem of balancing the stock losses due to total mortality against the stock gains from individual growth. The aim of such analyses is to determine the optimum size and age at which to begin harvesting the species. Growth over-fishing is where the fish are being caught before they have time to reach this optimal size.

The second type of over-fishing, the form particularly relevant to this chapter, is **recruitment over-fishing**. This occurs when a stock is fished so hard that the stock size is reduced below the level at which it, as a population, can produce enough new recruits to replace those dying (either naturally or otherwise). Obviously, such a set of circumstances could not continue for long and, sadly, recruitment over-fishing is usually a precursor to stock collapse.

Growth over-fishing is not difficult to detect. The data requirements for detection are a growth curve and an estimate of the age-structure of the fishery. With this information, one could conduct a yield-per-recruit analysis and determine how close the fishery is to the theoretical optimum. There can be complicating factors, such as whether the fishery really is at an equilibrium and which criterion to choose when selecting the optimum mean size at capture (i.e., which target fishing mortality to use, $F_{0.1}$ is now very common). However, the methods are well established and in common use. Sadly, the same cannot be said about the detection of recruitment over-fishing, which could require a determination of the relation between mature or spawning stock size and recruitment levels. This has proven to be a difficult task for very many fisheries.

9.1.3 The Existence of a Stock Recruitment Relationship

There is a commonly held mistaken belief that the number of recruits to a fishery is usually independent of the adult stock size over most of the observed range of stock sizes. This can be a dangerous mistake (Gulland, 1983). It suggests that scientists and managers can ignore stock-recruitment relationships unless there is clear evidence that recruitment is not independent of stock size. The notion of no stock-recruitment relationship existing derives from data on such relationships appearing to be very scattered with no obvious pattern (Fig. 9.1).

The fallacy of no relationship existing between stock size and subsequent recruitment originated because people made the invalid conclusion that because they could not observe a significant stock-recruitment relationship then one did not exist. If, in fact, such relationships did not exist then stock collapses would be less common. Hilborn and Walters (1992, p. 241) were explicit: "While recruitment may be largely independent of stock size as a fishery develops, experience has shown that most fisheries will reach a point where recruitment begins to drop due to over-fishing."

In a recent controversy there are strong arguments given for a relationship between recruitment and spawning stock biomass (Francis, 1997; Gilbert, 1997; Hilborn, 1997; Myers, 1997).

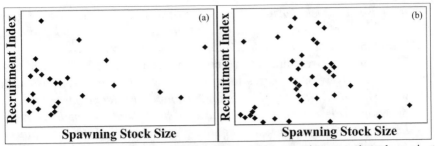

Figure 9.1 Stock-recruitment relationships with recruitment plotted against spawning adult stock size. The scales on the two plots differ. Plot (a) is for plaice (after Beverton, 1962) and plot (b) is for Georges Bank haddock (after Overholtz *et al.*, 1986). In both cases, a flat line provides an excellent fit.

9.2   STOCK-RECRUITMENT BIOLOGY

9.2.1 Properties of "Good" Stock-Recruitment Relationships

A good introduction to the biological processes behind the stock-recruitment relationships which we are going to consider is given by Cushing (1988) who provides an overview of the sources of egg and larval mortality along with good examples and a bibliography on the subject. There is an enormous literature on the biology of stock-recruitment relations and their modifiers. A great variety of influences, both biological and physical, have been recorded as affecting the outcome of recruitment. We will not be considering the biological details of any real species except to point out that the relation between stock size and resulting recruitment is not deterministic and there can be a number of forms of feedback affecting the outcome. We will primarily be considering how best to model stock-recruitment relationships from fisheries data. Various mathematical descriptions of stock-recruitment relationships have been suggested but we will only consider those by Beverton and Holt, Ricker, and Deriso-Schnute.

Ricker (1975) listed four properties of average stock-recruitment relationships that he considered desirable:

1) A stock-recruitment curve should pass through the origin; that is, when stock size is zero there should be no recruitment. *This assumes the observations being considered relate to the total stock and that there is no "recruitment" made up of immigrants.*

2) Recruitment should not fall to zero at high stock densities. *This is not a necessary condition but whilst declines in recruitment levels with*

increases in stock densities have been observed, declines to zero have not. Even for a population at equilibrium at maximum stock biomass, recruitment should still match natural mortality levels.

3) The rate of recruitment (recruits-per-spawner) should decrease continuously with increases in parental stock. *This is only reasonable when positive density-dependent mechanisms (compensatory) are operating (for example an increase in stock leads to an increase in larval mortality). But if negative density-dependent mechanisms (depensatory) are operating (for example predator saturation, and Allee effects – Begon and Mortimer, 1986) then this may not always hold.*

4) Recruitment must exceed parental stock over some part of the range of possible parental stocks. *Strictly, this is only true for species spawning once before dying (e.g., salmon). For longer lived, multi-spawning species, this should be interpreted as recruitment must be high enough over existing stock sizes to more than replace losses due to annual natural mortality.*

Hilborn and Walters (1992) suggested two other general properties that they considered associated with good stock-recruitment relationships:

5) The average spawning stock-recruitment curve should be continuous with no sharp changes over small changes of stock size. *They are referring to continuity, such that average recruitment should vary smoothly with stock size.*

6) The average stock-recruitment relationship is constant over time. *This is stationarity, where the relationship does not change significantly through time. This assumption seems likely to be false in systems where the ecosystem, of which the exploited population is a part, changes markedly.*

9.2.2 Data Requirements - Spawning Stock

There is potential for confusion over terms when we refer to the spawning stock biomass in discussions of stock-recruitment relationships. What is being considered is not necessarily a biomass but a measure of the reproductive productivity of the mature population. The optimal measurement of spawning stock is the number of eggs spawned (Rothschild and Fogarty, 1989). This measure of reproductive capability may be estimated from the average fecundity by age and the proportion of each age. Alternatively, one could estimate the number of mature females and multiply by the average fecundity, or use total biomass of mature individuals, or even an index of abundance for the population in the year of spawning. It is important to note that all of these methods would have a degree of uncertainty about the estimated values. Of the four methods listed, ignoring measurement errors, the methods were described in descending order of reliability (thus, an index of abundance would be the least reliable). By using measures other than egg production, some of the assumptions of the recruitment/stock relationship may be broken (Rothschild and Fogarty, 1989). The uncertainty associated with each indicator of spawning stock implies that the accuracy of the x-axis spawning stock size values are suspect even before attempting to estimate recruitment. This calls into question analyses where the independent variable is supposed to be measured without error.

9.2.3 Data Requirements - Recruitment

Generally, in a fisheries sense, recruitment often refers to the life-stage that first becomes vulnerable to fishing gear. But for purposes of stock-recruitment relationships, recruitment may be defined as the population still alive at any given time after the egg stage. In some fortunate fisheries (large fish in freshwater) the number of recruits can be counted directly as they pass through artificial weirs but in almost all fisheries only indices of relative abundance are possible (possibly from traps or trawl surveys for juveniles).

Once again, there will be errors in estimation as well as natural variation in recruitment levels from year to year. These levels of variation and error in the estimates are very important for the assessment of reliability of the final stock-recruitment relationship derived. If the estimates are not especially reliable, then even if a deterministic stock-recruitment relationship exists between the spawning stock size and subsequent recruitment, it may be difficult to identify or recognise. Such variation may be confused with environmentally induced recruitment variability and it would be difficult to distinguish the two (see later).

Before we investigate the possible effects of uncertainty, we will first consider the form of a number of equations that have become part of the toolkit of fisheries modellers for describing stock-recruitment relationships.

9.3 BEVERTON-HOLT RECRUITMENT MODEL

9.3.1 The Equations

The Beverton-Holt model of the spawning stock-recruitment relationship was devised to incorporate density-dependent survival rates reflecting intra-cohort competition for critical resources (Figs. 9.2; Appendix 9.1)

$$R = \frac{S}{\alpha + \beta S} e^{\varepsilon} \qquad (9.1)$$

where R is the recruitment, S is the measure of spawning stock size, and α and β are parameters of the Beverton-Holt relationship. The e^{ε} indicates that the residual errors between the relationship and observed data are expected to be lognormal. The β value determines the asymptotic limit (= $1/\beta$), while the differing values of α are inversely related to the rapidity with which each curve attains the asymptote, thus determining the relative steepness near the origin (the smaller the value of α the quicker the recruitment reaches a maximum). As with all stock-recruitment equations this is an average relationship and the scatter about the curve is as important as the curve itself.

There are a number of different formulations used for the Beverton-Holt stock-recruitment relationship. Perhaps the most commonly seen is

$$R = \frac{aS}{b + S} e^{\varepsilon} \qquad (9.2)$$

which is a restructuring of Eq. 9.1 so that R is recruitment, S is the spawning stock, a is the maximum number of recruits produced (the asymptote = $1/\beta$), and b is the spawning stock (=α/β) needed to produce, on average, recruitment equal to half maximum ($a/2$).

It is clear that the initial steepness of the Beverton-Holt curve (Fig. 9.2), along with the asymptotic value, captures the important aspects of the behaviour of the equation. The asymptote is given by the value of the parameter a, while the initial steepness is approximated by the value of ($a/b = 1/\alpha$) which happens when S is very small (see Eq. 9.2; Example Box 9.1).

9.3.2 Biological Assumptions/Implications

The Beverton-Holt model of stock-recruitment derives from the balance between density-independent and density-dependent juvenile mortality (see Appendix 9.1). This linear relationship implies that the larger the spawning stock the faster the juveniles will die. There is an inverse relationship between the average number of recruits per spawner and the spawning stock size.

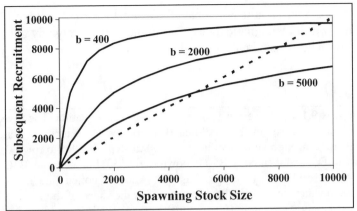

Figure 9.2 A comparison of Beverton-Holt curves relating recruitment against spawning stock size using Eq. 9.2 with different values of *a* and *b*. In all cases the value of *a* is 10,000, which is thus the maximum recruitment possible. Note that the steepness at the origin alters considerably as *b* alters (because a recruitment of 5000 = *a*/2, occurs at a spawning stock size of *b*). The straight, dotted line on the diagram is the line of replacement, so curves to the left of this represent growing populations while curves to the right represent shrinking populations. In practice, only species that have multiple years of spawning could be to the right of the dotted line (Example Box 9.1).

The idea of attempting to interpret the parameters of Eq. 9.2 draws attention to the differences between an equation as a theory and an equation as a summary description of a natural process. Essentially, Eqs. 9.1 and 9.2 attempt to describe the continuous reduction through mortality in the numbers of recruits from their initial egg-production numbers as spawning stock size increases (see Appendix 9.1).

Each of the parameters can be interpreted in terms of the observable world. However, despite the possibility of giving a real interpretation to the parameters, this description of recruitment is too simple to have great explanatory power and generally should be considered simply as a convenient mathematical description of the stock-recruitment relationship. All this means is that just because it may be possible to fit the equation to real data does not imply that the population concerned really has a stable, asymptotic limit to the number of recruits its population can produce.

Historically, Beverton and Holt (1957) introduced their curve because it had a simplistic interpretation, which meant it could be derived from first principles. But it was also mathematically tractable, which was important to them with the requirement at the time to use analytical methods. In fact, its continued use appears to stem a great deal from inertia and tradition. You should note that if we are to treat the Beverton-Holt curve simply as a mathematical description then effectively any curve, with the good properties listed earlier, could be used.

Example Box 9.1 The Beverton-Holt stock-recruitment equation. The rows of the *a* and *b* parameters extend from column B to column D. Extend column A, the spawning stock size, down to a value of 10000 in steps of 500 (to row 30). Copy B5 across to D5, and then copy B5:D5 down to row 30. Plot columns B, C, and D against A, as solid lines to mimic Fig. 9.2. Add the dotted line if you wish. Modify the *b* values to observe how this modifies the shape of the curve. Clearly, with an extremely steep curve (*b* very small) the relationship resembles a straight line from very low to very high biomass values. Change the equation in B5:B30 to become =((B$1*$A5)/(B$2+$A5))*loginv(rand(),0,B$3), to see the effect of random error on recruitment. Alter the line for column B to a scatter of points with no connecting line. Give C1:C2 the same values as in B1:B2. Change the value of the errors in B3 to 0.5 and press F9 a few times. What impact does that have on the apparent shape of the Beverton-Holt stock-recruitment curve?

	A	B	C	D
1	*a*	10000	10000	10000
2	*b*	5000	2000	400
3	error	0.000001		
4	Spawn	Recruit 1	Rec 2	Rec 3
5	1	=(B$1*$A5)/(B$2+$A5)	=(C$1*$A5)/(C$2+$A5)	24.9
6	10	=(B$1*$A6)/(B$2+$A6)	=(C$1*$A6)/(C$2+$A6)	243.9
7	100	=(B$1*$A7)/(B$2+$A7)	=(C$1*$A7)/(C$2+$A7)	2000.0
8	200	Copy Down to Row 30	Copy Down to Row 30	3333.3
9	300	566.0	1304.3	4285.7
10	400	740.7	1666.7	5000.0
11	500	909.1	2000.0	5555.6
12	1000	1666.7	3333.3	7142.9
13	1500	2307.7	4285.7	7894.7
14	2000	2857.1	5000.0	8333.3

9.4   RICKER MODEL

9.4.1 The Equation

As with the Beverton-Holt model, there are a number of different formulations of the Ricker equation, but one that is commonly used is

$$R = aSe^{-bS}e^{\varepsilon} \qquad (9.3)$$

where R is the recruitment from S the spawning stock, a is the recruits-per-spawner at low stock levels, and b relates to the rate of decrease of recruits-per-spawner as S increases. The e^{ε} indicates that the residual errors between the relationship and observed data are expected to be lognormal. Note that the parameters a and b are very different from those in the Beverton-Holt equation. This equation does not attain an asymptote but instead exhibits a decline in recruitment levels at higher stock levels (Fig. 9.3; Appendix 9.1; Example Box 9.2). It has been criticized for this detail (Hall, 1988) as being a theoretical input with no empirical support. But there has been argument on both sides and our coming discussion on uncertainty due to measurement errors may also illuminate the matter.

The Ricker model of stock-recruitment differs from that by Beverton-Holt in that the density-dependent mortality term for eggs and juvenile stages relates to the total stock size and not only to the cohort size.

9.4.2 Biological Assumptions/Implications

Various mechanisms have been suggested for generating this form of density-dependence (dependent upon total stock size and not just the cohort size). These include the cannibalism of the juveniles by the adults (hence stock density is more important than cohort density), density-dependent transmission of disease, damage by spawning adults of each other's spawning sites (occurs primarily in rivers with fish like salmon), and finally, there could be density-dependent growth combined with size-dependent predation. Each of these mechanisms can lead to different interpretations of the parameters of the Ricker curve.

Once again the distinction between whether the equation should be interpreted as a theoretical or explanatory statement about the observable world instead of just a convenient empirical description of the average recruitment becomes important. In addition, while the parameters can certainly be given a real world interpretation, the equations still tend to be overly simplistic and are best regarded as an empirical description rather than as an explanation of events.

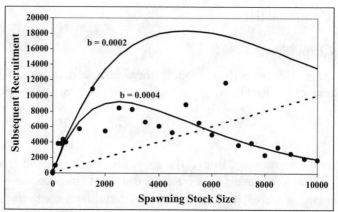

Figure 9.3   Three Ricker stock-recruitment curves based on Eq. 9.3. Each of the data series has an a of 10, with the b values indicated. The separate points are the same as the $b = 0.0004$ curve except they have lognormal error with a standard deviation of 0.5 associated with it. Note that the b value mainly influences the degree of recruitment decline with increasing stock and has little effect on initial steepness (Example Box 9.2).

Example Box 9.2 The Ricker stock-recruitment relationship. Extend column A, the spawning stock size, down to a value of 10000 in steps of 500 (to row 30). Copy B5 across to D5, and then copy B5:D5 down to row 30. To mimic Fig. 9.3, plot columns C and D against A, as solid lines and add column A as a scatter of points. Modify the a and b values in D1 and D2, to observe how this modifies the shape of the curve (leave B1:B2 alone). Change the error standard deviation value in B3 and press F9 a few times. Note how the scatter of points does not always mimic the deterministic curve.

	A	B	C	D
1	a	=C1	10	10
2	b	=C2	0.0004	0.0002
3	**error**	0.0001	0.0001	0.0001
4	**Spawn**	**Recruit**	**Rec 2**	**Rec 3**
5	1	=(b$1*$a5*exp(-b$2*$a5))*loginv(rand(),0, b$3)	10.0	10.0
6	10	=(b$1*$a6*exp(-b$2*$a6))*loginv(rand(),0, b$3)	99.6	99.8
7	100	Copy Down to Row 30	960.8	980.2
8	200	3566.3	1846.2	1921.6
9	300	2419.6	2660.8	2825.3
10	400	5203.6	3408.6	3692.5
11	500	3240.6	4093.7	4524.2

9.5 DERISO'S GENERALIZED MODEL

9.5.1 The Equations

The Beverton-Holt and Ricker stock-recruitment curves are special cases of a more general model proposed by Deriso (1980). Schnute (1985) restructured Deriso's equation to produce an even more flexible version with even greater flexibility

$$R = \alpha S \left(1 - \beta \gamma S\right)^{1/\gamma} \tag{9.4}$$

where R is recruitment and S is the spawning stock as before, with the three parameters, α, β, and γ. By modifying the value of γ different special cases can be produced (Schnute, 1985)

$$
\begin{array}{llll}
\gamma = -\infty & : & R = \alpha S & \\
\gamma = -1 & : & R = \alpha S / (1 + \beta S) & \\
\gamma \to 0 & : & R = \alpha S e^{-\beta S} & \\
\gamma = 1 & : & R = \alpha S (1 - \beta S) &
\end{array}
\tag{9.5}
$$

The first form in Eq. 9.5 is a simple, constant productivity model where α recruits are produced for each unit of stock; this is density-independent recruitment. The same result follows from setting $\beta = 0$ in Eq. 9.4. The next three cases correspond to the standard stock-recruitment relationships of Beverton-Holt (1957), Ricker (1954, 1958), and Schaefer (1954), respectively. The Beverton-Holt equation is yet another version of their model but it has the same properties (asymptotic). The arrow in the Ricker equivalent merely means "approaches", thus as γ approaches zero the equation becomes equivalent to the Ricker model. Finally, the Schaefer equivalent is really a form of the logistic equation, which is equivalent to the number of recruits per spawner declining linearly with increasing spawning stock. Mathematically, the Schaefer equation could lead to negative recruitment if spawning stock managed to rise above the level that could generate the theoretical maximum recruitment. Parameters α and β should always be positive although γ can have either sign. The curve always passes through the origin but its shape will depend upon the balance between the three parameters (Fig. 9.4; Example Box 9.3).

There are some mathematically unstable properties to this model; consider the implication of $\gamma = 0$ which would lead to a mathematical singularity (divide by zero). The parameter limits should rather be $\gamma \to 0$, from either the negative or the positive direction.

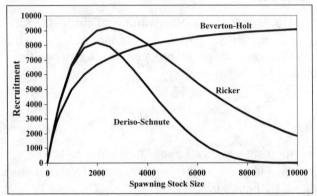

Figure 9.4 An array of stock-recruitment curves using the Deriso-Schnute equation (Eq. 9.4), the Ricker equation (Eq. 9.5), and a version of the Beverton-Holt equation (Eq. 9.5). In all cases $\alpha = 10$. For the Deriso-Schnute and Ricker curves, $\beta = 0.0004$ but for the Beverton-Holt equivalent, $\beta = 0.001$. For the particular Deriso-Schnute curve illustrated $\gamma = 0.25$. If the γ is set $=0.000001$ (approaches zero) the curve becomes very similar to the Ricker curve illustrated. If the β is set to 0.001 and the γ is set to -1, the Deriso-Schnute curve will sit on the Beverton-Holt curve (Example Box 9.3).

Example Box 9.3 The Deriso-Schnute generalized stock-recruitment model. In B5 put =B$1*A5*(1-B$2*B$3*A5)^(1/B$3), in C5 put as shown below, in D5 put =($D$1*A5)/(1+$D$2*A5). Extend column A down to 10000 in steps of 500. Copy B5:E5 down to row 30. To mimic Fig. 9.4, plot columns B, C, and D against A. Slowly change B3 towards zero and inspect the graph and the differences in column E. Set B2 = D2, set B3 = -1, and E5 to =D5-B5 (and copy down). How close to the Beverton-Holt in column D is the result?

	A	B	C	D	E
1	alpha	10	10	10	
2	beta	0.0004	0.0004	0.001	
3	gamma	0.25			
4	Spawn	Deriso-Schnute	Ricker	Bev-Holt	Diff
5	1	10.00	=C$1*A5*exp(-C$2*A5)	9.99	=C5-B5
6	10	99.60	=C$1*A6*exp(-C$2*A6)	99.01	=C6-B6
7	100	960.60	=C$1*A7*exp(-C$2*A7)	909.09	=C7-B7
8	200	1844.74	Copy Down to Row 30	1666.67	1.496
9	300	2655.88	2660.76	2307.69	4.883

9.6   RESIDUAL ERROR STRUCTURE

The stock-recruitment curves we have considered are all average expected curves. Observations concerning the stock-recruitment relationship would have values that stray above and below the expected average curve (the residual errors from the curve). Consideration of the form these residuals take in wild populations has led to the conclusion that observed variations should be distributed lognormally and reflects the possibility of spawning stocks giving rise to occasional very large recruitment levels.

Hilborn and Walters (1992) recommend that a lognormal distribution of residuals about the average stock-recruitment relationship be used unless there is evidence to the contrary. This will generally mean the data or the parameters for each model will have to be log-transformed before being fitted to data (Fig. 9.5; Example Box 9.4).

There is also a theoretical justification for using the lognormal distribution to describe the residual errors around stock-recruitment relationships. The stock-recruitment process can be considered to be the outcome of a whole series of successful survivorships from the egg stage to recruit. The overall survivorship is simply the product of all of these separate events

$$S = S_1 S_2 S_3 \ldots\ldots\ldots\ldots S_n \tag{9.6}$$

where S is the total survivorship over n life stages, and the S_i are the survivorships (probabilities) through life history stage i. To convert the product into a summation we can take logs on both sides

$$Ln(S) = \sum Ln(S_i) \tag{9.7}$$

The central limit theorem states that the sum of a long series of independent, identically distributed random variables ($Ln[S_i]$ in this example) will have a distribution that approaches the normal distribution as n increases. Thus, if each S_i is an independent random variable, and there are no particular stages that dominate the sum, the overall survivorship should be log-normally distributed, represented as

$$R_i = \bar{R} \ e^{N(0,\sigma_i^2)} \tag{9.8}$$

where the $N(0, \sigma_i^2)$ is a normally distributed random variate with mean of zero and a standard deviation of σ_i. The lognormal distribution has two properties of interest in this context. It occasionally gives rise to very high recruitment values (skewed tail out towards high values) and the amount of variation will be proportional to the average recruitment. We would thus expect to see higher variation at high levels of recruitment although this may also be related to the geographical distribution of the species Myers (1991).

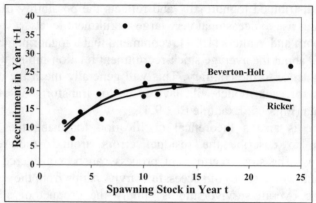

Figure 9.5 Two different stock recruitment models fitted to data from Penn and Caputi (1986) concerning tiger prawns (*Penaeus esculentus*) from Exmouth Gulf, Western Australia. The data points refer to the 13 years between 1970/71 to 1983/84 (Table 9.3; Example Box 9.4).

Table 9.1 Stock-recruitment data, after Penn and Caputi (1986), for Exmouth Gulf tiger prawns (*Penaeus esculentus*), including relative rainfall as an index of cyclonic activity (Fig. 9.5; Example Box 9.4). Sorted on index of spawning.

Year	Spawning Stock Index	Recruitment Index	Cyclone Index January	February
82-83	2.4	11.6	0	0
81-82	3.2	7.1	85	28
83-84	3.9	14.3	0	54
71-72	5.7	19.1	0	1
80-81	6	12.4	18	19
79-80	7.4	19.7	14	41
74-75	8.2	37.5	0	213
73-74	10	18.5	102	22
76-77	10.1	22.1	2	1
77-78	10.4	26.9	4	10
78-79	11.3	19.2	0	0
72-73	12.8	21	1	5
70-71	18	9.9	353	19
75-76	24	26.8	23	38

In order to fit a stock-recruitment model with lognormal residual errors to raw data, it is best to log-transform the equations to normalize the error structure. Then one can use ordinary least squares or normally distributed likelihoods. Thus, with the Ricker equation (Eq. 9.3), we divide through by S and transform using natural logarithms

$$Ln\left(\frac{R}{S}\right) = Ln(a) - bS + \varepsilon \qquad (9.9)$$

where the right hand side provides the expected value of $Ln(R/S)$. This can be compared with the observed value to provide a residual that can be used in a least squares determination of the optimal values of the parameters a and b (see Example Box 9.4). Equation 9.9 has the form of a linear relation and the parameters could be determined using a linear regression rather than using a non-linear technique.

Example Box 9.4 Fitting the Ricker and Beverton-Holt stock recruitment models to Exmouth Gulf tiger prawn data (after Penn and Caputi, 1986). Rather than log-transform the data we will log-transform the parameters as per Eqs. 9.9 and 9.10. Thus, copy the data from Table 9.1 into columns B and C (down to row 18). In D5 put =Ln($D$1)-($D$2*B5), in E5 put =(Ln(C5/B5)-D5)^2, and in F5 put =$D$1*B5*exp(-$D$2*B5), copy D5:F5 down to row 18. Then in G5 put =Ln($G$1)-Ln($G$2+B5), in H5 put =(Ln(C5/B5)-G5)^2, and in I5 put =$G$1*B5/($G$2+B5), and copy G5:I5 down to row 18. The sum of squared residuals for the Ricker fit is in E3; put =sum(E5:E18), that for the Beverton-Holt fit is in H3; put =sum(H5:H18). Mimic Fig. 9.5 by plotting column C against B, as a scatter of points. Add columns F and I as solid lines. Fit the two curves by minimizing E3 through changing D1:D2, and minimizing H3 through changing G1:G2. Which curve provides the smallest sum of squared residuals? While this is the best fitting curve does it really differ from the alternative by very much? Which curve would be least conservative?

	A	B	C	D	E	F	G	H	I
1			a	4.0		a	25.0		
2			b	0.1		b	3.0		
3	Year	Spawn	Recruit	Ssq_R	3.6336		Ssq_BH	1.7552	
4	Date	Si	Ri	Ricker	$(O-E)^2$	E(R)	BevHolt	$(O-E)^2$	E(B-H)
5	82-83	2.4	11.6	2.022	0.184	7.552	1.532	0.0019	11.111
6	81-82	3.2	7.1	2.229	0.073	9.295	1.394	0.3569	12.903
7	83-84	3.9	14.3	2.357	0.092	10.562	1.287	0.0001	14.13

Similarly with the Beverton-Holt model of stock-recruitment (Eq. 9.2) we divide through by S and then transform using natural logarithms

$$\text{Ln}\left(\frac{R}{S}\right) = \text{Ln}(a) - \text{Ln}(b+S) + \varepsilon \qquad (9.10)$$

where, once again, the right hand side provides the expected value of Ln(R/S), which can be used in a least squares determination of the optimal values of the parameters a and b (see Example Box 9.4). Equation 9.10 does not have a linear form and so the use of a non-linear solving method is required.

9.7   THE IMPACT OF MEASUREMENT ERRORS

9.7.1 Appearance over Reality

In Section 9.1, we briefly discussed the fallacy that because stock-recruitment relationships are poorly defined they can reasonably be ignored. This idea originated with inadequate data being used to describe the stock-recruitment relationship. Such problems were especially significant when indices of catch per unit effort of juveniles are compared with spawning biomass estimates. The resulting scatterplots gave the appearance of a random or flat distribution of recruitment relative to stock size. This is what suggested (although it was taken to be an implication) that there was no relation between spawning stock size and subsequent recruitment. Using catch per unit effort as an index of spawning stock size is generally the same as estimating stock size with large errors. We can summarize the problem to be where large estimation errors make recruitment appear to be independent of spawning stocks.

This whole problem is another instance of the "errors in variables" problem where the independent variable in a relationship (the x-axis) cannot be measured or estimated without a large component of error. Therefore, because this violates all the assumptions of such analyses, it should not be used as an independent variable in correlation and regression analyses.

9.7.2 Observation Errors Obscuring Relationships

Walters and Ludwig (1981) and Ludwig and Walters (1981) carried out some simulations in which they modelled the impact of observation or measurement errors on estimating the spawning stock size. They assumed that the spawning stock size was measured with some error ε, lognormally distributed with mean zero, and standard deviation σ_ε. Taking into account bias in the estimates of spawning stock size, the estimated spawning stock size is given by

$$\hat{S} = Se^\varepsilon \tag{9.11}$$

Similarly, for observing recruitment levels

$$\hat{R} = Re^\varepsilon \tag{9.12}$$

Using these equations, even small levels of error can transform an underlying linear stock-recruitment relationship into one showing little relation between spawning stock and recruitment (Fig. 9.6; Example Box 9.5).

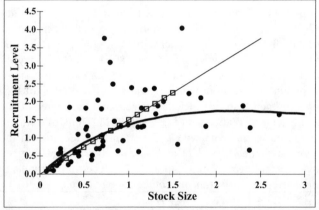

Figure 9.6 Comparison of an hypothetical linear stock-recruitment relationship (squares and fine straight line) with the observed relationship (filled circles and curve) with Eqs. 9.11 and 9.12, plus σ_ε of 0.5 in both cases. The solid curve is the best fitting Ricker curve fitted to the resulting error ridden data (Example Box 9.5).

The inference to draw from this simple demonstration is that the uncertainty surrounding estimates of spawning stock size and subsequent recruitment can badly obscure the detection of any underlying stock-recruitment relationship. In part, how one then proceeds will depend upon the use to which the stock-recruitment relationship is to be put. In simulation models, one can define the desired model structure, but when fitting a stock-assessment model the form of the stock-recruitment relationship may be important. Typically, however, annual recruitment levels are estimated directly from the model (either directly as recruitment levels or as residual errors around some deterministic stock-recruitment equation). Thus, the precise form of the relationship may not be critical.

Example Box 9.5 Influence of measurement error on detection of any underlying stock-recruitment relationship. As in Fig. 9.6 the underlying relationship is linear (R = 1.5 x S) as depicted in columns A and B. Column A contains values ranging from 0.1 to 1.5 in steps of 0.1, with four replicates of each value (i.e. down to row 64). To include lognormal measurement error to both sets of observations, in C5 put =A5*loginv(rand(),0,$B$1), and put =B5*loginv(rand(),0,$B$2) into D5. E5 is as shown. This column of transformed data is to facilitate the calculation of the Ricker curve relating to the data in columns C and D. Copy C5:E5 down to row 64. Plot column B against A, as empty squares and add the line defined by F1:G2 [copy the selection, select the graph, paste special/tick New Series and Categories (X values) in first column] to mimic Fig. 9.6. Add column D against C on the same graph in the same way as above, as a set of points. To obtain the Ricker curve using a linear regression on the transformed data, select D1:E1 and put =linest(E5:E64,C5:C64,true,true), using <Ctrl><Shift><Enter> to enter the array function (check the help for this function). Sexp and Rexp provide the back transformed data for drawing the Ricker curve onto the graph. Fill column F with values from 0.1 down to 3, then put =F5*exp($E$1+$D$1*F5) into G5 and copy down. Add column G against F as a curve to the graph (to fully mimic Fig. 9.6). Press F9 to obtain new random numbers and new dispersions of the "observed" values along with their respective Ricker curve. Alter the degree of variability and observe the impact (see Walters and Ludwig, 1981). Think of a way to compare a Ricker curve fitted to this data with a Beverton-Holt recruitment model. It will require copying columns C and D and pasting their values somewhere before analysis. Instead of a linear relationship, install a Beverton-Holt relationship. How does the initial steepness affect the outcomes? See Example Box 9.1.

	A	B	C	D	E	F	G
1	Sigma1	0.5		-0.4821	0.88987	0	0
2	Sigma2	0.5		Gradient	Intercept	2.5	=F2*1.5
3	Actual Values		Observed with error				
4	Sp	Rec	ObsSp	ObsRec	Ln(R/S)	Sexp	Rexp
5	0.1	=A5*1.5	0.056	0.587	=ln(D5/C5)	0.100	0.232
6	0.1	=A6*1.5	0.271	0.117	=ln(D6/D6)	0.200	0.442
7	0.1	=A7*1.5	0.079	0.177	=ln(D7/C7)	0.300	0.632
8	0.1	0.15	0.087	0.095	0.086	0.400	0.803
9	0.2	0.3	0.200	0.473	0.863	0.500	0.957
10	0.2	0.3	0.136	0.110	-0.209	0.600	1.094
11	0.2	0.3	0.159	0.307	0.656	0.700	1.216
12	0.2	0.3	0.155	0.244	0.451	0.800	1.324
13	0.3	0.45	0.430	0.330	-0.262	0.900	1.420

Clearly, observation errors may be introducing a great deal of uncertainty into any assessment we make. What can be done about this problem is less clear. Ludwig and Walters (1981) and Walters and Ludwig (1981) both suggest bias corrections for use when attempting to estimate the underlying model. It is debateable whether there would be sufficient information in such data to be able to distinguish between the Beverton-Holt and the Ricker curve.

9.8   ENVIRONMENTAL INFLUENCES

It is often stated that the effects of the physical environment may influence recruitment success, and this is certainly the case. However, clear examples of taking account of environmental effects in stock recruitment relationships are not common. Penn and Caputi (1986) provide an excellent example where environmental effects are explicitly recognized and included in a Ricker stock-recruitment relationship (Fig.9.7).

Penn and Caputi (1986) obtained a series of indices of spawning stock (derived from a detailed study of catch effort data and research cruises) plus the recruitment in the following year. This was described by the log-transformed Ricker equation

$$Ln(R_{t+1}) = Ln(a) + Ln(S_t) - (bS_t) + \varepsilon \qquad (9.13)$$

When this is fitted to the available data the parameters $a = 4.339$ and $b = 0.0761$ lead to the smooth curve in Fig. 9.7. In order to account for the monthly rainfall (a cyclone index) Penn and Caputi (1986) added further terms to the Ricker model (Table 9.1; Example Box 9.6)

$$R_{t+1} = aS_t e^{-bS} e^{cJ_t} e^{dF_t} e^{\varepsilon} \qquad (9.14)$$

where c and d are the new parameters relating, respectively, to the January cyclone index J_t in year t, and the February cyclone index F_t (see Table 9.1). On log-transformation this gives

$$Ln(R_{t+1}) = Ln(a) + Ln(S_t) - (bS_t) + (cJ_t) + (dF_t) + \varepsilon \qquad (9.15)$$

which can be solved for the optimal parameters using multiple regression or some minimizer. When this is done $a = 3.769$, $b = 0.0575$, $c = -0.00029$, and $d = 0.0027$, which gives rise to the dashed line in Fig. 9.7. Clearly, the fit of the line accounting for the cyclone index is a much better fit than without. The rainfall during the two months, induces different effects on recruitment (one increasing recruitment, the other decreasing success), this is indicated by the parameters c and d being of opposite sign. In January, the recruits are still in shallow water and are vulnerable to storm actions, in February, the recruits are in deeper water and the increased turbidity is thought to increase survivorship. By adding this change, Eq. 9.14 is clearly no longer a general equation but is certainly more informative about the Exmouth Gulf fishery. Ideally, the relationships between the cyclone indices and the stock-recruitment relationship should be tested experimentally but the dramatic improvement in fit certainly suggests the hypothesis of cyclones affecting survivorship would be worth testing. This relates back to how one might design a model.

A second look at the data and model makes it clear that we have added two extra parameters to account for three data points that have very large residuals. Using four parameters to describe 14 data points it is not surprising that a reasonable quality of fit was obtained. Clearly more data in the time series would be needed and independent tests of the assumptions behind the two parameters relating to the cyclonic indices would be required before one could conclude that the modified model would lead to better predictions in the future than the average or general model (Example Box 9.6).

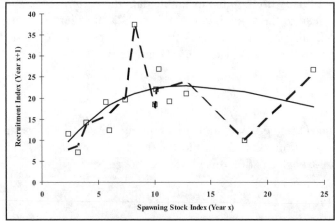

Figure 9.7 The Ricker stock-recruitment relationship for tiger prawns in the Exmouth Gulf as described by Penn and Caputi (1986). The open squares are the data points, the smooth curve is the best fitting standard Ricker curve, and the dashed line is the same curve but with the effects of rainfall in January and February included (Eq. 9.14; Table 9.1; Example Box 9.6).

9.9 RECRUITMENT IN AGE-STRUCTURED MODELS

9.9.1 Strategies for Including Stock-Recruitment Relationships

A standard problem in fisheries stock assessment is to model an age-structured population through time. Data in the form of a time series of the relative catch-at-age needs to be available along with some index of total relative abundance (e.g., catch-rates and/or biomass estimates).

Example Box 9.6 The influence of environmental factors on recruitment in tiger prawns in Exmouth Gulf, Australia (after Penn and Caputi, 1986). Copy the four columns of data from Table 9.1 into columns A, B, C, and D. In E8 put the modified Ricker =Ln($B$1)+Ln(A8)-($B$2*A8)+($B$3*C8)+($B$4*D8). The squared residuals are shown as in F8. The predicted value for the Ricker curve is as shown in G8; note the back transformation. The initial transformation is required to normalize the residual errors. Plot columns B against A as a scatterplot. Add column G as a solid line to mimic Fig 9.6. To fit the Ricker curve only, as shown, put zero into the January and February cells, B3:B4, and use the solver to minimize B5 by altering B1:B2. To completely mimic Fig. 9.6, save the values from column G into H and add those to the plot as a line. Then, use the solver again to minimize B5 but this time by altering B1:B4. Note the relative values attributed to January and February. Which is likely to increase recruitment and which to decrease it? Compare the two lines on the graph. Is the improvement worth the alteration? Using the AIC to compare the two models, which one is to be preferred? [AIC = $nLn(\sigma^2) + 2K$, where n is the number of observation =14, K is the number of parameters =2 or 4, and σ^2 is the sum of squared residuals (B5) divided by n, $=\Sigma\varepsilon^2/n$]

	A	B	C	D	E	F	G
1	a	4.50974					
2	b	0.075674					
3	Jan – c	0.0					
4	Feb –d	0.0					
5	SSQ	=sum(F8:F21)					
6	Spawn	Recruit	Cyclone Index				
7	Si	Ri	Jan	Feb	Ricker	(O-E)2	Model
8	2.4	11.6	0	0	2.2001	=(Ln(B8)-E8)^2	=exp(E8)
9	3.2	7.1	85	28	2.4272	=(Ln(B8)-E9)^2	=exp(E9)
10	3.9	14.3	0	54	2.5721	0.0078	13.093
11	5.7	19.1	0	1	2.8154	0.018	16.699
12	6	12.4	18	19	2.844	0.1064	17.184
13	7.4	19.7	14	41	2.9477	0.0011	19.063
14	8.2	37.5	0	213	2.9899	0.4026	19.883

The two main age-structured approaches (Megrey, 1989) are to use some version of Virtual Population Analysis (VPA) or some form of statistical Catch at Age Analysis (CAGEAN). Both of these approaches attempt to determine the historical recruitment levels that would be consistent with the proportional age-structure as observed by current sampling. Thus, a large number of five-year-old fish present in one year, implies that recruitment was relatively successful five years previously. Similarly, relatively low numbers of three year olds (taking selectivity of gear into account), would indicate a relatively poor recruitment three years previously. The statistical catch-at-age methodology (see later) would entail varying the hypothesized recruitment values during the search for a maximum likelihood or minimum least squares.

Recruitment can be implemented either deterministically or stochastically in such a CAGEAN model. If one used a deterministic relationship, one of the stock-recruitment equations would be included in the model. Most models, however, would tend to use a stochastic representation of recruitment. For this, one could either estimate the (assumed) annual recruitment level directly or, alternatively, one could assume an underlying deterministic stock-recruitment relationship and estimate a set of residual errors around that. The two approaches produce the same result in terms of the predicted/expected values of recruitment in each year. A major advantage of including a deterministic stock-recruitment relationship and working with residuals around that is that in years where there are no data for predicting the level of recruitment, the deterministic recruitment level could be used to fill in the gap in knowledge.

9.9.2 Steepness

In the discussion of the Beverton-Holt and Ricker stock-recruitment equations, reference was made to the importance of the initial steepness of the curves as representing importance aspects of the behaviour of each curve. With the Beverton-Holt model, if the steepness were very high, the asymptote would be reached at relatively small spawning stock sizes. With the addition of measurement error, any relationship might be difficult to distinguish from a scattered straight line. Under such circumstances, it might be said that environmental variation was more important than spawning biomass in the determination of recruitment levels. Alternatively, if the steepness were relatively low, a stock-recruitment relationship might be more apparent. If, in Example Box 9.5, you replaced the original underlying linear relationship with a Beverton-Holt equation, the impact of steepness could have been determined easily. With a shallow, less steep curve, environmental variation still has effects, but at low spawning stock sizes there could not be high levels of recruitment. Obscuring the relationship would be more difficult and spawning stock size cannot be ignored (see also Example Box 9.1).

The steepness of any underlying stock-recruitment relationship can either be included or investigated by including a stock-recruitment equation in one's model. Apart from determining the annual recruitment levels one could also be capturing other ideas/beliefs concerning the species concerned in relation to whether recruitment year-class strength is dependent upon either environmental factors or spawning biomass. The common option is to include the deterministic stock-recruitment relationship but to search for the residual errors around this to obtain the optimum fit. Of course, this would only be possible for those years where there was sufficient data available for estimating the recruitment residuals. If the model in which the recruitment relationship was included was attempting to describe events in years where there was no information concerning recruitment, then the stock-recruitment relationship would provide estimates of mean expected levels that could fill the gap.

9.9.3 Beverton-Holt Redefined

By assuming that the unfished or virgin population had attained a stable age distribution, Francis (1992) was able to provide a biologically meaningful re-parameterization of the Beverton-Holt stock-recruitment model in terms of the steepness of the stock-recruitment curve, h, the initial recruitment, R_0, and the given virgin biomass, B_0. The steepness parameter h is defined by a consideration of the deterministic number of recruits arising when the mature biomass is reduced to 20% of its virgin level, thus

$$R_0 = \frac{A_0}{\alpha + \beta A_0} \qquad (9.16)$$

and, by definition

$$hR_0 = \frac{0.2 A_0}{\alpha + 0.2\beta A_0} \qquad (9.17)$$

where the α and β are the Beverton-Holt parameters. A_0 is the total mature biomass per recruit from the stable age distribution found in a virgin population. The "per-recruit" part is important because this permits us to determine a relationship between R_0 and B_0 independently of Eq. 9.1 (see Appendix 9.3). The 0.2 is present because hR_0 is defined as the constant number of recruits that occur at 20% of the virgin mature biomass. The stable age distribution derives from a constant recruitment level, R_0, exposed to a constant level of natural mortality leading to a standard exponential decline on numbers at age. If natural mortality is low then a plus group may be needed (Example Box 9.7; Eq. 9.18)

$$n_{0,i} = \begin{cases} R_0 e^{-Mi} & i < t_{max} \\ R_0 e^{-Mt_{max}} / (1 - e^{-M}) & i = t_{max} \end{cases} \tag{9.18}$$

where $n_{0,i}$ is the virgin number of fish per recruit of age i and t_{max} is the maximum age being modelled. The t_{max} class acts as a plus-group and hence the necessity for the division by $1-e^{-M}$ (to provide the sum of an exponential series, see Chapter 11). The biomass A_0 corresponds to the stock biomass that would give rise to a constant recruitment level of one. Thus, at a biomass of A_0, distributed across a stable age distribution, the resulting recruitment level would be $R_0 = 1$. A_0 acts like a scaling factor in the recruitment equations by providing the link between R_0 and B_0 (Example Box 9.7)

$$A_0 = \left(\sum_m n_{0,i} w_i \right) e^{-0.5M} \tag{9.19}$$

where m is the age at maturity (assumed to equal age of recruitment to the fishery), $n_{0,i}$ is the virgin number of animals per recruit of age i, and w_i is the weight of an animal of age i. The $e^{-0.5M}$ is half the natural mortality imposed upon the population. This imposition implies that half the mortality for the year will have occurred before reproduction occurs (i.e., A_0 is the mid-year per-recruit, mature biomass). This suggests spawning must occur at least half way through the year. Without this optional term, spawning could occur at either the beginning or end of each year depending on whether natural mortality occurs before or after spawning.

A_0 acts as a scaling factor, because a stable age distribution will arise in the unfished population given any constant recruitment level. The magnitude of A_0 will be scaled by the estimated virgin biomass, but its value, relative to the constant recruitment needed to maintain the stable age-distribution, will remain the same (Example Box 9.7; Eq. 9.22). Given the mature biomass per recruit, A_0, Francis (1992) was able to provide definitions of α and β as used in Eq. 9.1

$$\alpha = \frac{B_0(1-h)}{4hR_0} \quad \text{and} \quad \beta = \frac{5h-1}{4hR_0} \tag{9.20}$$

Alternatively, if we want to use Eq. 9.2 we could use (Appendix 9.3)

$$a = \frac{4hR_0}{5h-1} \qquad \text{and} \qquad b = \frac{B_0(1-h)}{5h-1} \qquad (9.21)$$

Example Box 9.7 The calculation of the deterministic Beverton-Holt stock recruitment relationship, given the growth properties and natural mortality of a species. Ages in column A extend from 0 to 20 down to row 29. The value in B9 is the constant recruitment value and the rest of column B generates the stable age distribution. In B10 put =B9*exp(-$B$4). Column C is the von Bertalanffy equation, put =$B$1*(1-exp(-$B$2*(A9-$B$3))) into C9. The **a** and **b** values for males and females (B6:C7), define the weight at length relationship. Put =B$6*$C9^B$7 into D9 and copy across to E9. Fill F9 as shown. Select B9:F9 and copy down to row 29. In B29, put =B28*exp(-$B$4)/(1-exp(-$B$4)), to act as the plus group (see Eq. 9.18 and the later chapter on age-structured models). Put the equations shown into F2:F7. Note that Ro is determined by a rearrangement of Eq. 9.22. Notice that, by definition, Ao should relate to mature biomass per recruit. However, unlike F7, the Ao value in F6 is not divided by the number of recruits. If the constant recruitment level in B9 is altered from 1, Ao in F6 also alters. However, the actual mature biomass per recruit in F7 always remains the same. Alter the value in B9 and see what happens to F6 and F7. See the effect of altering the growth parameters and the natural mortality. See Example Box 9.8 for how to link these calculations into a workable model.

	A	B	C	D	E	F
1	Linf	152.5			h	0.75
2	K	0.15			Alpha	=F5*(1-F1)/(4*F1*F4)
3	t0	0			Beta	=(5*F1-1)/(4*F1*F4)
4	M	0.15			Ro	=F5/F6
5		Females	Males		Bo	=F6
6	a	0.000271	0.000285		Ao	=sum(F12:F29)*exp(-0.5*B4)
7	b	3.135	3.114		Ao/Rec	=F6/B9
8	Age	Nage	Length	Wt Fem	Wt Mal	Biomass
9	0	1.00	0.0	0.0	0.0	=(B9/2)*(D9+E9)/1000
10	1	0.86	21.2	3.9	3.9	=(B10/2)*(D10+E10)/1000
11	2	0.74	39.5	27.5	26.8	=(B11/2)*(D11+E11)/1000
12	3	0.64	55.3	78.6	76.0	0.0493
13	4	0.55	68.8	156.3	150.4	0.0842
14	5	0.47	80.5	255.3	244.8	0.1181

At the virgin biomass per recruit, $B_0 = A_0$, the R_0, virgin recruitment parameter, is directly related to the virgin mature, or recruited, biomass, B_0

$$B_0 = R_0 \left(\sum_m n_{0,j} w_j \right) = R_0 A_0 \qquad (9.22)$$

By determining A_0, from a constant recruitment level of one, the recruitment levels from realistic B_0 levels can be obtained by applying Eq. 9.22. When these equations are being used in a stock assessment model, it does not matter whether the model is fitted by varying R_0 or B_0. Given A_0, which is derivable from growth and mortality characteristics (Eq. 9.19), the other two parameters can be obtained from each other. In order to include this relationship into a model, one needs to provide parameters describing growth, natural mortality, and h. From these it is first necessary to estimate A_0. This provides us with the link between B_0 and R_0, so that if we provide an estimate of B_0, during a model fitting or simulation process, each of the stock recruitment parameters can be determined and the relationship is defined (Example Box 9.8).

Example Box 9.8 The extension of Example Box 9.7 to permit the plotting of the Beverton-Holt stock-recruitment curves derived from the growth parameters, natural mortality, and steepness, h. Extend Example Box 9.7 by copying F1:F6 across into G1:G6. Replace G6 with =F7 to keep the correct virgin biomass per recruit. Give a particular value to Bo, such as the 2478 shown and see the impact on the two parameters Alpha and Beta. In G9 and below set out a series of stock biomass levels (steps of 200 below row 12 will lead to 3800 in row 29). In H9 put =G9/($G$2+$G$3*G9), which is the Beverton-Holt Eq. 9.1. In I9 put =$I$2*G9/($I$3+G9), which is Eq. 9.2. Select H9:I9 and copy down to the extent of column G. Note that they generate identical numbers. Plot column H against column G as a solid line. Note how this changes when the Bo value is altered. Clearly, given a Bo value, the expected recruitment for any calculated stock biomass could then be calculated.

	E	F	G	H	I
1	h	0.75	0.75		
2	Alpha	0.225162	=G5*(1-G1)/(4*G1*G4)	a	=1/G3
3	Beta	0.916667	=(5*G1-1)/(4*G1*G4)	b	=G2/G3
4	Ro	1	=G5/G6		
5	Bo	2.7019	2478		
6	Ao	2.7019	=F7		
7	Ao/Rec	2.70194			
8	Wt Mal	Biomass	Sp_Biom	Recruit	Recruit
9	0.0	0.000	10	42.5	42.5
10	3.9	0.003	100	307.6	307.6
11	26.8	0.020	200	470.5	470.5
12	76.0	0.049	400	640.0	640.0

When fitting an age-structured model the algorithm is often to select a B_0 value, generate the expected mature biomass, calculate the deterministic recruitment level, and finally, estimate the additional residual error needed to fit each of the recruitments.

9.10 CONCLUDING REMARKS

The review given here is extremely brief relative to the amount of work extant on stock-recruitment relationships in fished populations. Nevertheless, sufficient material is given to capture the requirements for including a stock-recruitment relationship into a model. Keep in mind that there is usually a great deal of uncertainty over the form of the stock-recruitment relationship, so there is little benefit to strictly adhering to one equation over another. Reducing uncertainty over recruitment levels will almost always make modelling a population more convincing. However, empirical estimates of recruitment are very difficult to obtain, so they remain the Achilles heel as well as the Holy Grail in fisheries.

APPENDIX 9.1

DERIVATION OF BEVERTON-HOLT EQUATIONS

Beverton-Holt

Beverton and Holt (1957, p. 48-49) considered intra-specific competition to be the main cause of the density-dependent effects. Their model of stock size and subsequent recruitment was devised to incorporate density-dependent survival rates (μ) thought to reflect intra-cohort competition for critical resources. As stated by Beverton and Holt (1957, p. 48): "The simplest assumption we can make, and that which is in best agreement with data from many population studies ..., is that the mortality coefficient is itself linearly related to the population density..." If there are density-independent (μ_1 - intercept) and density-dependent (μ_2 - gradient) mortality terms, then, in a manner exactly analogous to that described for simple population models in Chapter 2, the simple exponential model becomes linear

$$\frac{dN}{dt} = -\left(\mu_{r,1} + \mu_{r,2} N\right) N \qquad (A9.1)$$

where the r subscript denotes a particular stage/period during the pre-recruit phase. Beverton and Holt go on to point out that mortality rates during the various early stages of a fishes life can vary greatly with age so the parameters μ_1 and μ_2 would not be expected to remain constant during the pre-recruit stages in the life-cycle (the different r periods). They continue by demonstrating that the expected recruitment from an initial number of eggs is the product of the effect of each of these separate r periods. By implication, any particular stage in the spawning cycle can be substituted for the number of eggs (*e.g.* spawning biomass can be converted to eggs simply through multiplying by average fecundity). It is strongly recommended that reader consult Beverton and Holt's (1957, p48-49) original proof to see their demonstration of this relationship. As a general rule, such classic texts in fisheries science are well worth the effort it might take to read them. Thus, Beverton and Holt concluded with

$$R = \frac{1}{\left[\alpha + (\beta/E)\right]} \tag{A9.2}$$

which is one version of the Beverton-Holt recruitment model, where R is recruitment, E is the number of eggs produced by the spawning biomass, and the α and β are the recruitment parameters. It would be more common to relate recruitment levels directly to spawning stock biomass (Quinn and Deriso, 1999) and this can be achieved by including an inverse average fecundity term into the β and substituting S for the E (some additional algebra gives rise to a more commonly seen version).

$$R = \frac{1}{\left[\alpha + (\beta/S)\right]} = \frac{S}{\alpha S + \beta S/S} = \frac{S}{\beta + \alpha S} \tag{A9.3}$$

It is common practice, though possibly confusing, to alter the parameter usage and replace the β with an α, and the α with a β, as in Eq. 9.1.

APPENDIX 9.2

DERIVATION OF THE RICKER EQUATIONS

The Ricker model of stock-recruitment differs from that by Beverton-Holt in that the density-dependent mortality term for eggs and juvenile stages relates to the stock size and not only to the cohort size. We would thus have (Hilborn and Walters, 1992)

$$\frac{dN}{dt} = -(q + pS)N \qquad (A9.8)$$

which is equivalent to Eq. 9.15 except inside the brackets on the left-hand side we have S (spawning stock size) instead of N the cohort size. In it q is the instantaneous, density-independent mortality rate, pS is the density-dependent, instantaneous mortality rate for the cohort N. Solving this for any time t

$$N_t = N_0 e^{-pSt} e^{-qt} \qquad (A9.9)$$

where N_0 is the initial cohort size ($= fS$ where f is the average number of eggs per spawner), and N_t is the number of recruits at time t, thus

$$R = fSe^{-ptS} e^{-qt} \qquad (A9.10)$$

Now, e^{-qt} is the density-independent mortality rate, if this is multiplied by the fecundity and re-labelled a, and if we combine p and t into a new parameter b then Eq. A9.10 becomes the familiar

$$R = aSe^{-bS} \qquad (A9.11)$$

APPENDIX 9.3

DERIVING THE BEVERTON-HOLT PARAMETERS

Francis (1992) provides definitions of the Beverton-Holt parameters in terms of the more biologically meaningful terms relating to steepness (h), virgin mature biomass (B_0), and virgin recruitment (R_0). He does this for the recruitment equation

$$R_i = \frac{S_{i-1}}{\alpha + \beta S_{i-1}} \qquad (A9.12)$$

where R_i is the recruitment in year i, S_{i-1} is the spawning stock size in the year prior to i, and α and β are the usual Beverton-Holt parameters (see Eq. 9.1). In fact, it is easier to derive the definitions from the common alternative form of the Beverton-Holt equation ($R = aS/[b+S]$), which is what we will do here. If we assume that recruitment in the virgin stock derives from the virgin biomass we have

$$R_0 = \frac{aB_0}{b+B_0} \tag{A9.13}$$

steepness is defined as the recruitment obtained at 20% of virgin biomass

$$hR_0 = \frac{0.2aB_0}{b+0.2B_0} \tag{A9.14}$$

substituting Eq. A9.13 into Eq. A9.14 leads to

$$h = \frac{(0.2aB_0)(b+B_0)}{(b+0.2B_0)(aB_0)} = \frac{0.2(b+B_0)}{(b+0.2B_0)} \tag{A9.15}$$

multiplying through and exchanging terms leads to

$$hb - 0.2b = 0.2B_0 - 0.2hB_0 \tag{A9.16}$$

multiplying both sides by 5 and simplifying leads to

$$b(5h-1) = B_0(1-h) \tag{A9.17}$$

and therefore

$$b = \frac{B_0(1-h)}{5h-1} \tag{A9.18}$$

Reverting to Eq. A9.13 we can see that

$$R_0 = \frac{aB_0}{\dfrac{B_0(1-h)}{5h-1} + B_0} \tag{A9.19}$$

which multiplies through to become

$$\frac{R_0B_0(1-h)}{5h-1} + R_0B_0 = aB_0 \tag{A9.20}$$

dividing through by B_0 and multiplying the second R_0 by $5h$-1 allows the simplification

$$a = \frac{R_0 - hR_0 + 5hR_0 - R_0}{5h - 1} = \frac{4hR_0}{5h - 1} \qquad \text{(A9.21)}$$

Remembering that $\alpha = b/a$ and $\beta = 1/a$, we finish with

$$\alpha = \frac{B_0(1-h)}{4hR_0} \qquad \text{and} \qquad \beta = \frac{5h-1}{4hR_0} \qquad \text{(A9.22)}$$

as in Francis (1992). This has re-defined the parameters in terms of h, B_0, and R_0. However, this means we cannot use Eq. A9.13 to define the relationship between B_0 and R_0, because they are used in the generation of the α and β values used in the equation. The solution is to assume that the virgin population had a stable age distribution. The mature biomass generated per recruit from the stable age-distribution (A_0), therefore, defines the relationship required between B_0 and R_0.

10

Surplus-Production Models

10.1 INTRODUCTION

10.1.1 Stock Assessment Modelling Options

Surplus-production models are the simplest analytical method available that provides for a full fish stock assessment. Described in the 1950s (Schaefer, 1954, 1957), they are relatively simple to apply partly because they pool the overall effects of recruitment, growth, and mortality (all aspects of production) into a single production function. The stock is considered solely as undifferentiated biomass, that is, age- and size-structure, along with sexual and other differences, are ignored. The minimum data needed to estimate parameters for such models are time-series of an index of relative abundance and of associated catch data. The index of stock abundance is most often catch-per-unit-effort but could be some fishery independent abundance index (e.g., from trawl surveys, acoustic surveys) or both could be used.

To conduct a formal stock assessment it is necessary, somehow, to model the dynamic behaviour of the exploited stock. One objective is to describe how the stock has responded to varied fishing pressure. By studying the impacts on a stock of different levels of fishing intensity it is possible to assess its productivity. If statistics are collected, the process of fishing a stock can provide information about how the stock responds to perturbations (the extra mortality, above natural mortality, imposed by fishing). If a reduction in the stock size cannot be detected reliably (i.e., catch rates or survey results are hyper-stable relative to stock size), then stock assessment will be difficult, unreliable, or even impossible.

Given the necessary data, stock dynamics may be modelled using relatively simple surplus-production models, also known as biomass-dynamic models (Hilborn and Walters, 1992). A common alternative would be to use the more complex and data demanding age-structured models (e.g., cohort analysis, virtual population analysis, or statistical catch-at-age; see Chapter 11; Quinn and Deriso, 1999). A less common alternative, which tends to be used with those species that are difficult or impossible to

279

age, would be to use a length-based assessment model (Sullivan *et al.*, 1990; Sullivan, 1992). Because age-based models follow identifiable cohorts they suffer from fewer problems (given good data, which is not necessarily easy to obtain). Fisheries scientists usually try to collect the data required to produce an age-structured model in preference to the simpler data requirements of a surplus-production model. Ludwig and Walters (1985, 1989) have shown, however, that this is not always the best strategy, as stock production models may produce answers just as useful and sometimes better for management than those produced by age-structured models, at a fraction of the cost.

In a discussion of model selection, Hilborn and Walters (1992) suggest adopting a pragmatic approach. Assuming the data are available, they imply that one should apply both surplus-production and age-structured methods, which, because they are fundamentally different, will provide a test of relative performance. They state, "If biomass dynamic methods provide a different answer than age-structured methods, then the scientist should try to understand why they are different and analyze the management implications of the different predictions, rather than concentrating on deciding which method is correct" (Hilborn and Walters, 1992, p. 329). Surplus-production models, therefore, can be useful, and in this chapter we will be examining their use and properties in some detail.

Despite occasional recent use (Saila *et al.*, 1979; Punt, 1994), the use of surplus-production models went out of fashion in the 1980s. This was possibly because early on in their development it was necessary to assume the stocks being assessed were in equilibrium and this often led to over-optimistic conclusions that were insupportable in the longer term. Hilborn (1979) analyzed many such situations and demonstrated that the data used were often too homogeneous; they lacked contrast and hence were uninformative about the dynamics of the populations concerned. For the data to lack contrast means that fishing catch and effort information is only available for a limited range of stock abundance levels. However, a lack of contrast in this way can also lead to inconclusive results from age-structured models. There should also be concern that uncertainty (bias or lack of precision) in the observed abundance estimates exacerbates this problem by reducing the information content of the data used in relation to the actual stock size.

10.1.2 Surplus-Production

Surplus-production, as the name implies, relates to the production from a stock beyond that required to replace losses due to natural mortality. Production, in this case, is the sum of new recruitment and the growth of individuals already in the population minus those dying naturally. Early ideas on surplus-production were discussed briefly in Chapter 2 when the logistic population model was introduced. Many of the intuitions regarding surplus-production in fisheries derive from the logistic and we will revisit them here in order that their strengths and weaknesses can be illustrated.

   Schaefer (1954) first applied the logistic curve as a description of the production of Pacific halibut and later to eastern Pacific yellowfin tuna (Schaefer, 1957). Until then, no simple method of assessing a fished stock was available. Using the logistic as a foundation, Schaefer (1954, 1957) demonstrated a theoretical link between stock size and expected catch rates. This all related back to the expected level of surplus-production produced by particular stock sizes. Thus, given a known stock biomass the total production could be predicted thus

$$B_{t+1} = B_t + rB_t\left(1 - \frac{B_t}{K}\right) \tag{10.1}$$

where B_t is the stock biomass at time t, and r is the population growth rate and K is the maximum population size for growth to be positive. Both r and K are parameters of the logistic equation (see Chapter 2). A property of this equation is that the maximum production occurs at K/2 (Fig. 10.1; Example Box 10.1).

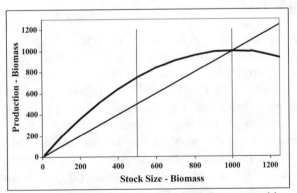

Figure 10.1 Production curve for the discrete logistic curve with r = 1 and K = 1000. The right hand vertical line indicates the carrying capacity while the central line illustrates the point of maximum production at K/2. The diagonal line is the line of replacement (Example Box 10.1).

   Irrespective of the stock size, it should be possible to take the excess production, above the equilibrium line of replacement, and leave the stock in the condition it was in before production and harvesting (Fig. 10.1). An obvious management strategy deriving from this theory would be to bring the stock to a size that would maximize the surplus-production and hence the potential yield. This supports the intuition that it is necessary to fish a stock down in size so that it becomes more productive. There are, of course, many problems with this simplistic view of fisheries production. It assumes that the population is in equilibrium with all of its inputs and outputs (a poor assumption). It also implies that, while one can fish inefficiently one should not be able to crash a fish stock through overfishing.

One problem that was easily solved was the fact that the logistic curve generated a symmetrical production curve, which was felt to be overly constraining. Pella and Tomlinson (1969) solved this by introducing an asymmetry term p, which modifies the logistic as follows:

$$B_{t+1} = B_t + \frac{r}{p} B_t \left(1 - \left(\frac{B_t}{K} \right)^p \right) \tag{10.2}$$

Equation 10.2 is from Polacheck *et al.* (1993). With this model, the stock size at which maximum production occurs is not necessarily at K/2 (Example Box 10.1).

10.2 EQUILIBRIUM METHODS

Wherever possible, equilibrium methods should be avoided in fisheries assessments. With surplus-production models, if a fish stock is in decline, then equilibrium methods consistently overestimate the sustainable yield. Their use in the past undoubtedly contributed to a number of major fishery collapses (Boerema and Gulland, 1973; Larkin, 1977). While equilibrium methods are no longer recommended, how they were used will be demonstrated so that the literature may be more easily understood.

Equilibrium methods are best described using the Schaefer or Pella-Tomlinson model of stock dynamics; this is just Eq. 10.2 minus any catch

$$B_{t+1} = B_t + \frac{r}{p} B_t \left(1 - \left(\frac{B_t}{K} \right)^p \right) - C_t \tag{10.3}$$

where C_t is the total catch in year t, if p = 1, this is equivalent to the Schaefer model (Eq. 10.1).

Example Box 10.1 Different production curves using an asymmetric production equation (Eq. 10.2). Name cell B1 as ri, cell B2 as K, and cell B3 as p. Extend the biomass levels in column A in steps of 100 down to 1300 in row 20. Column B contains the equivalent of Eq. 10.2, while column C indicates the production above replacement levels. Plot column B against A, adding a diagonal line as in Fig. 10.1. Select D2:E3, copy, and paste-special these data as a new series on the same graph to illustrate the stock size at maximum production. Using the solver, maximize C5 by modifying A5. This will automatically reposition the line of maximum production. Modify r and p to see their relative effects. When p = 1, the equation is equivalent to the simpler Schaefer equation. The effect of the asymmetry parameter is to make the effects of density-dependent regulation non-linear (see Chapter 2).

	A	B	C	D	E
1	ri	1			
2	K	1000		=A5	0
3	p	1		=D2	1200
4					
5	500	=A5+(ri/p)*A5*(1-(A5/K)^p)	=B5-A5		
6	**Biomass**	**Production**	**Surplus**		
7	10	=A7+(ri/p)*A7*(1-(A7/K)^p)	=B7-A7		
8	100	=A8+(ri/p)*A8*(1-(A8/K)^p)	=B8-A8		
9	200	Copy down to Row 20	Copy Down		
10	300	510	210.0		
11	400	640	240.0		
12	500	750	250.0		

Equilibrium methods rely on the assumption that for each level of fishing effort there is an equilibrium sustainable yield. The stock is assumed to be at some equilibrium level of biomass producing a certain quantity of surplus-production. If the fishing regime is changed the stock is assumed to move immediately to a different stable biomass with its associated surplus-production. This is patently wrong as it ignores the difference in standing crop between the two different biomass levels and the time it takes the system to respond to changed conditions. At heart, the assumption is that the yield taken is always surplus-production from a population in equilibrium. From this assumption it is possible to estimate the maximum sustainable yield (MSY) and the associated effort that will give rise to the MSY (E_{MSY}) given the appropriate biomass (B_{MSY}). It is necessary to assume that the rate of change of biomass is zero for all years i.e., $B_{t+1} = B_t$ = constant, and that Eq. 10.4 is exact. One requires a set of data to fit this model, and in the case of fisheries data, this is usually a time series of catch-rates, I_t. In fact, because of the assumption of equilibrium, the time series nature of the data is ignored (another of the flaws in the methodology). Real contrast in the data is required (meaning that information on catch rates from widely different effort and stock size levels provide the best information for fitting this sort of model). To connect the population dynamics model (Eq. 10.3) to reality we connect the catch rates to the stock biomass, B, and q, the catchability coefficient (= proportion of the total stock taken by one unit of effort). Given that C is the catch and E is the associated effort, then

$$I = \frac{C}{E} = qB \qquad (10.4)$$

Note the lack of any t subscripts. This is to emphasize that the time series nature of the data is ignored in this method. At equilibrium, B_{t+1} will equal B_t, and so the year when each data point was generated becomes irrelevant. We can solve Eq. 10.3 for C after assuming $B_{t+1} = B_t$, and after substituting

$$\frac{C}{(qE)} = B \qquad (10.5)$$

for B in Eq. 10.3 to give

$$C = \frac{rC}{pqE}\left[1 - \left(\frac{C}{qEK}\right)^p\right] \qquad (10.6)$$

which, in turn, can be solved for C/E or I (see Appendix 10.1), so that

$$I = \frac{C}{E} = \left((qK)^p - \frac{pq^{p+1}K^pE}{r}\right)^{\frac{1}{p}} \qquad (10.7)$$

If we re-parameterize the constants, by defining $(qK)^p$ to be a new parameter a, and the second term, $(pq^{p+1}K^p)/r$, to be a new parameter b, this would lead to the form (Polacheck *et al.*, 1993)

$$C/E = (a - bE)^{1/p} \qquad (10.8)$$

and therefore, the equilibrium catch is

$$C = E(a - bE)^{1/p} \qquad (10.9)$$

The sum of squares estimates of the parameters a, b, and p can then be obtained by minimizing the quantity

$$\sum (I - \hat{I})^2 \qquad (10.10)$$

where $(C/E) = I$ is the observed catch-rate from which we subtract the predicted or expected catch per unit effort from the model (denoted by the $\wedge$ symbol). If we assume $p = 1$, the model simplifies to the original Schaefer stock production model. Estimates for the two parameters a and b can then be obtained using standard linear regression techniques (Fig. 10.2; Example Box 10.2).

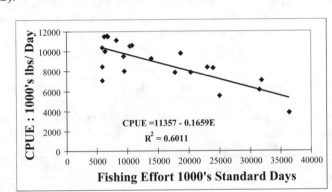

Figure 10.2 Relationship between CPUE and effort for eastern tropical Pacific ocean yellowfin tuna. Data from Schaefer (1957), see Table 10.1. Given p = 1 and the parameters a (11357) and b (0.1659) permits the calculation of an equilibrium MSY and E_{MSY} (194366258 lbs. and 34228 class four clipper days, respectively).

Referring to Eq. 10.8 makes it clear why plotting CPUE against effort was such a popular analytical tool in the past (Fig. 10.2). When $p = 1$, $a = qK$, and $b = (q^2K)/r$, then differentiating Eq. 10.9 with respect to E gives

$$E_{MSY} = \frac{a}{2b} = \frac{r}{2q}, \quad \text{if } p <> 1 \qquad E_{MSY} = \frac{pa}{b(p+1)} = \frac{r}{q(p+1)} \qquad (10.11)$$

substituting Eq. 10.11 into Eq. 10.6 gives

$$MSY = \frac{(a/2)^2}{b} = \frac{rK}{4},$$ (10.12)

and if $p \diamond 1$ $$MSY = \frac{p}{b}\left(\frac{a}{p+1}\right)^{\frac{p+1}{p}} = \frac{rK}{(p+1)^{\frac{(p+1)}{p}}}$$ (10.13)

The best way to visualize these analyses is to plot the expected equilibrium catches and the observed catches (as in Fig. 10.3) against the imposed effort. The outcome of this equilibrium analysis can appear dangerously convincing but this is to be resisted as it usually overestimates the safe catch levels (Example Box 10.2). Such analyses can be conducted with remarkable speed, but these should be considered as examples where the approximations and assumptions used (especially the assumption of equilibrium) mean the analyses cannot be used validly.

Table 10.1 Total Eastern Pacific catch of yellowfin tuna, catch per standard day fishing, and calculated relative fishing effort (data from Schaefer, 1957). Catch is in 1,000s of pounds, effort is in standardized class 4 clipper days, and catch rates are in pounds per class 4 clipper day. This data is used in Example Boxes 10.2 and 10.3 and is illustrated in Figs. 10.3 and 10.5.

Year	Catch	Effort	CPUE	Year	Catch	Effort	CPUE
1934	60913	5879	10361	1945	89194	9377	9512
1935	72294	6295	11484	1946	129701	13958	9292
1936	78353	6771	11572	1947	160134	20381	7857
1937	91522	8233	11116	1948	200340	23984	8353
1938	78288	6830	11462	1949	192458	23013	8363
1939	110417	10488	10528	1950	224810	31856	7057
1940	114590	10801	10609	1951	183685	18726	9809
1941	76841	9584	8018	1952	192234	31529	6097
1942	41965	5961	7040	1953	138918	36423	3814
1943	50058	5930	8441	1954	138623	24995	5546
1944	64094	6397	10019	1955	140581	17806	7895

Surplus-production models no longer need the assumption of equilibrium to enable them to be fitted to fisheries data. The non-equilibrium approach to fitting the models means they are better able to represent the dynamics of fished populations. Nevertheless, the inherent simplicity of surplus-production models means there are limits to how far their development may be taken. We will investigate some recent developments after a discussion of the different methods that can be used to fit non-equilibrium surplus-production models.

Example Box 10.2 An equilibrium surplus-production model fitted to Peruvian anchovy data (after Pitcher and Hart, 1982). The equations used in B4 and B5 are Eqs. 10.11, and 10.13. Select B1:C1, type =linest(D7:D15,C7:C15, true, true), then press <Ctrl><Shift><Enter> to enter the array function for linear regression (see Excel Help). The gradient is in B1 and the intercept in C1. Column F is Eq. 10.9, but for the Schaefer model (i.e., p = 1). Plot the regression data (columns D against C) and add a linear trend-line, showing the equation (*cf.* Fig 10.2). In a separate graph, plot column F against E as a line. Add columns C vs. B as a scatter of points to mimic the type of graph shown in Fig. 10.3 (select and copy them, select the graph, paste-special as a new series, edit the series so that column C is on the x-axis and column B on the y-axis). Replace the data in columns A, B, and C with the eastern Pacific yellowfin tuna data from Schaefer (1957), found in Table 10.1. Alter the linest function in B1:C1, and the effort values in column E appropriately. Do the calculated values for the a, b, MSY, and E_{MSY} match those given in Fig. 10.2?

	A	B	C	D	E	F
1	Linest	-0.01409	0.782203			
2	a	=C1				
3	b	=B1				
4	MSY	=((B2/2)^2)/-B3				
5	Emsy	=B2/(-2*B3)				
6	Year	Catch	Effort	CPUE	E	Equil C
7	60	6.00	10.1	0.594	2	=($B$2+$B$3*E7)*E7
8	61	8.40	12.0	0.700	6	=($B$2+$B$3*E8)*E8
9	62	10.10	21.0	0.481	10	6.413
10	63	10.60	24.0	0.442	14	8.188
11	64	11.00	28.0	0.393	18	9.513
12	65	9.00	25.0	0.360	22	10.387
13	66	9.20	22.0	0.418	26	10.809
14	67	10.75	21.5	0.500	30	10.781
15	70	12.70	30.0	0.423	34	10.301

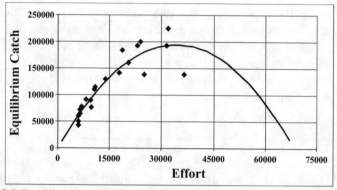

Figure 10.3 Equilibrium analysis of eastern tropical Pacific ocean yellowfin tuna; data from Schaefer (1957). The expected equilibrium catches for particular effort levels (1000s of class 4 clipper days) are represented by the curves solid line. The observed catch levels (1000s of lbs) are the scattered points. The Maximum Sustainable Yield (MSY) is the top of the curve and the optimum effort (E_{MSY}) is directly below the MSY peak of equilibrium catch ($C = (a - bE)E$; see Example Box 10.2).

10.3 SURPLUS-PRODUCTION MODELS

10.3.1 Russell's Formulation

Surplus-production models relate directly to Russell's (1931) verbal formulation of stock dynamics and, in difference equation or discrete form, have the general structure

$$B_{t+1} = B_t + f(B_t) - C_t \qquad (10.14)$$

with

$$\hat{I}_t = \frac{C_t}{E_t} = qB_t \qquad (10.15)$$

where B_{t+1} = exploitable biomass at the end of year t or the beginning of year $t+1$, B_t = exploitable biomass at the start of year t, $f(B_t)$ = production of biomass, as a function of biomass at the start of year t, C_t = biomass caught during year t; I_t = an index of relative abundance for year t (often CPUE but could be trawl survey results), and q = catchability coefficient. The ^ symbol denotes a value estimated from the model. The function describing the production of biomass in any year can take many forms; here are three which all derive from a form of the logistic equation, this is especially clear in the Schaefer model

$$f(B_t) = rB_t\left(1 - \frac{B_t}{K}\right)$$ classic Schaefer (1954) form   (10.16)

$$f(B_t) = Ln(K)rB_t\left[1 - \left(\frac{Ln(B_t)}{Ln(K)}\right)\right]$$ modified Fox (1970) form   (10.17)

$$f(B_t) = \frac{r}{p}B_t\left[1 - \left(\frac{B_t}{K}\right)^p\right]$$ modified Pella and Tomlinson (1969)   (10.18)

where r is a growth rate parameter (derived from the intrinsic rate of natural increase), K is the virgin biomass (B_0) or the average biomass level prior to exploitation (derived from the idea of carrying capacity). If $p = 1$, then Eq. 10.18 is equivalent to Eq. 10.16. As p tends to zero, Eq. 10.18 becomes equal to Eq. 10.17. The usual manner of writing the Fox model is without the Ln(K) at the front of the production equation (Eq. 10.17). However, without its inclusion the two equations may take on an equivalent form as p tends to zero, but they do not produce exactly the same value for production. The form of Eq. 10.18 was given by Polacheck *et al.* (1993) and alternative forms for the Pella-Tomlinson model exist (the most common alternative using an asymmetry parameter m-1 instead of the p parameter).

The constant p permits asymmetry in the surplus-production curve (Fig. 10.4). The linear density-dependent effects built into the logistic become non-linear with the addition of the p parameter. Note that by modifying p the absolute level of production alters (Fig. 10.4; Example Box 10.1) which implies that adding the asymmetry term has altered the interpretation that can be placed upon the other parameters. For example, as p decreases below a value of one, and all other parameters do not change, the absolute production increases. This increase could be offset during the model fitting process by a decrease in the value of r. The point is that it is not possible to directly compare the parameter values of the three different models [although if the Ln(K) term is included in the Fox model then Eq. 10.18 and Eq. 10.17 are directly comparable if p is very close to zero].

Equation 10.15 captures how the expectations from the model are compared against reality. It constitutes a very strong assumption, which is that catch rates are linearly related to stock biomass. The catchability coefficient, q, is often referred to as a nuisance parameter whose job is simply one of scaling the model to match the catch rates. Assessment becomes very difficult without some time series of relative abundance estimates. Surplus-production methods tend to be used when large amounts of good quality data are not available. The assumption that catch rates relate to the stock biomass must be considered carefully when such analyses are conducted.

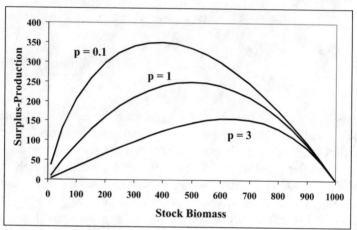

Figure 10.4 Influence of the parameter p (with values 0.1, 1, and 3) on the discrete Pella-Tomlinson version of the biomass dynamic model. When p = 1, the equation is equivalent to the Schaefer model and thus has a symmetrical production curve around the mid-point in biomass. Values of p < 1 skew the curve to the left and values >1 skew it to the right.

10.3.2 Alternative Fitting Methodology

In the development of fisheries methods a number of contrasting algorithms or strategies have been used for fitting stock-production models to observed data on catches and catch-rates. These methods differ with regard to whether or not they assume the population to be in equilibrium with the exploitation rate. Secondly they differ with regard to where the residual errors are attributed between the model and the data.

The earliest method used estimated two equilibrium management targets, the MSY and E_{MSY}. For this method to be valid the fishery was assumed to be in equilibrium. It was recognized, however, that fish stocks are rarely in equilibrium. An early solution to this problem was to use a weighted average of a number of years fishing effort for each year instead of just observed effort for that year. Unfortunately, if this is done the interpretation of the relationship between catch and effort becomes problematical. While effort averaging certainly improved the analyses in a limited way it is an *ad hoc* solution that can be thought of as spreading an invalid assumption of equilibrium across a number of years. Here, it will not be pursued further.

Process error estimators were then developed, which assumed all observations, such as catch rates, were made without error (Eq. 10.15 was without error), and that all error was in the equation describing changes in population size (Eq. 10.14 was imprecise). Thus, in any one year, to obtain a close match between the predictions of the model and the data, residual errors would need to be added to some or all of the parameters (Polacheck *et al.*, 1993).

Observation error estimators have been developed which assume all residual errors are in the catch rate or biomass observations (Eq. 10.15) and that the equation describing the time series of biomass values (Eq. 10.14) is deterministic and without error. In this chapter we will focus our efforts on this method of fitting surplus-production models.

More recently, attempts have been made to create estimators which use both forms of error, most notably using a technique borrowed from control engineering termed the Kalman filter (Meinhold and Singpurwalla, 1983). However, currently, there are no generally available methods that can make estimates where both types of error are being modelled without constraints (Quinn and Deriso, 1999).

Process error assumes that the observations used to fit the model are made without error, thus, all error occurs in the predicted change in population size (i.e., observations are made without error but the model does not exactly mimic reality). Observation error is the inverse of process error i.e., $B_{t+1} = \hat{B}_{t+1}$, with no error term (once again, the ^, or caret symbol, denotes the expected or estimated parameter value), so that observations are made with error but the model exactly describes the population dynamics. Observation errors imply

$$C_t / E_t = qB_t e^{\varepsilon_t}$$

where C_t is catch and E_t is the fishing effort in year t. The term e^ε represents lognormal residual errors. Alternatively, in this model process errors would imply

$$r = \hat{r} + \varepsilon_r \quad \text{or} \quad K = \hat{K} + \varepsilon_K$$

the form of the residual errors here could be other than normally distributed.

The approach of observation error estimation is the method now most commonly recommended because simulations have demonstrated that it can more closely reflect the circumstances underlying the observations (Hilborn and Walters, 1992; Polacheck *et al.*, 1993; Punt, 1990, 1995). Generally, if both types of error are present (seems generally likely), then so far, it has been found to be more efficient to assume only the presence of observation error (Ludwig and Walters, 1989). There has been some work attempting to model some restricted types of fisheries data using both kinds of residual error via the Kalman filter (Sullivan, 1992; Reed and Simons, 1996). But this remains generally intractable. One way in which both types of error could be modelled would arise if there were an estimate of the ratio of the respective variances of the two processes (Ludwig *et al.*, 1988). In practice, it is extremely difficult to work with both forms of error in an estimation model.

10.4 OBSERVATION ERROR ESTIMATES

10.4.1 Outline of Method

If we assume that the equation describing the stock dynamics is deterministic (i.e., zero process residual error), then all residual errors are assumed to occur in the relationship between stock biomass and the index of relative abundance. The stock biomass time-series can therefore be estimated by projecting forward the initial biomass, B_0, at the start of the catch series, using the selected biomass dynamic model and the historic annual catches (Example Box 10.3; Eq. 10.19)

$$B_{t+1} = B_t + \frac{r}{p} B_t \left(1 - \left(\frac{B_t}{K} \right)^p \right) - C_t \tag{10.19}$$

Using Eq. 10.19, and given the observed C_t's, the parameters r, K, and p, and an initial starting biomass B_0, a series of expected B_t's, can be produced. In the following example, the asymmetry parameter p is set $=1$, so the model simplifies to the Schaefer model. The discrete version of the surplus-production model is used to produce the predicted series of B_t's, and these, given a q value, are used to produce a predicted series of CPUE values (C/E = qB; see Eq. 10.20) that can be compared with those observed, using either maximum likelihood or least squares.

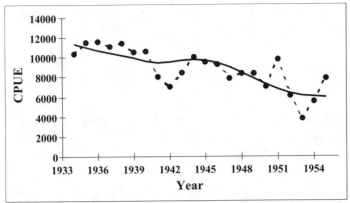

Figure 10.5 Observed index of relative abundance (dotted line) vs. year, with best fit predicted index of relative abundance superimposed (smooth line) for the eastern tropical Pacific ocean yellowfin tuna. Data from Schaefer (1957; Table 10.1). The model fitted is the Schaefer surplus-production model (Eq. 10.19 with p=1). Because the index of relative abundance (CPUE) is only qB_t, where q is a constant, the thick line is also a representation of the time series of predicted stock biomass levels (Example Box 10.3).

In summary, one makes predictions about the deterministic trajectory of the system from a hypothesized set of parameters and initial starting conditions, and then compares the observed values with the predictions (Fig. 10.5). Connecting the deterministic series of biomass levels to observed catch rates is implemented using:

$$\hat{I}_t = \frac{\hat{C}_t}{E_t} = qB_t e^\varepsilon \quad \text{or} \quad \hat{I}_t = q\frac{(B_{t+1} + B_t)}{2}e^\varepsilon \qquad (10.20)$$

where the e^ε indicates that the residual errors are assumed to be log-normally distributed (a standard assumption with catch rate data). Taking the average of two biomass levels relates to using the average biomass at the start and end of year t so that the catches relate to the biomass more realistically. In effect, the expected catch rates are related to mid-year biomass. In the process of fitting the observed data to the model, it will be necessary to log-transform both the observed and the predicted catch rates to normalize the residual errors.

10.4.2 In Theory and Practice

Assuming that the error in Eq. 10.20 is multiplicative and lognormal with a constant variance [i.e., $I_t = qB_t e^\varepsilon$ where $\varepsilon = N(0; \sigma^2)$], then the estimates of the model parameters (B_0, r, q, and K) are obtained by maximizing the appropriate likelihood function

$$L\left(\text{data}|B_0,r,K,q\right) = \frac{1}{\sqrt{2\pi}\hat{\sigma}}\prod_t e^{\frac{-\left(\text{Ln}\,I_t-\text{Ln}\,\hat{I}_t\right)^2}{2\hat{\sigma}^2}} \qquad (10.21)$$

where $L(\text{data}|B_0,r,K,q)$ is the likelihood of the data given the parameters, the product is over all years (t) for which CPUE data are available and, where (Neter *et al.*, 1996, p. 34)

$$\hat{\sigma}^2 = \sum_t \frac{\left(\text{Ln}\,I_t - \text{Ln}\,\hat{I}_t\right)^2}{n} \qquad (10.22)$$

and n is the number of observations. An estimate of q, which maximizes Eq. 10.21, is given by the geometric average of the time series of individual q estimates (see Appendix 10.2 for the derivation) thus

$$\hat{q} = e^{\frac{1}{n}\sum \text{Ln}\left(\frac{I_t}{\hat{B}_t}\right)} \qquad (10.23)$$

Alternatively, one could estimate the q value directly but the value thus determined should differ from that derived from Eq. 10.23 by only a small fraction. Such closed form estimates, as in Eq. 10.23, are valuable because, on average, the model tends to be quicker to fit and more robust when it has fewer directly estimated parameters.

   With a model as indicated in Eqs. 10.19, 10.20, and 10.23 (with $p=1$), then the actual fitting process would be as follows: given a time-series of catches (C_t), and guesses at r, K, and B_0, the model produces a series of expected biomass values $\hat{B}_t$. Then, given a catchability coefficient q (see Eq. 10.23), the $\hat{B}_t$ are used to produce a series of expected catch rates $C/E = \hat{I} = q\hat{B}_t$ (Eq. 10.20) and these are compared with the observed catch rates (Eq. 10.21; in fact Eq. 10.25).

   There is more than one method available for determining B_0. It can be set equal to K (i.e., $B_0/K = 1$); or it can be estimated by $\hat{B}_0 = \hat{I}_0/q$ (forces the first point of the fitted curve to coincide with first point of abundance index, one degree of freedom is lost), or B_0 can be estimated directly as a separate parameter. These alternatives can give very different results, and which method is used might depend upon whether fishing had occurred before records of catch were available. Punt (1990) found, using simulations, that even with situations where B_0/K was substantially different from unity, estimation performance was better when B_0/K was set at unity than when attempts were made to estimate B_0 separately. However, there are situations, especially in fisheries for shorter lived, more recruitment driven species, where a direct estimation of B_0 tends to be more efficient (Haddon, 1998).

In a similar example to the one illustrated here, Hilborn and Walters (1992) used a least squares criterion of fit using normal random residual errors between the observed CPUE, I_t, and the expected CPUE, $\hat{I}_t$.

$$\sum \left(\frac{C_t}{E_t} - \frac{\hat{C}_t}{E_t} \right)^2 = \sum \left(I_t - \hat{I}_t \right)^2 \qquad (10.24)$$

An obvious alternative is to use a log-likelihood approach (Polacheck et al., 1993). In addition, instead of using normal random errors (implied in Eq. 10.24) one uses lognormal residuals because we are dealing with catch rate data. Equation 10.21 can be converted to a log-likelihood and greatly simplified (see Appendix 10.3 and Example Box 10.3) so that

$$LL = -\frac{n}{2} \left(Ln(2\pi) + 2Ln(\hat{\sigma}) + 1 \right) \qquad (10.25)$$

where LL refers to log-likelihood, n is the number of observed catch rates, and σ is defined in Eq. 10.22.

10.4.3 Model Outputs

A general objective of fisheries modelling is to generate outputs in terms useful to fisheries management. There are many outputs possible from most fishery models but focus tends to be placed on those that can act as fishery performance indicators or that inform about selected limit thresholds. The two classical performance indicators, that derive from surplus-production modelling, are the maximum sustainable yield (MSY) and the effort, E_{MSY}, that, given $B_{MSY} = K/2$ (for the logistic), should lead to the MSY (as in Eqs. 10.11 and 10.13)

$$E_{MSY} = \frac{r}{2q} \qquad\qquad MSY = \frac{rK}{4} \qquad (10.26)$$

Example Box 10.3 A non-equilibrium, Schaefer surplus-production model fitted to Schaefer's (1957) original eastern Pacific yellowfin tuna data (Table 10.1). Copy the data into columns A, B, and C, down to row 31. In D11 put Eq. 10.19 as =max(D10+$B$1*D10*(1-(D10/$B$2))-B11,100). The max function ensures that the stock biomass cannot go extinct when using the solver. In E10 put =((D10+D11)/2)*$B$4, the mid-year biomass in a given year multiplied by the estimate of q (in E31 put =D31*$B$4). Put =(C10-E10)^2 into F10 to obtain the squared normal residual errors. In G10 put =Ln(C10/D10), to generate the contributions to the q estimate. In H10 put =(Ln(C10)-Ln(E10))^2 to generate the residuals for lognormal random errors. Select D10:H10 and copy down to row 32. Complete the worksheet by putting =exp(average($G$10:$G$31)) into B4 to calculate the closed form of q (see Eq. 10.23). Count the number of observations by putting =count(C10:C31) into B6, and, finally, to estimate the standard deviation of the residuals put =sqrt(sum(H10:H31)/B6) into B7 (See Eq. 10.22). Plot columns C and E against column A (use different colours). The parameters shown are close to optimum. Use the solver to maximize the log-likelihood in E7 by changing B1:B3. Compare the results obtained by minimizing the sum of squared residuals in E8. They should be the same. How different are the answer's when you minimize E6, which contains normal random residual errors instead of lognormal? How stable is the answer? Start the model from different starting points (e.g., ri, Ki, B0 = 0.05, 1500000, 1500000; or 0.5, 3500000, 1000000). In the solver options, turn on automatic scaling and increase the required precision and decrease the convergence limits. Try removing the max function from column D. Try estimating q directly (just vary the values of B1:B4 when using the solver). Put the p parameter into B5. Alter column D to match Eq 10.19 and E3 and E4 to match Eqs. 10.11 and 10.12. How does permitting an asymmetric production curve affect the results? Are the results biologically sensible?

	A	B	C	D	E	F	G	H
1	ri	0.17830		Bcurr	=D32			
2	Ki	2525547		Bcurr/K	=E1/B2			
3	B0	2748326		MSY	=B1*B2/4			
4	qi	0.00436		Emsy	=B1/(2*B4)			
5	p		1	F$_{Final}$	=B32/E32			
6	n	22		SSQ_I	=sum(F11:F32)			
7	Sigma	0.1726		LogLik	=-(B6/2)*(Ln(2*pi())+2*Ln(B7)+1)			
8				SSQ_Ln(I)	=sum(H11:H32)			
9	Year	Catch	O(I)	ExpectB	E,(I)	SSQ_I	Expt_q	Ln(I-I)
10	1934	60913	10361	=$B$3	11763.4	1966371.8	-5.542	0.0161
11	1935	72294	11484	2549745	11330.2	23763.4	-5.403	0.0002
12	1936	78353	11572	2467036	10943.8	394501.1	-5.362	0.0031
13	1937	91522	11116	2385705	10585.9	281500.8	-5.369	0.0024

In addition, we could consider such outputs as the current estimated biomass, the ratio of the current biomass with K or B_0, and possibly include an estimate of fishing-mortality rate, F (see Example Box 10.3). A number of alternative methods exist for calculating various model parameters and outputs.

The instantaneous fishing mortality rate can be estimated in two ways, either as a conversion of the annual exploitation rate (catch/biomass) to an instantaneous fishing mortality rate

$$F_t = -Ln\left(1 - \frac{C_t}{(B_t + B_{t+1})/2}\right) \tag{10.27}$$

where F_t is the instantaneous fishing mortality rate in year t, C_t is the catch in year t, and $(B_t + B_{t+1})/2$ is the mid-year biomass for year t (as in Eq. 10.20 and Example Box 10.3). Alternatively, we could use the standard catch equation so that instantaneous fishing mortality relates to expected effort and the catchability coefficient, q

$$F_t = qE_t = q\frac{C_t}{C_t/E_t} = q\frac{C_t}{\hat{I}_t} \tag{10.28}$$

Given Eq. 10.28, and that $E_{MSY} = r/2q$ (from Eq. 10.26) we can see that the instantaneous fishing mortality rate at MSY, F_{MSY}, would be

$$F_{MSY} = qE_{MSY} = q\frac{r}{2q} = \frac{r}{2} \tag{10.29}$$

Prager (1994) described many extensions to standard surplus-production models and one of these was to point out that $F_{0.1}$ (see Chapter 2) is approximately 90% of F_{MSY}. Thus, it would be simple to include both of these potential management targets in the outputs from the model.

The real world interpretation of management targets is not always straightforward. An equilibrium is now assumed to be unlikely in a fished population, so the interpretation of the MSY is more like an average, long-term expected potential yield if the stock is fished optimally. The E_{MSY} is only the effort that should give rise to the MSY if the stock biomass is at B_{MSY}, the biomass needed to generate the maximum surplus-production. Clearly, a fishery could be managed by limiting effort to E_{MSY}, but if the stock biomass is depleted, then the average long-term yield will not result.

In fact, the E_{MSY} effort level may be too high to permit stock rebuilding.

Few of these potential management outputs are of value without some idea of the uncertainty around their values. It would also be very useful to be able to project the models into the future to provide a risk assessment of alternative management strategies.

10.5 BEYOND SIMPLE MODELS

10.5.1 Introduction

Prager (1994) described and reviewed a range of extensions to simple surplus-production models. He gave detailed descriptions of some of the fundamental equations relating to the population dynamics and to some of the management targets ($F_{0.1}$ has already been mentioned). He also discussed the handling of multiple data series and missing data, along with suggestions for dealing with changing catchability through time. He briefly described how to estimate the uncertainty around parameters using bootstrap procedures, and an algorithm, that was an extension of the bootstrap, for using surplus-production models in projections. Projections are necessary for conducting risk assessments of different proposed management options.

Prager's (1994) paper is recommended reading, but we will still consider some of these subjects here so as to give more detail and make further extensions. We will focus on the more general surplus-production model described first by Pella-Tomlinson but developed by Polacheck *et al.* (1993) (Eq. 10.19 and following equations, but with p constrained to =0.00000001 instead of =1; sections 10.3.1 and 10.4.1).

10.5.2 Changes in Catchability

One major assumption in the use of surplus-production models is that the relationship between catch rates and stock biomass is constant ($C/E = qB$). This relationship implies that the catchability coefficient, q, remains constant through time. In fact, because fishers tend to be good at what they do, there tend to be continual improvements to fishing gear and fishing practices such that the effectiveness of each unit of effort increases through time. This effort creep is often considered in terms of changes in fishing power brought about, for example, by introducing new gear such as radar, coloured echo sounders, and Global Positioning System (GPS) receivers and plotters (Brown *et al.*, 1995). By using General Linear Models to compare the catch rates of vessels that had adopted GPS and related plotters in different years, Robins *et al.* (1998) found that vessels in the

Australian northern prawn fishery obtained a 4% advantage with the introduction of GPS, and this figure grew to 7% if a plotter was also installed. Over the subsequent two years there were further improvements of between 2 and 3% per year (i.e., learning was a factor). Overall, once the complete fleet had adopted the technology (a matter or 3 to 4 years), the increase in fishing power accorded to this alteration alone was 12%. Multiplying the units of effort by 1.12 is a possibility but such an approach would make the units of effort confusing. For example, if effort were in hours fished, then it would become necessary to refer to effort as hours standardized relative to some reference year (100 hours in 1996 might be 112 1994-hours). Instead, perhaps the simplest interpretation to place on increases in fishing power is to consider them as changes to the catchability coefficient. In numerical terms, because $C = qEB$, it does not matter whether the E or the q changes.

Clearly, the assumption that q is a constant is rather an over-simplification. Prager (1994) pointed out that if it were suspected that the catchability coefficient had changed rather suddenly, then the non-equilibrium model could be applied as if there were two time-series of catches and catch-rates. The same parameters (r, K, B_0, and perhaps p) would apply to each time-series and would be fitted together. However, there would need to be as many q parameters as there were separate time-series, and these would need to be fitted separately. Alternatively, two or more sets of closed-form calculations could be produced, but if the number of observations in each time-series becomes very low then the closed-form calculations may become sub-optimal and direct estimation might be more robust. Each suspected major change in catchability would entail the addition of a further parameter. Naturally, as the number of parameters increases one would expect the quality of model fit to improve. Prager (1994) suggests using an F-ratio test to compare the simple models with the more complex. This would be equivalent to using a likelihood ratio test.

Prager (1994) also considers a linear increase in catchability through time. This would be equivalent to a constant absolute improvement each year

$$q_t = q_0 + q_{add} \times t \qquad (10.30)$$

where the t subscript denotes the particular year, 0 to n-1, q_0 is the catchability in the first year, and q_{add} is the constant increase added to the catchability each year. Prager (1994) suggested this could be parameterized by estimating the first and last year's q and interpolating for the intervening years. Perhaps this would be most easily implemented by using Eq. 10.30, directly estimating the q_0 for the first year and then the q_{add} that provides the best fit. Alternatively, a closed form estimate of q_0 and q_{add} can be generated by implementing the appropriate regression analysis (see Appendix 10.2 for the derivation).

In some fisheries it has been suggested that there is a constant proportional increase in catchability each year. For example, in the Australian northern tiger prawn fishery the annual proportional increase in the effectiveness of effort formally accepted by managers and Industry (for purposes of discussing effort reduction targets) is 5% per annum (Pownall, 1994). Thus, instead of Eq. 10.30, we would need

$$q_t = q_0 \times q_{inc}^t \qquad (10.31)$$

where q_t is the catchability in year t and q_0 is the catchability in the first year. In year 0, the q_{inc} would be raised to the power zero and hence equal 1. For a 5% per annum increase, q_{inc} would $= 1.05$. As with the additive form of catchability increase, closed form estimates of q_0 and q_{inc} can be obtained if we log-transform the Eq. 10.31 to give it the form of a linear regression (see Appendix 10.2 for the derivation)

$$Ln(q_t) = Ln(q_0) + t \times Ln(q_{inc}) \qquad (10.32)$$

10.5.3 The Limits of Production Modelling

We will consider the Australian northern tiger prawn fishery and illustrate some of the variations possible when implementing non-equilibrium surplus-production models. This fishery extends across the top of Australia from the Gulf of Carpentaria to the west of Joseph Bonaparte Gulf (Pownall, 1994). The fishery has been operating for over 30 years with significant tiger prawn catches since 1970 (Table 10.2; Fig. 10.6). Management is via input controls (being a mixture of limited entry, closed seasons, closed areas, and gear controls), and stock assessment uses a relatively complex model (Wang and Die, 1996). Nevertheless, because effort creep has been identified as a major issue in this fishery, it provides an opportunity to apply the techniques described above involving the estimation of effort creep levels.

The dynamics of the model are described by the same non-equilibrium model as used with the Schaefer data (Eq. 10.19) along with the relationship between catch rates and biomass illustrated by Eq. 10.20. The estimation of q will involve either the constant catchability (Eq. 10.23) or the additive or the multiplicative incremental increases in q, as in Eqs.

10.31 and 10.32. The precise relationship between catch rates and stock biomass is

$$\hat{I}_t = \frac{\hat{C}_t}{E_t} = qB_t e^\varepsilon \qquad (10.33)$$

In this case the biomass is not averaged across two years as the prawns are almost annual in their life-cycle with very few animals surviving from one year to the next.

When fitting such an array of options it is obviously best to start simple and progress to the more complex.

Table 10.2 Published catch, in tonnes, and effort, in fishing days, for the northern Australian tiger prawn fishery (from Pownall, 1994, and AFMA, 1999). Data are for both the brown (*Penaeus esculentus*) and grooved tiger prawns (*P. semisulcatus*) combined. Catch rates for each year can be determined by dividing the catch by the effort for each year.

Year	Catch	Effort	Year	Catch	Effort	Year	Catch	Effort
1970	1138	5818	1980	5124	30594	1990	3550	25525
1971	1183	6057	1981	5559	31895	1991	3987	20744
1972	1380	7380	1982	4891	32956	1992	3084	21789
1973	1672	7362	1983	5751	34551	1993	2515	16019
1974	666	3439	1984	4525	32447	1994	3162	18592
1975	973	6010	1985	3592	26516	1995	4125	16834
1976	1118	6660	1986	2682	26669	1996	2311	16635
1977	2900	11673	1987	3617	22478	1997	2694	15385
1978	3599	18749	1988	3458	26264	1998	3250	18003
1979	4218	17791	1989	3173	27036			

10.6 UNCERTAINTY OF PARAMETER ESTIMATES

10.6.1 Likelihood Profiles

Polacheck *et al.* (1993) used the log-likelihood criterion, even though it provides the same estimates as the least squares estimates (as long as the *I* values are log-transformed first so as to keep the same error-structures). They did this because they also suggested using Venzon and Moolgavkar's (1988) approximate likelihood profile method to produce confidence intervals around the parameter estimates. The methodology behind this was discussed in the parameter estimation chapter (Section 3.4.14). For single parameters the results are essentially the same as standardizing the log-likelihoods so they all add to one and then finding the confidence intervals by using the parameter limits which contain 95% or 99% (or whatever confidence interval chosen) of the likelihood curve. Polacheck *et al.* (1993) found that the likelihood profiles obtained when using observation error estimators were much smaller than those deriving from process error estimators.

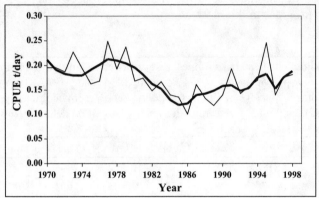

Figure 10.6 Observed catch per unit effort data from the northern Australian tiger prawn fishery (see Table 10.2) as the fine irregular line with an optimal model fit as the thicker smoother line (Example Box 10.4). The optimal model was constrained to be equivalent to the modified Fox model ($p = 0.0000001$; Eq. 10.17).

Example Box 10.4 Non-equilibrium surplus-production model of the northern Australian tiger prawn fishery (Table 10.2; Fig. 10.6). The row numbers are omitted to save space but they start in row 1. The manipulations in C1:C4 assist the solver by keeping the values to vary similar in value, alternatively logs could be used. Name C1 as 're', C2 as 'K', and C4 as 'p'. Select B5:C5 and put =linest(F9:F37,I9:I37,true,false) using, <Ctrl><Shift><Enter> to enter the array function (*cf.* Eq. 10.32). The closed form estimates of q and $qinc$ are in B6:C6. Column I contains numbers 0 to 28 representing t-1 in the closed form calculations. In B7 put =-(F6/2)*(Ln(2*pi())+2*Ln(F7)+1), which is Eq. 10.25. Enter the data into columns A, B, and C from Table 10.2 (CE_Obs is Catch/Effort). D9 is B0 (=C3). D10 is =max(D9+D9*(re/p)*(1-(D9/K)^p)-B9,100), which is Eq. 10.33 with the max function preventing the possibility of negative biomass (which would halt the solver). G9 is =(Ln(C9)-Ln(E9))^2, which represents the lognormal squared residuals. Select E9:G9 and copy down to row 10. Select D10:H10

Example Box 10.4 [cont.] and copy down to row 37. Plot columns C and E against A to mimic Fig. 10.6. Optimize the model fit by maximizing B7 by varying B1:B3, leaving the *p* value as it is. Save the parameter and model output values somewhere else on the sheet. Now solve by varying B1:B4 to see the impact on the *p* value when it is free to vary. Are the values for *qinc, p,* and B0/K reasonable? Would the shape of the production curve make biological sense (*c.f.* Fig. 10.1)?   Alter the structure of the sheet to estimate a constant *q* (put =exp(average(F9:F37)) into C6, and copy H9 down to row 37). Re-solve, either for B1:B3 or B1:B4. How big an impact would this have on the model outputs? Because the multiplicative effects of *qinc* lead to an exponential impact on catchability, the results are very sensitive to the model of *q* used. Try implementing the additive model of *q* and *qadd.* When solving this model, it may be necessary to search carefully for the optimal solution. Try solving for individual parameters and moving towards the correct solution if you lose the optimum. Try varying the parameters to see how sensitive the solution is to the starting position. Can you find any false minima. If you do, does altering the options in the solver alter the solution? Is adding a constraint that *p* be greater than zero a good idea? Try the different possibilities listed for generating B0 (independent fit, =K, =CE/q). How much do they affect the results? Plot the predicted biomass history for the stock. Are things looking good for the northern Australian tiger prawn fishery according to this model? Compare the solution obtained for the maximum likelihood with that obtained from the summed squared residuals. Implement normally distributed residuals (put =(C9-E9)^2 into G9 and copy down), and solve by minimizing these. How big a difference does the residual error structure make?

A	B	C	D	E	F	G	H	I
r	32.965	=B1/100		**B98**	=D37			
K	27.3301	=B2*1000		**B98/K**	=F1/K			
B0	42.1005	=B3*1000		**MSY**	=(re*K)/((p+1)^((p+1)/p))			
p	1	=B4/1E9		**B0/K**	=C3/K			
	0.0763	-12.2195	Ln(q0)	**Ssq**	=sum(G9:G37)			
qinc	=exp(B5)	=exp(C5)	q0	**n**	=count(G9:G37)			
LL	17.2676			**Sigma**	=sqrt(average(G9:G37))			
Year	Catch	CE_Obs	PredB	Pred_I	Pred_q	SSQ	q	Yr
1970	1138	0.1956	=C3	=D9*H9	=Ln(C9/D9)	0.0072	=$C$6	0
1971	1183	0.1953	35774	0.1905	-12.118	0.0006	=H9*$B$6	1
1972	1380	0.1870	31593	0.1816	-12.037	0.0009	0.0000058	2
1973	1672	0.2271	28840	0.1789	-11.752	0.0569	0.0000062	3
1974	666	0.1937	26767	0.1792	-11.837	0.0060	0.0000067	4
1975	973	0.1619	26377	0.1906	-12.001	0.0267	0.0000072	5

A maximum of two parameters can be visualized at one time. To determine the likelihood profile confidence intervals, as shown in Chapter 3, involves subtracting 5.9915/2 (= $\chi^2/2$, for 2 degrees of freedom) from the maximum likelihood and searching for the parameter combinations that generate this likelihood (Fig. 10.7; Example Box 10.5).

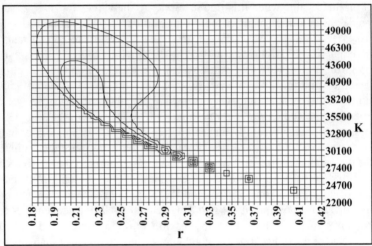

Figure 10.7 Approximate 95% confidence intervals from the two-dimensional likelihood profile, for the r and K parameters in the non-equilibrium surplus-production model for the northern Australian tiger prawn fishery (as in Example Box 10.4). The optimum solution was at r = 0.32965 and K = 27,330. The target log-likelihood was 14.273, which is the outer curve (Example Box 10.5). The gaps in the tail are unreal, the resolution of the graph is insufficient to show the detail of connected contours.

10.6.2 Bootstrap Confidence Intervals and Estimates of Bias

There are disadvantages to using likelihood profiles. These include the complexity of implementing the method when there are many parameters involved (parameter correlation usually increases the width of confidence intervals), and likelihood profiles give no indication of any bias in the parameter estimates and bias can be an important aspect of uncertainty. Fortunately, alternative methods exist for fitting confidence intervals around parameter estimates. A common approach is to use a bootstrap strategy as introduced in Chapter 6. This approach resamples the residuals from the optimum fit to generate new bootstrap samples of the observed time series. The model is fitted to many of these bootstrap samples and the outputs stored so that percentile confidence intervals can be determined, as is usual with bootstrap methods. The confidence intervals generated can be asymmetric and synthesize the effects of all the parameters varying at once. A further advantage of the bootstrap strategy is that one can estimate whether or not the parameter estimates are biased.

Example Box 10.5 The generation of confidence intervals using likelihood profiles. First construct the table of likelihoods from the array of values for the two parameters of interest, in this case, r and K. The sixth row relates to the K value and column U7 downwards, to the r parameter. Copy W6 across to column BB, then copy U8 down to row 59. Cell U6 points to the Likelihood value in B7. Select U6:BB59 (or however big you have made the table), then choose the Data/Table menu item. In the dialog box the row input cell is $B$2 and the column input cell is $B$1. This should complete the Excel table. To plot this as a contour surface plot it is best to convert the borders to text so that the axes are automatically labeled. In V66 put =text(V6*1000,"#0") and copy across to column BB. In U67 put =text(U7/100,"#0.000") and copy down to row 119 to match the table above. In V67 put =V7 and copy down and then copy the column across to regenerate the table. Select the whole table including the borders (U66:BB119) and generate the contour surface plot by selecting the appropriate menu item (Insert/Chart/Surface). The target log-likelihood for the approximate 95 % confidence intervals for two parameters would be the optimum $\pm 5.99/2$ (=B7-chiinv(0.05,2)), which in this case = 14.273. Double click on the legend and alter the scale to a minimum of 13.273 and a major unit (tick value) of 1.0, which will lead to a graph similar to Fig. 10.7. Drag the graph up near to U1:U4 and experiment with the table ranges and the scale shown. Be wary of selecting a major unit which is too small else the poor hardworking hardware can reach its limits and you will need to crash out of the program. Do NOT run the solver if a table is active unless you have a very great deal of time to spare (deleting the core of the table – V7:BB59, will fix that problem).

	T	U	V	W	X
1	K	22			
2		0.9			
3	r	17.5			
4		0.48			
5					
6		=B7	=U1	=V6+$U$2	26.2
7		=U3	=table(B2,B1)	=table(B2,B1)	=table(B2,B1)
8		=U7+$U$4	=table(B2,B1)	=table(B2,B1)	=table(B2,B1)

Surplus-production models are fitted to time-series of relative abundance indices (CPUE). To obtain confidence intervals and bias estimates using bootstrap procedures it is important that the time-series nature of the data is not disrupted. So that the time-series nature of the CPUE data is retained one should not bootstrap the raw data but instead bootstrap the residuals between the observed and expected values; i.e., randomly sample from the original best-fit residuals, with replacement, to generate a new vector of bootstrapped residuals. This vector of bootstrapped residuals is combined with the optimum vector of expected CPUE data to obtain each new bootstrap sample of CPUE data (Prager, 1994; Haddon, 1998).

If we had been using normal, additive errors, then the residuals would simply be (observed CPUE - expected CPUE), and after bootstrapping, these would each have been added to the sequence of original expected CPUE value (Eq. 10.34)

$$I_t^* = \hat{I}_t + \left(I_t - \hat{I}_t\right)^* \equiv \hat{I}_t + \varepsilon^* \tag{10.34}$$

where I_t^* is the bootstrapped CPUE value which equals the Expected CPUE value, $\hat{I}_t$, plus a bootstrapped normal residual $(I_t - \hat{I}_t)^*$, which is equivalent to combining each expected catch/effort value with a bootstrapped error or residual term. In this way a new time-series of bootstrapped "Observed"-CPUE data are generated to which the model may be refitted and the parameters may be re-estimated. However, with lognormal, multiplicative errors (see Eq. 10.33) we must use the ratio of the CPUE values (observed / expected) to calculate the residuals that are to be bootstrap sampled. To obtain the bootstrapped CPUE values the residuals are multiplied with their respective original expected CPUE values (Eq. 10.35)

$$I_t^* = \hat{I}_t \times \left(\frac{I_t}{\hat{I}_t}\right)^* \tag{10.35}$$

Confidence intervals can be estimated by generating 1000s of bootstrap samples, refitting the model, and generating an equal number of parameter estimates (these could include outcomes such as MSY). The central 95% of these (the 0.025 and 0.975 percentile values of the sorted estimates) would represent the bootstrap, percentile confidence intervals (Example Box 10.6).

If we wished to take into account any bias in the parameter estimates, we would do best to calculate bias-corrected percentile confidence intervals (Efron and Tibshirani, 1993). Percentile confidence intervals are determined by using the 25th and 975th ordinal values (out of 1,000 replicates for 95% intervals). Bias-correction leads one to use different percentile values depending on whether the parameter estimates are positively or negatively biased. The procedure begins by determining what proportion (LT) of the bootstrap replicates are less than the original optimal fit estimate of the parameter or output of interest and this value is transformed via the inverse cumulative standard normal distribution (Φ^{-1})

$$z = \Phi^{-1}(LT)$$ (10.36)

A convenient way to do this would be to use the z=NORMINV(LT,0,1) in Excel, where the mean of 0 and the standard deviation of 1 imply the standard normal distribution. This z value is used in the cumulative standard normal distribution (Φ) to calculate the appropriate percentile to use instead of the standard 25th and 975th

$$P_{lower} = \Phi(2z - 1.96)$$
$$P_{upper} = \Phi(2z + 1.96)$$ (10.37)

where Φ is the cumulative standard normal distribution function, conveniently provided by NORMDIST(P_{index}, 0, 1, 0.05) where the 0 and 1 define the normal distribution to be standard, and the 0.05 and 1.96 reflect the normal values required for 95% intervals. With bias-corrected confidence intervals, if LT were 0.5, then z would be zero and we would, of course, obtain the 25th and 975th percentiles (Haddon, 1998). However, if, for example, LT were 0.459, then z would be -0.10295, which would lead us to use the 16th and 960th percentiles (note they are no longer necessarily symmetrical around the median).

The percentile confidence intervals around the parameter estimates for the northern Australian tiger prawn fishery tend to be narrower than the likelihood profile intervals (Table 10.3). However, all of these confidence intervals are only approximate and only capture the variability inherent in the data, ignoring other sources of variability [these would include: the simplicity of the model failing to capture the full dynamics of the population (for example, there are two species of tiger prawns that are lumped in the catch and catch rate information), and the short time series of fisheries data not capturing the full range of environmental variation possible].

Example Box 10.6 Bootstrap implementation for the surplus-production model from Example Box 10.4. The procedure replaces the observed catch rates with a bootstrap sample; it is therefore prudent to store a copy of the original catch rate data into column J. Select E9:E37, copy as values into K9 (Paste Special/Values). Given the optimum fit put =C9/E9 into L9 and copy down L37. These are the residuals around the catch rates calculated as (observed/expected). An easy error is to forget to convert these to values. Select L9:L37, copy and paste as values onto themselves so they are ready for bootstrap sampling. The resampling is conducted in column M using the offset function. The trunc(rand()*29)+1 will provide a random integer between 1 and 29. Copy M9 down to row 37. The bootstrap sample is generated in column N by multiplying the selected residual with the expected catch rate in that year. To conduct the bootstrap select N9:N37 and copy as values into C9 and re-solve for an optimum to provide a single set of bootstrap estimates, which will need to be stored as values somewhere else on the worksheet. The cells in column P have been arranged to make this copying and storage a simpler process. Of course, to do the bootstrapping in a sensible way one needs to write a macro to do the necessary copying, solving, and storing of results. Cells L1:L3 and M1:M3 all relate to a suitable macro to do the bootstrapping. Before constructing the macro, carry out the bootstrap a few times manually. Under the solver options be sure to provide generous time and iteration limits, and refine the precision and convergence criteria. For this problem the other options that seem to provide stable answers are Estimate: Tangent, Derivatives: Forward, and Search: Conjugate, but do try the alternatives. Do you think it is a good idea to always start the search from the original optimum? To provide a convenient source of the optimum value is why they were copied into L1:L3 before bootstrapping started. Do you ever have to run the solver twice to find a stable optimum? It would be a good idea to run it twice for each bootstrap sample, giving the solution a slight shift before the second solve (the reason for M1:M3). Add a few of the bootstrap samples to the plot of the observed and expected catch rates (retaining the original observed values). How closely do they compare? Create a macro to do the bootstrapping. Start with only a few bootstraps to see how long it takes and whether there are problems. Then set it going for 1000 bootstraps. See the macro below details that cannot be recorded.

	K	L	M	N	O	P
1	r	32.969	=B1+B1/100		r	=C1
2	K	27.328	=B2+B2/100		K	=C2
3	B0Est	42.106	=B3+B3/100		B0Est	=C3
4					p	=C4
5					q0	=C6
6					qinc	=B6
7					Bcurr	=F1
8	Exply	Resid	Resample Residuals	Bootstrap	Bcurr/K	=F2
9	0.2117	0.9242	=offset($L$8,trunc(rand()*29)+1,0)	=K9*M9	MSY	=F3
10	0.1897	1.0295	=offset($L$8,trunc(rand()*29)+1,0)	=K10*M10	B0/K	=F4
11	0.1812	1.0320	0.83075	0.209113	LogLike	=B7

Example Box 10.6. [cont.] Note the solver is used twice and that the predicted biomass time series and the bootstrap sample are all stored along with the results. The times will only be indicative but will determine the approximate time taken for 1000 replicates. You may wish to place labels across the columns starting at S13 with 'r'. All bootstrap analyses should include the original optimum solution (put it in S14:AC14). The bootstrap estimates may be tabulated and percentile intervals calculated (Example Box 10.7, Fig. 10.8).

```
Option Explicit
Sub Do_Bootstrap()
Dim i As Integer
Dim start As Double, endtime As Double
start = Timer
Application.ScreenUpdating = False
For i = 1 To 10                                   ' Set the number of bootstraps
    Range("N9:N37").Select
    Selection.Copy
    Range("C9").Select
    Selection.PasteSpecial Paste:=xlValues
    Range("L1:L3").Select                          ' Paste original optimum solution
    Selection.Copy
    Range("B1").Select
    Selection.PasteSpecial Paste:=xlValues
    SolverOk SetCell:="$B$7", MaxMinVal:=1, ValueOf:="0", ByChange:="$B$1:$B$3"
    SolverSolve (True)
    Range("M1:M3").Select                          ' optimum + 1%
    Selection.Copy
    Range("B1").Select
    Selection.PasteSpecial Paste:=xlValues
    SolverOk SetCell:="$B$7", MaxMinVal:=1, ValueOf:="0", ByChange:="$B$1:$B$3"
    SolverSolve (True)
    Range("P1:P11").Select                         ' Save the answers
    Selection.Copy
    ActiveCell.Offset(13 + i, 3).Range("A1").Select
    Selection.PasteSpecial Paste:=xlValues, Transpose:=True
    Range("D9:D37").Select                         ' Save the biomass values
    Selection.Copy
    ActiveCell.Offset(5 + i, 26).Range("A1").Select
    Selection.PasteSpecial Paste:=xlValues, Transpose:=True
    Range("C9:C37").Select                         ' Save the bootstrap data
    Selection.Copy
    ActiveCell.Offset(5 + i, 56).Range("A1").Select
    Selection.PasteSpecial Paste:=xlValues, Transpose:=True
    Application.CutCopyMode = False
Next i
endtime = Timer
Application.ScreenUpdating = True
MsgBox Format(endtime - start, "##0.000"), vbOKOnly, "Time Taken"
End Sub
```

Table 10.3 Comparison of different approximate confidence intervals for the two parameters r and K, from the non-equilibrium surplus-production model for the northern Australian tiger prawn fishery (Example Boxes 10.4 10.5, 10.6, and 10.7). Note that the first order bias corrected bootstrap percentile intervals are closer to the likelihood profile intervals than the straight percentile intervals. These intervals are only approximate and are likely to be underestimates, it is therefore tempting to adopt the widest as providing the best estimates, or having the greatest chance of covering the true value.

	r	r	r	K	K	K
Interval Type	Lower95	Average	Upper95	Lower95	Average	Upper95
Likelihood Profile	0.1825	0.32965	0.4050	23,800	27,330	50,400
Bootstrap Percentile	0.2303	0.32965	0.4194	23,475	27,330	38,820
BC_1 Percentile	0.1954	0.32965	0.4098	24,095	27,330	44,905

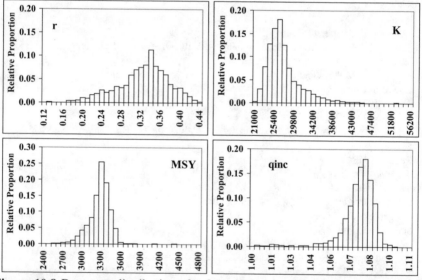

Figure 10.8 Bootstrap distributions for four of the parameters and model outputs from Example Box 10.6, for tiger prawns until 1998. By comparing the panels for the r and K parameters and Table 10.3, the first order bias correction can be seen to shift the confidence intervals in the direction of the skew of the distribution of bootstrap values (Example Box 10.7). The lower two panels represent values of interest to management. The MSY can be interpreted as the long term average yield expected from the stock when it is at its optimum size. The value of *qinc* suggests the level of effort creep and this indicates the urgency of any measures to limit effort in the fishery.

Example Box 10.7 Calculation of the bootstrap percentile confidence intervals and the first-order bias-corrected percentile intervals from the results of the bootstrapping from Example Box 10.6. The macro shown in Example Box 10.6 will deposit the bootstrap estimates in cells S15:CI1014 (assuming 1000 replicates). If more than 1000 are calculated, then the ranges in the various functions listed here will need to be extended. This assumes that the optimum answers were all selected and pasted into row 14 (hence S4 and T4). S1:S3 are just the standard bootstrap percentile intervals and the bootstrap estimate of the parameter. Generally the median in S5 (could be =median(S14:S1014)) will be closer to the optimum model fit when there is any bias. S6 is the count of bootstrap estimates that were less than the optimum value in S4, which must be converted to text for the countif function to work. The many decimal places are to obtain robust answers across the range of values experienced by the different parameters. S7 is Eq. 10.36 and S8:S9 is Eq. 10.37, which are used in the percentile estimates of S10:S11. S12 is an estimate of the bias shown in the bootstrap estimates. Select S1:S12 and copy across to column BF. The formatting (number of decimal places will need adjustment in each column). Store these equations in a separate worksheet and delete the originals if you wish to do more bootstrapping (otherwise they slow the calculations a great deal). Your row 15 will differ from that shown here because the bootstrap replicates will differ. Construct a bootstrap for the eastern Pacific yellowfin tuna (Example Box 10.3). How certain is the non-equilibrium estimate of the average long-term yield, MSY? The bias-correction is converting the confidence intervals from being centred on the average to being centred more on the median of the distribution.

	R	S	T	U	~	X
1	U95%	=percentile(S14:S1014,0.975)		64250.4		1.09289
2	Average	=average(S14:S1014)		43169.6		1.07631
3	L95%	=percentile(S14:S1014,0.05)		31376.5		1.05388
4	Optimum	=S14	=T14	42105.6		1.07931
5	Median	=percentile(S14:S1014,0.5)		42512.1		1.07902
6	LT_mean	=countif(S14:S1014,"<" & text(S4,"#0.0000000"))				359
7	Z	=norminv(S6/1000,0,1)		-0.60828		-0.91727
8	Pupper	=normdist(2*S7+1.96,0,1,0.05)		0.77139		0.54992
9	Plower	=normdist(2*S7-1.96,0,1,0.05)		0.00075		0.00007
10	U95%	=percentile(S14:S1014,S8)		48094.6		1.07992
11	L95%	=percentile(S14:S1014,S9)		22299.0		0.99272
12	%Bias	=100*(S2-S4)/S4				
13		r	K	B0Est		Qinc
14		0.32969	27327.694	42105.588		1.079306
15		0.210056	40102.191	45006.252		1.039695

The confidence intervals produced by both likelihood profile and bootstrap methods will often be asymmetric and will vary greatly between parameters (Fig. 10.8). The predicted biomass estimates can be treated in exactly the same way as the parameter estimates so that the estimated history of the stock biomass can be illustrated (Fig. 10.9). The confidence intervals are relatively wide around B_0, which would be typical of the uncertainty surrounding this parameter (Fig. 10.9). In this fishery, the situation is even more complicated in that the fishery was only developing over the years 1970 to 1975, so it is difficult to be certain as to how meaningful the early catch rates are in terms of stock biomass. Note the asymmetry of the confidence intervals. The lower bound is much closer to the average than the upper. This makes sense as there must be a certain minimum present to sustain the history of the catches that have been taken from the fishery. In this case, bias correction makes only a slight difference (from Example Box 10.7, try plotting the two data series for comparison). Clearly, the model outputs are consistent with the stock being in a relatively depressed state. Over the last 10-12 years the stock has been in a relatively low level, below the size at which we might expect it to be maximally productive.

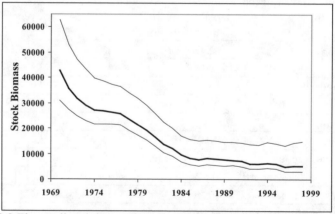

Figure 10.9 The predicted time series of stock biomass for the northern Australian tiger prawn fishery (Example Boxes 10.6 and 10.7). Most of the severe decline is driven by the continually increasing catchability coefficient. If q is increasing and catch rates stay stable, then this actually implies that the stock biomass must be declining. This is one reason why un-standardized catch rates are dangerous as a basis for managing a fish stock.

None of the parameter estimates and model outputs in the tiger prawn fishery are especially certain. It would almost always be a good strategy to obtain further information from the fishery rather than attempting to estimate peripheral parameters within the model. Systematic changes in catchability can have a huge effect and using such a model form should not be adopted routinely. In the case of the tiger prawns, if the changes to the catchability could be determined empirically by determining the timing of novel gear changes and introductions, and their relative effect, this would have the potential for providing greater certainty. Of greatest value would be to obtain a direct and relatively precise estimate of abundance in more than one year. This could be used to anchor the model to reality along with the catch rates. More than one data series can be included in the likelihood equation. Each series might best be weighted in accord with their relative precision (standard error or coefficient of variation for each estimate, or some other estimate of their relative precision).

10.7 RISK ASSESSMENT PROJECTIONS

10.7.1 Introduction

Invariably, there will be many sources of error and uncertainty that are not accounted for in the model. Determining the uncertainty in an analysis only tells us that we need to be careful when attempting to interpret the model outcomes, it cannot inform resource managers about the risk level associated with a particular management option. To answer such questions a risk assessment is required.

Risk assessment implies projecting the population dynamics model into the future under the constraint of different management options (for example, a particular catch or effort regime, or different open and closed areas, etc.). Given the selected catch or effort we need to be able to model the projected recruitment levels in a stochastic manner, with the variability of that recruitment reflecting the stock dynamics observed in the available time series of data. The projected recruitments would be offset against the catches and the trajectory of the stock biomass through time could thus be generated. The problem, when using surplus-production models is to generate these stochastic recruitments.

10.7.2 Bootstrap Projections

Prager (1994) suggested that because surplus-production models imply a recruitment function, they could be used to make projections based upon hypothetical catch or effort allocations. In the standard operation of surplus-production models the stock biomass is projected forwarded under the constraint of the time series of catches and catch rates (e.g., Eq. 10.19). To do this for a risk assessment would be simply to extend this stock biomass projection beyond the years for which data is available. The projected catches or efforts (which, given the catchability and stock biomass would imply catches) would be dictated by proposed management options. The stock biomass projections are deterministic so a mechanism for introducing the required stochastic element is still required. Prager's (1994) suggestion for varying the population projections, was to conduct a bootstrap analysis and project each bootstrap forward to obtain a risk assessment at the same time as a determination of the level of uncertainty in the analysis. This mechanism uses the variation inherent in each bootstrap sample to represent the variation likely to occur in the stock dynamics of the species concerned. The variation is represented by the residuals between the observed and predicted catch rates, which are assumed to relate back to stock biomass via the catchability coefficient (C/E = qB).

10.7.3 Projections with Set Catches

Many fisheries are now managed through output controls in the form of a total allowable catch (TAC). In such fisheries, a vital management control is to set the TAC at a level consistent with stock sustainability and, often, with optimizing production. Investigation of the implications of setting different catch levels is relatively simple with the surplus-production models described in this chapter. If the stock dynamics are assumed to be described by the deterministic equation

$$B_{t+1} = B_t + \frac{r}{p} B_t \left(1 - \left(\frac{B_t}{K} \right)^p \right) - C_t \qquad (10.38)$$

or any production function from which a catch is subtracted, then forward projection only requires those catches to be defined and the projections can be implemented (Example Box 10.8).

Exactly what characteristic of the population to consider in the projections can vary depending on circumstances and what would be most informative. For example, a common performance indicator would be to determine whether the predicted stock biomass in any given projection year is greater than a selected reference year. If many replicate projections are generated then, in any year, the proportion in which the stock biomass is greater than that in the reference year, can represent the probability that the modelled stock will have increased in size in that year (cf. Figs. 10.10 and 10.11). Alternatively, if there is a risk of stock collapse, then the number of replicates under a given set of management constraints, that led to collapse (defined as some low biomass level) can also be collated and graphed.

10.7.3 Projections with Set Effort

The northern Australian tiger prawns is an input controlled fishery so the management controls used to constrain the projections will involve considering the impact of different effort levels. There is also the problem of effort creep to attend to in the risk assessment. In theory, it should be possible to constrain the annual proportional increase in the effectiveness of effort. The presently accepted level is 5% per annum but it is recognized that this level cannot continue so alternative, lower levels will also need consideration. The best strategy is to conduct a grid analysis, running the projections for each of the selected effort level and all the levels of the annual increment in catchability ($qinc$) that are to be considered (Figs. 10.10 and 10.11; Example Box 10.8).

Given a recommended effort level the model can still be projected into the future by calculating the catch implied by the stock biomass, the catchability in that year, and the effort imposed

$$C_t = q_t E_t B_t \qquad (10.39)$$

Of course, the catches in each projected year are likely to vary from year to year. It is harder to make the stock crash with a constant effort management scenario than with a constant catch level. This is because as stock biomass declines, so does the catch from a certain effort. But a constant catch will be taken until there is no biomass left to take (until an infinite effort is implied).

10.8 PRACTICAL CONSIDERATIONS

10.8.1 Introduction

Despite only having minimal data requirements, surplus-production models purport to provide an assessment of the state of a given fishery at a particular time. The assumption that the stock and its dynamics can be described purely in terms of its biomass exposes this class of models to some problems peculiar to themselves. Because stock biomass can be either recruitment or standing crop these possibilities can be confounded in the model outputs.

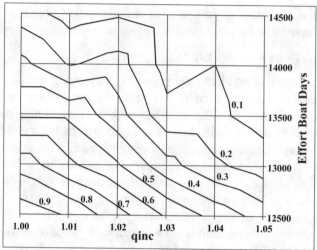

Figure 10.10 The contours are the probability of stock biomass being greater in the year 2002 than it was in 1998. The contours are not particularly smooth because each intersection is only represented by 100 bootstrap projections, using a fixed effort strategy in the northern Australian tiger prawn fishery. It is clear that positive stock growth, relative to 1998, will only have a greater than 50% chance of occurring if effort creep (*qinc*) is kept below 3.5% per annum, and effort is less than 13500 fishing days.

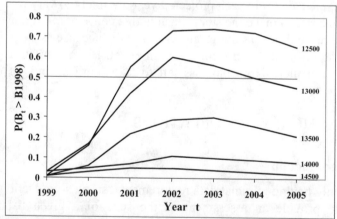

Figure 10.11 An alternative view of risk assessment output for the northern Australian tiger prawn fishery. The vertical axis is the probability of the stock biomass in a projected year being greater than the stock biomass in 1998. Each curve relates to a given effort level depicted at the end of each line. The annual increase in catchability was set at 2% (*qinc* = 1.02) for all series. The fine horizontal line is at 50%, indicating the desirability of fixing effort at less than 13000 fishing days to encourage stock rebuilding if qinc = 1.02.

If the available data exhibit a steady decline in catch rate through time, with no contrast in the ranges of effort imposed over the different stock biomass levels, then the data has little information with respect to the form taken by the stock biomass. The model can have trouble distinguishing between a stock with a high population growth rate and low standing crop and a stock with almost no production but an enormous standing crop from which all the catches have been taken. All the model can do is present outputs that are consistent with the input data and both of these interpretations can provide an internally consistent description of events. As long as one is aware of the possibility of the catch rate data having little useful information about the stock biomass, then care can be taken in the interpretation of the results from surplus-production models.

Equilibrium surplus-production models will almost always provide apparently workable management advice whereas non-equilibrium models can lead to the conclusion that the data provide no information. This latter situation may not appear to be useful but it is far better to know that the information one has is non-informative than to follow model results blindly.

10.8.2 Fitting the Models

Many options for the implementation of surplus-production models have been described in this chapter and it is unlikely that only a single set of options will suit a particular problem. If the results of the modelling are similar irrespective of the options selected, then at least there is a consistent story to tell. If different options lead to different results then how they differ should inform us about the value of the data and the relative sensitivity to the different parameters. The point is that, wherever possible, multiple options should be used.

In the case of the northern Australian tiger prawns, the constant proportional increase in the catchability coefficient ($qinc$) has an overwhelming impact on the assessment provided by the surplus-production model. This parameter has important management implications. If it is assumed equal to one (i.e., constant catchability, then the stock does not appear to be as depleted. These simple assessment models can emphasize where research effort should be focused to improve an assessment.

Where there are sufficient years data available, it is worthwhile conducting hind-casting trials to compare the assessment results for different series of years. Thus, if data is available from 1970 to 1998, then tabulating any changes derived from comparing years 1970-1995, with

1970-1996, with 1970-1997, and so on, can provide insight into the stability of the model and its outputs.

Example Box 10.8 Additions to the bootstrapping procedure, implemented in Example Boxes 10.6 and 10.7, to permit risk assessment projections based upon setting the future catch or effort levels. Save the workbook from Example Box 10.6 under a new name and make the following changes. The worksheet below reflects the original data and optimal fit, before any bootstrapping has occurred. In this fishery we do not assume the catchability will stay constant but will need to trial different catch or effort levels against different levels of qinc (hence H1:H2). To use allocated catch levels put =$H$1 into B38 and copy down to row 44. To use allocated effort levels put =D38*H38*$H$1 into B38 and copy down to row 44 (this is just C = qEB; Eq. 10.39). Select D37:E37 and copy down to row 44. Copy H37 down to row 44. These simple extensions will generate the population projection for a single instance of the combination of catch or effort level and qinc. To generate replicates we need to modify the macro controlling the bootstrap to save only the data we are interested in during the projections (see below). The bootstrap percentile limits can be used to describe the projections as with the usual analysis but the projections permit many other questions to be answered. Beside the columns of projected stock biomass we can add columns determining whether, for example, the biomass in the given year is greater than in some reference year. Thus, if only year 1985-2005 are stored according to the macro below, then in AN14 we could put =if(AG14>$AF14,1,0), where AG14 is the predicted biomass in 1999 and AF14 is that in 1998, the selected reference year. If AN14 is copied across to column AT and down to however many replicates were run, we can sum the columns relating to each of the projected years and ask about the probability of the stock being larger than the reference year under the catch or effort and qinc regime adopted (Figs. 10.10 and 10.11). Any reference year can be chosen. In this way, risk averse management strategies can be developed.

	A	B	C	D	E	F	G	H
1					**Future Catch or Effort**			14500
2						**Managed qinc**		1.02
~								
8	**Year**	**Catch**	**Obs_CE**	**Biomass**	**Pred_CE**	**Pred_q**	**Ln(I-I)**	**q**
~								
36	1997	2694	0.1686	4431.5	0.1749	-10.139	0.0000	0.00002167
37	1998	3250	0.2142	4395.4	0.1872	-10.100	0.0013	0.00002287
38	1999	=D38*H38*$H$1		3793.4	0.1648			=H37*$H$2
39	2000	=D39*H39*$H$1		3873.3	0.1716			=H38*$H$2
40	2001	=D40*H40*$H$1		3879.6	0.1754			=H39*$H$2
41	2002	2563		3833.9	0.1768			0.00004610
42	2003	2559		3753.4	0.1765			0.00004703
43	2004	2539		3650.8	0.1751			0.00004797
44	2005	2507		3534.5	0.1729			0.00004893

Example Box 10.8 [cont.]

```
Option Explicit            ' To run the macro first Set values in H1 and H2

Sub Projection()                  ' macro to conduct projections
Dim i As Integer
Calculate
Application.ScreenUpdating = False
For i = 1 To 100                       ' Number of projections; change to suite
   Range("N9:N37").Select              ' Copy and paste the bootstrap sample
   Selection.Copy
   Range("C9").Select
   Selection.PasteSpecial Paste:=xlValues
   Range("L1:L3").Select              ' Paste original optimum solution
   Selection.Copy
   Range("B1").Select
   Selection.PasteSpecial Paste:=xlValues
   SolverOk SetCell:="$B$7", MaxMinVal:=1, ValueOf:="0", ByChange:="$B$1:$B$3"
   SolverSolve (True)
   Range("M1:M3").Select              ' optimum + 1%
   Selection.Copy
   Range("B1").Select
   Selection.PasteSpecial Paste:=xlValues
   SolverOk SetCell:="$B$7", MaxMinVal:=1, ValueOf:="0", ByChange:="$B$1:$B$3"
   SolverSolve (True)
   Range("D24:D44").Select            ' Save biomass values from 1985 to 2005
   Selection.Copy
   ActiveCell.Offset(-10 + i, 15).Range("A1").Select
   Selection.PasteSpecial Paste:=xlValues, Transpose:=True
   Application.CutCopyMode = False
   If (i Mod 10) = 0 Then             ' Keep a check on progress.
     Range("L6").Value = i
     Application.ScreenUpdating = True
     Application.ScreenUpdating = False
   End If
 Next i
Application.ScreenUpdating = True
Beep                                  ' Announce completion
End Sub
```

10.9 CONCLUSIONS

Now that surplus-production models have moved away from their equilibrium-based origins they provide a useful tool in the assessment of stocks for which there is only limited information available. Their simplifying assumption imply that any conclusions drawn from their outputs should be treated with caution. Nevertheless, given the constraints of only considering the total stock biomass, they can provide insights as to the relative performance of the stock through time.

Surplus-production models now have surprising flexibility and can be used in risk assessment and to produce management advice that goes well beyond the old traditional performance indicator notions of MSY and E_{MSY}.

Surplus-production models have now been developed to a point where even if more information is available and more complex and realistic models can be implemented, it would be sensible to implement a simpler model if only to act as a contrast.

APPENDIX 10.1 DERIVATION OF EQUILIBRIUM BASED STOCK-PRODUCTION

The steps between Eq. 10.3 and Eq. 10.9 are as follows

$$B^* = B^* + \frac{r}{p}B^*\left(1-\left(\frac{B^*}{K}\right)^p\right)-C^* \qquad (A10.1)$$

where the * superscript denotes an equilibrium level. The left hand B^* and first B^* on the right hand side can be cancelled and the C^* can be moved across

$$C^* = \frac{r}{p}B^*\left(1-\left(\frac{B^*}{K}\right)^p\right) \qquad (A10.2)$$

substituting $B^* = C^*/qE^*$ (Eq. 10.5), by assuming equilibrium at all times, we obtain

$$C^* = \frac{rC^*}{pqE^*}\left[1-\left(\frac{C^*}{qE^*K}\right)^p\right] \qquad (A10.3)$$

$$pqE^* = \frac{rC^*}{C^*}\left(1-\left(\frac{C^*}{qKE^*}\right)^p\right) \qquad (A10.4)$$

$$\frac{pqE^*}{r} = 1-\left(\frac{C^*}{qKE^*}\right)^p \qquad (A10.5)$$

$$\left(\frac{C^*}{qKE^*}\right)^p = 1 - \frac{pqE^*}{r} \qquad (A10.6)$$

$$\frac{(C^*)^p}{(qK)^p (E^*)^p} = 1 - \frac{pqE^*}{r} \qquad (A10.7)$$

$$\frac{(C^*)^p}{(E^*)^p} = (qK)^p - \frac{pqq^p K^p E^*}{r} \qquad (A10.8)$$

$$\left(\frac{C^*}{E^*}\right)^p = (qK)^p - \frac{pq^{p+1} K^p E^*}{r} \qquad (A10.9)$$

and finally

$$\frac{C^*}{E^*} = \left[(qK)^p - \frac{pq^{p+1} K^p E^*}{r}\right]^{\frac{1}{p}} \qquad (A10.10)$$

If we re-parameterize by defining $(qK)^p$ to be a new parameter a, and the second term: $(pq^{p+1})/r$ to be the new parameter b, this would lead to the form

$$\frac{C^*}{E^*} = \left(a - bE^*\right)^{\frac{1}{p}} \qquad (A10.11)$$

If $p = 1$, Eq. A10.11 collapses to $C/E = (a - bE)$.

APPENDIX 10.2 THE CLOSED FORM OF THE ESTIMATE OF THE CATCHABILITY COEFFICIENT

Version 1: Constant q

Derivation of the statement that we can directly estimate the value of q, which relates to the maximum likelihood fit of the model, by using the geometric average of the time series of q estimates $I_t / \hat{B}_t$

$$\hat{q} = e^{\frac{1}{n}\sum \ln\left(\frac{I_t}{\hat{B}_t}\right)} \qquad (A10.12)$$

By definition we have: $$\hat{I}_t = \hat{q}_t \hat{B}_t \qquad\qquad (A10.13)$$

where $\hat{I}_t$ is the expected CPUE in a given year t, $\hat{q}_t$ is the expected catchability coefficient in year t, and $\hat{B}_t$ is the predicted biomass in year t. However, the assumption is that the catchability coefficient is a constant and each $\hat{q}_t$ is only an estimate of the overall $\hat{q}$. We can either directly estimate this expected catchability coefficient by using non-linear estimation or we can modify Eq. A10.13 to include observed data instead of purely expected values. In this way we can generate the expected catchability coefficient; this is known as a closed form of the equation.

Using observation errors which are lognormal, multiplicative, and with a constant variance, we could fit the model using the sum of squared residuals criterion. The model residuals are related to the observed data in the usual way for log-normal errors

$$I_t = \hat{I}_t e^\varepsilon \qquad \text{or} \qquad \frac{I_t}{e^\varepsilon} = \hat{I}_t \qquad\qquad (A10.14)$$

where I_t is the observed CPUE in a given year t. In order to obtain the closed form of Eq. A10.13, we can substitute the right hand version of Eq. A10.14 into Eq. A10.13 to include the observed CPUE values instead of the expected values

$$\frac{I_t}{e^\varepsilon} = \hat{q}_t \hat{B}_t \qquad\qquad (A10.15)$$

which is equivalent to

$$I_t = \hat{q}_t \hat{B}_t e^\varepsilon \qquad \text{or} \qquad \frac{I_t}{\hat{B}_t} = \hat{q}_t e^\varepsilon \qquad\qquad (A10.16)$$

and log-transforming this gives

$$Ln\left(\frac{I_t}{\hat{B}_t}\right) = Ln\left(\hat{q}_t\right) + \varepsilon \qquad\qquad (A10.17)$$

The value of $\hat{q}$ that minimizes the residuals, ε, in Eq A10.17 (remember because of the normalized error term this would be the same as maximizing the likelihood) would be the value which minimized the residuals of Eq. A10.17 for all the observed values of catch-effort (I_t) and biomass B_t. If there are n observations, then the best estimate of the log of the constant, $\hat{q}$, is simply the mean of the t estimates from the set of observed catch-effort values with associated expected biomass values

$$Ln\left(\hat{q}\right) = \frac{\sum_{t=1}^{n} Ln\left(\hat{q}_t\right)}{n} = \frac{\sum Ln\left(\frac{I_t}{\hat{B}_t}\right)}{n} \qquad (A10.18)$$

To obtain the expected value of q we clearly need to anti-log the outcome of Eq. A10.18 which is, in fact, the geometric mean of the original estimates of q_t

$$\hat{q} = e^{\frac{1}{n}\sum Ln\left(\hat{q}_t\right)} = e^{\frac{1}{n}\sum Ln\left(\frac{I_t}{\hat{B}_t}\right)} \qquad (A10.19)$$

Version 2: Additive Increment to Catchability

In the case where the catchability is assumed to increase by a constant absolute amount each year, the q value for each year, q_t can be determined using a simple linear equation

$$q_t = q_0 + t \times q_{add} \qquad (A10.20)$$

where q_t is the catchability in year t, q_0 is the catchability in the first year for which data is available (time zero), and q_{add} is the constant absolute amount by which the catchability is incremented each year. Estimation of the two parameters involves finding the gradient, q_{add}, and intercept, q_1, of a linear regression between q_t and time t, where t ranges from 0 to n-1 years (a total of n years).

In the model, for each year, the implied estimate of q_t is obtained by dividing each observed catch rate (I_t) by the estimated biomass for that year

$$\hat{q}_t = \frac{I_t}{\hat{B}_t} \qquad (A10.21)$$

In the maximum likelihood fit one would have a time series of expected catchability coefficients, which would be described in the model by Eq. A10.20 or A10.21. Equation A10.20 has the form of a linear regression, so the equations to find the closed form parameter estimates are thus

$$q_{add} = \frac{\sum\limits_{0}^{n-1}\left((t-\bar{t}) \left[\left(\dfrac{I_t}{B_t} \right) - \left(\sum \left(\dfrac{I_t}{B_t} \right) \right) / n \right] \right)}{\sum (t-\bar{t})^2}$$ (A10.22)

and

$$q_0 = \frac{\sum \left(\dfrac{I_t}{B_t} \right)}{n} - (q_{add})\bar{t}$$ (A10.23)

where n is the number of years of data, t-bar is the mean of the t values representing the 0 to n-1 years of data (i.e., with 9 years of data t-bar would equal 4.0, i.e., the mean of 0…8).

By estimating these parameters using the closed form the number of parameters directly estimated by the fitting procedure is reduced, which simplifies the procedure and speeds the process.

Version 3: Constant Proportional Increase - q_{inc}

In the case where the catchability is assumed to increase annually by a fixed proportion then the q value for each year q_t is determined as in exponential growth or compound interest

$$q_t = q_0 \times qinc^t$$ (A10.24)

which, when log-transformed takes the form

$$Ln(q_t) = Ln(q_0) + t \times Ln(q_{inc})$$ (A10.25)

where t ranges from 0 to n-1 years. In the final maximum likelihood fit one would have a time series of expected catchability coefficients which would be described in the model by the equation A10.24 or A10.25 (*cf.* Fig. A10.1). Thus, the estimation of the two parameters involves finding the gradient ($Ln(q_{inc})$) and intercept ($Ln(q_0)$) of a linear regression between $Ln(q_t)$ and time t, where time t ranges from 0 to n-1 years.

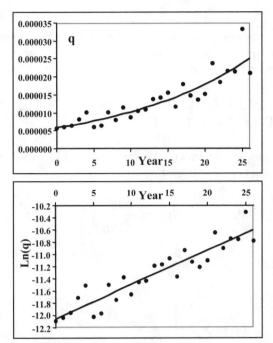

Figure A10.1 The top panel is the final distribution of expected catchability coefficients with the fitted curve of form Eq. A10.24. In order to estimate the two parameters q_0 and q_{inc} instead of a simple geometric average of the expected q values we have to fit the curve. By log-transforming each q_t value and plotting these against the number of times the q_{inc} is applied to the starting value a straight line is obtained as in the lower panel. The straight line is defined by Eq. A10.25, so the two parameters may be determined by anti-logging the two parameters from the linear regression. Data is a bootstrap sample from the northern Australian tiger prawn fishery (see Example Box 10.6).

The closed form equations are thus

$$Ln(q_{inc}) = \frac{\sum_{0}^{n-1}\left((t-\bar{t})\left[Ln\left(\frac{I_t}{B_t}\right) - \left(\sum Ln\left(\frac{I_t}{B_t}\right)\right)/n\right]\right)}{\sum(t-\bar{t})^2} \tag{A10.26}$$

and

$$Ln(q_0) = \frac{\sum_{t=0}^{n-1} Ln\left(\frac{I_t}{B_t}\right)}{n} - Ln(q_{inc})\bar{t} \tag{A10.27}$$

where n is the number of years of data, t-bar is the mean number of years of data. The final parameter estimates are determined by anti-logging the values from Eqs. A10.26 and A10.27.

APPENDIX 10.3 SIMPLIFICATION OF THE MAXIMUM LIKELIHOOD ESTIMATOR

Showing the simplification of the maximum likelihood estimator for lognormal random errors. Given Eq. 10.21

$$L\left(data|B_0,r,K,q\right) = \frac{1}{\sqrt{2\pi}\hat{\sigma}}\prod_t e^{\frac{-\left(Ln\,I_t-Ln\,\hat{I}_t\right)^2}{2\hat{\sigma}^2}} \tag{A10.28}$$

we can convert this to a log-likelihood

$$LL = \sum Ln\left[\frac{1}{\sqrt{2\pi}\hat{\sigma}}e^{\frac{-\left[\left(Ln(I_t)-Ln(\hat{I}_t)\right)^2\right]}{2\hat{\sigma}^2}}\right] \tag{A10.29}$$

Simplifying this by removing constants from the summation and cancelling the Ln and e, we obtain

$$LL = nLn\left(\frac{1}{\sqrt{2\pi}\hat{\sigma}}\right) + \frac{1}{2\hat{\sigma}^2}\sum\left[-\left[\left(Ln(I_t)-Ln(\hat{I}_t)\right)^2\right]\right] \tag{A10.30}$$

where the maximum likelihood estimator of the standard deviation σ is given by

$$\hat{\sigma} = \sqrt{\frac{\sum\left[\left(Ln(I_t)-Ln(\hat{I}_t)\right)^2\right]}{n}} \tag{A10.31}$$

Note the division by n instead of n-1 to give the maximum likelihood estimate (Neter *et al.*, 1996). Given Eq. A10.3, we can simplify Eq. A10.30 much further by substituting one into the other

$$LL = nLn\left(\frac{1}{\sqrt{2\pi}\hat{\sigma}}\right) + \frac{-\sum\left(Ln(I_t) - Ln\left(\hat{I}_t\right)\right)^2}{\dfrac{2\sum\left(Ln(I_t) - Ln\left(\hat{I}_t\right)\right)^2}{n}} \qquad \text{(A10.32)}$$

Both of the terms can be further simplified

$$LL = nLn\left(\left[\sqrt{2\pi}\hat{\sigma}\right]^{-1}\right) + \left(\frac{-1}{2/n}\right) \qquad \text{(A10.33)}$$

which simplifies again to become

$$LL = -nLn\left(\sqrt{2\pi}\hat{\sigma}\right) - \frac{n}{2} \qquad \text{(A10.34)}$$

A little algebra leads to an alternative final version in A10.38

$$LL = -n\left[Ln\left([2\pi]^{\frac{1}{2}}\right) + Ln(\hat{\sigma})\right] - \frac{n}{2} \qquad \text{(A10.35)}$$

$$LL = -n\left[\frac{1}{2}Ln(2\pi) + Ln(\hat{\sigma})\right] - \frac{n}{2} \qquad \text{(A10.36)}$$

$$LL = -\frac{n}{2}\left(Ln(2\pi) + 2Ln(\hat{\sigma})\right) - \frac{n}{2} \qquad \text{(A10.37)}$$

$$LL = -\frac{n}{2}\left(Ln(2\pi) + 2Ln(\hat{\sigma}) + 1\right) \qquad \text{(A10.38)}$$

11

Age-Structured Models

11.1 TYPES OF MODELS

11.1.1 Introduction

Surplus-production models (Chapter 10) ignore sexual, size-, and age-based differences by treating a stock as undifferentiated biomass. Even a superficial consideration of the ecological differences to be found within and between the members of a population suggest that this assumption may be leaving out important influences on the population dynamics. An obvious example would be the time delays present in the dynamics of populations that have a number of years between biological recruitment as juveniles and sexual maturity. In a favourable year, leading to a strong year class, there will be a major increase in stock biomass but it may take a few years before that biomass starts to contribute to reproduction. By lumping growth, reproduction, and mortality into one production function, dynamic interactions between these processes are ignored.

The obvious solution is to differentiate a stock's biomass into component parts. To describe the dynamics of the population we will still need to account for recruitment, growth, and mortality, but these processes will either have to be dealt with separately or be included in however the stock biomass is sub-divided. Thus, we could generate production models of a stock in which the two sexes are differentiated. Similarly, one could generate a length-structured model of stock dynamics, as briefly discussed in the chapter on growth (Section 8.3.5; Sullivan *et al.*, 1990). The most commonly used option, however, is to subdivide the population into age-classes or cohorts and follow the dynamics of each cohort separately, combining them when inputs to the dynamics, such as recruitment, and outputs, such as yield, are being considered. Of course, to utilize age-structured methods it must be possible to age a species accurately, or at least with a known error rate. There are methods for accounting for ageing error but ideally the ageing should be highly accurate and with low bias.

The age-structured models to be discussed in this chapter are relatively complex and have more parts than the models presented in earlier chapters. Indeed, the example boxes will need to be generated in sections to account for the model complexity. In fact, rather than using Excel to work with these models it would usually be more efficient to adopt a programming language and write custom programs to conduct the model fitting. If a modeller wishes to utilize such complex models, then eventually the need for expertise in a programming language becomes essential. Which language is used is not an issue as there are active modellers using Pascal, Fortran, C, C++, Visual Basic, APL, and others. Fortran used to be the language of choice among fishery modellers (the new Fortran 90 and 95 is a nicely versatile language) but C++ is now gaining headway. Personally, I find C++ to be more of a computer programmer's language requiring a higher level of programming skill than that needed in, say, Pascal or Fortran. However, as with all software, the choice is up to particular users and generally all languages permit the necessary speed and complexity.

Despite this need for custom programs we will continue to use Excel in the examples in this chapter. We will only be considering relatively simple age-structured models but even so in one of the models we will attempt to estimate 29 parameters. This chapter only aims to introduce some of the more important ideas behind age-structured models. It does not investigate all of their intricacies and will only attempt to describe a fraction of the range of possibilities. Detailed reviews of age-structured models, such as those by Megrey (1989) and Quinn and Deriso (1999), provide a great deal more information and detail about the complexities and other developments possible with this class of models. To treat these models with the same detail as that given to the simpler models in this book would require a much larger book.

In this chapter we will first generalize the standard catch curve (Beverton and Holt, 1957) into a dynamic system of equations simulating the dynamics of an age-structured population. We will then provide a very brief introduction to cohort analysis (Gulland, 1965, cited in Megrey, 1989; Pope, 1972), followed by an introduction to statistical catch-at-age methods (Doubleday, 1976; Methot, 1989; 1990). In this latter section we will also consider the algorithms necessary to conduct bootstrap estimates of uncertainty around parameter estimates and model outputs.

We have already considered some of the fundamentals of the dynamics of cohorts in the coverage of yield-per-recruit in Chapter 2 and in the Monte Carlo simulation of catch-curves in Chapter 7. Following the dynamics of each cohort has the advantage that after each cohort has recruited, assuming there is no immigration or emigration, the numbers present in the population can only decline. Age-structured models are founded on the basis that a careful examination of this decline can provide information on the total mortality being experienced. If there is an estimate of the natural mortality then, clearly, deductions can be made concerning the exploitation rates of each cohort.

As with surplus-production models we need to have data on the total catch (fishing mortality can include discarding mortality) as weight, but we also need data on the numbers-at-age in the catch. Ideally, for each year of the fishery, there will be an estimate of the relative numbers caught in each age-class (e.g., Table 11.1). Effort information is also required to obtain an optimum fit in all of the models. Beyond this minimum there are many other forms of information that can be included in such stock assessment models. Such information might include estimates of annual recruitment, stock biomass estimates, catch rates, and the mean length and weight of animals in the catch.

Table 11.1 Number of North Sea plaice in each of nine age-classes, landed at Lowestoft per 100 hours fishing by British 1st class steam trawlers from 1929 to 1938. Data from Table 13.1 in Beverton and Holt (1957). The fishing year is April 1st to March 31st. The + symbol after the age indicates that a fish of t+ years is in its t+1th year, somewhere between t and t+1 years old. Note there are relatively strong year-classes, such as the one highlighted, which would have arisen in 1928/1929. Despite being standardized to the same levels of effort the 5+ group in 33/34 is larger than the 4+ group in 32/33. This suggests that the early ages are not fully selected (Fig. 11.1). Effort is in millions of steam trawler hours fishing (from Table 14.15 in Beverton and Holt, 1957).

Year\Age	2+	3+	4+	5+	6+	7+	8+	9+	10+	Effort
29/30	328	2120	2783	1128	370	768	237	112	48	5.81
30/31	223	2246	1938	1620	302	106	181	58	18	5.84
31/32	95	2898	3017	1150	591	116	100	82	33	4.97
32/33	77	606	4385	1186	231	138	42	21	51	4.91
33/34	50	489	1121	4738	456	106	80	27	18	5.19
34/35	44	475	1666	1538	2510	160	50	43	14	4.94
35/36	131	1373	1595	1587	1326	883	144	30	28	4.63
36/37	38	691	2862	1094	864	382	436	27	15	4.44
37/38	138	1293	1804	1810	426	390	163	228	26	4.39

A commonly observed phenomenon in age-structured data is the progression of year classes (Table 11.1; Fig. 11.1). This also provides evidence that the ageing of the animals concerned is at least consistent through time and also that the ring counts used to age the animals actually relate to yearly increments. Note also that the absolute numbers and proportions caught in each year do not necessarily decline through time as one might expect. If the availability of each age-class changes from year to year, for example, for reasons of selectivity, or through different fishing gear being used, or different people doing the fishing, then a simple progression of the cohorts will not be observed. The North Sea plaice data (Table 11.1) are standardized to a constant amount of the same kind of effort so, after the species is fully selected by the gear (age 5+ and older), the numbers always decline from one year and age to the next (Fig. 11.1).

Data on catch and on catch-at-age will provide information regarding the population dynamics. However, for stock assessment purposes, some index of relative abundance through time is required to strengthen the attachment of the model to changes in stock size through time or to provide robust estimates of fishing mortality. A suitable index could use catch rates or the effort imposed to obtain the standardized catches, or even fishery independent survey estimates. In summary, the minimum data requirements are the commercial catches, the catch-at-age as numbers, and effort or some index of relative abundance through time.

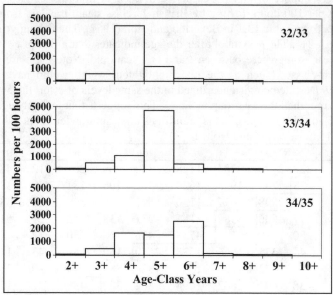

Figure 11.1 Progression of a relatively strong year class (recruited into the population as 0+ in 1928/1929) through the population of North Sea plaice over the three years from 1932/1933 to 1934/1935 (Table 11.1). The fishery is primarily imposed upon three to seven year old fish. Data from Beverton and Holt (1957).

11.1.2 Age-Structured Population Dynamics

As demonstrated in Chapters 2 and 7, if there is no emigration or immigration, then, after recruitment, the numbers in any cohort will decline exponentially through time with a rate of decline equal to the instantaneous total mortality rate

$$N_y = N_0 e^{-Zy} \tag{11.1}$$

where N_y is the numbers in year y, N_0 is the initial recruitment into the cohort, Z is the instantaneous rate of total mortality (M is natural mortality and F is fishing mortality; $Z = M + F$). A log-transformation leads to the familiar

$$\text{Ln}\left(N_y\right) = \text{Ln}\left(N_0\right) - Zy \qquad (11.2)$$

which has the form of a linear relationship and is the basis of the classical catch curve (Fig. 11.2; Beverton and Holt, 1957). There are two kinds of catch curve possible. The first samples all age-classes present in a particular year, assumes the system is in equilibrium, and treats the proportions of the different age-classes as if they were the product of a single cohort (Fig. 11.2). The major assumptions here are that there has been a constant recruitment level in all years, that all ages have been exposed to the same history of fishing mortality (fishing mortality is constant across years), and that after age 4+ all animals are fully selected and have the same catchability. One problem with using this snapshot approach is that standardizing the data collected to a constant amount of effort using a particular gear does not necessarily remove all occasions where an age class is more numerous than the previous age-class (e.g., age groups 8 and 9 in Fig. 11.2). This is a combination of a failure of all or some of the assumptions of this approach to catch curves and also may reflect the impact of sampling error. The assumption of constant recruitment is particularly unlikely, as demonstrated by the strong year classes evident in Table 11.1.

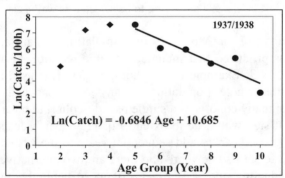

Figure 11.2 Natural logarithm of the average number of North Sea plaice in each age-group caught per 100 hours fishing by a 1st class steam trawler in 1937/1938 (data from Beverton and Holt, 1957). The negative gradient of the slope estimates the total mortality $Z = 0.685$. Age groups 2 to 4 are omitted from the regression because they are not fully selected by the fishing gear. The elevated value in the 9th age group derives from the strong year class that arose in 1928/29 (Table 11.1).

The second kind of catch curve follows the fate of single cohorts (Fig. 11.3). This approach is more difficult because it relies on standardizing the catch of the cohort being followed to a given amount of effort with a particular type of fishing gear. This is necessary because if catchability varied between years, then cohort numbers would not always decline in the steady fashion required by the catch curve methodology. Also, of course, many more years of comparable sampling are required to generate a single catch curve.

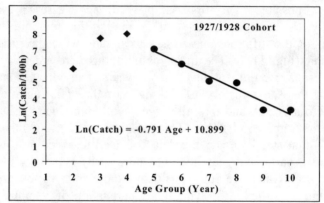

Figure 11.3 A catch curve following the average number of North Sea plaice from a single cohort caught per 100 hours fishing by a 1st class steam trawler over the years 1929 to 1938 (Table 11.1; data from Beverton and Holt, 1957). The negative gradient of the slope estimates the total mortality $Z = 0.791$. Age groups 2 to 4 are omitted from the regression because they are not fully selected by the fishing gear (despite, in this instance, age 4 appearing to be selected as well as later ages).

In Chapter 7, a Monte Carlo simulation of an age-structured population was produced that included variable recruitment and sampling error (Section 7.4.3; Example Boxes 7.6 and 7.7). This used the approach of sampling the whole population in any one year and treating the combination of many cohorts as a single pseudo-cohort to be analysed.

The next step would be to extend Eq. 11.1, which deals solely with ages within a given year, to add sequential years to the system. Obviously the system has to start somewhere and there are two alternatives commonly used when conducting stock assessments of age-structured populations.

The first is to assume the modelling starts at the beginning of exploitation and to ascribe an equilibrium age-structure to the starting population. The second is to directly estimate the starting numbers-at-age in the model (this has the obvious disadvantage of adding the same number of parameters as there are age-classes to the estimation problem). At equilibrium, in an un-exploited population, the age-structure would be the result of natural mortality acting alone upon the average virgin levels of recruitment. Thus, the relative numbers in each age-class in the initial year would be equivalent to the single year catch curve where all of the assumptions have been met

$$N_{a,1} = \begin{cases} R_{0,1} & a = 0 \\ N_{a-1,1}e^{-M} & 1 \le a \le a_{max} - 1 \\ N_{a_{max}-1,1}e^{-M} / (1 - e^{-M}) & a = a_{max} \end{cases} \quad (11.3)$$

where $N_{a,1}$ is the numbers of age a, in year 1, a_{max} is the maximum age modelled (the plus-group), M is, as usual, the instantaneous rate of natural mortality. Recruitment variability has been omitted from Eq. 11.3, but this will be addressed later. In a pre-exploitation population there is no fishing mortality. The final component of Eq. 11.3, where $a = a_{max}$, is referred to as the plus group because it combines ages a_{max} and all older ages that are not modelled explicitly. The inclusion of the $(1 - e^{-M})$ divisor forces the equation to be the sum of an exponential series (Example Box 11.1)

After initial conditions are defined (e.g., Example Box 11.1), the population will grow approximately in accord with some stock recruitment relationship and the mortality imposed naturally and via fishing pressure. In yearly steps, the numbers in each age-class in each year will depend on the numbers surviving from the preceding age-class from the preceding year

$$N_{a+1,y+1} = N_{a,y}e^{-(M+s_a F_y)} \quad (11.4)$$

which is the numbers at age a in year y, multiplied by the survivorship (e^{-Z}) after natural mortality M, and $s_a F_y$, the selectivity for each age a times the fully selected fishing mortality in year y. This is equivalent to the catch curve equation and could be transformed into a linear format. Not all age classes are necessarily fully selected, thus the fishing mortality term must be multiplied by the selectivity associated with the fishing gear for age a, s_a, we use a lower-case s for selectivity to leave the upper-case S for the survivorship term. We also need a term for the recruitment in each new year and this is assumed to be a function of the spawning biomass of the stock in the previous year y, B^s_y (Example Box 11.2), thus

$$N_{a,y+1} = \begin{cases} f(B^s_y) = R_{0,y+1} & a = 0 \\ N_{a-1,y}e^{-(M+s_{a-1}F_y)} & 1 \le a \le a_{max} - 1 \\ N_{a_{max}-1,y}e^{-(M+s_{a-1}F_y)} + N_{a_{max},y}e^{-(M+s_a F_y)} & a = a_{max} \end{cases} \quad (11.5)$$

where the plus group is modelled by applying the total mortality to the preceding age-class and the maximum age-class in the preceding year. These two cannot be combined prior to multiplication by the survivorship term if the selectivities of the two age classes are different (Example Box 11.3). Compare Eq. 11.5 with Eq. 11.3 to see the impact of including fishing mortality.

Example Box 11.1 An equilibrium age-structured simulation model illustrating the initial age-structure (using Eq. 11.3). The selectivities in row 11 are constants but an equation could be used. In B9 put =$E$1*(1-exp(-$E$2*(B12-$E$3))) and in B10 put =$E$4*B9^$E$5. Select B9:B10 and copy across to column K. Copy C13 across to column J, and in K13 put =$B$6*exp(-K12*M)/(1-exp(-M)) to obtain the plus group. Copy C7 across to column K and modify K7 to be =J7*exp(-M)/(1-exp(-M)) to obtain the per-recruit numbers at age. Copy C8 into B8 and into D8:K8 to obtain the biomass per recruit, needed for the estimation of the stock recruitment relationship. To test the plus-group calculation, record the value in K13 (133) and extend the ages across to age 25 in column AA, then copy J13 across to column AA and in K6 put =sum(K13:AA13). This summation should also equal 133. B13 is the first term, C13:J13 is the second, and K13 is the third term from Eq. 11.3 (see Example Boxes 11.2 and 11.3 to fill column O).

	A	B	C	D	E	~	J	K
1	Steepness	0.75		Linf	200			
2	Nat Mort	0.4		K	0.45			
3	Bzero	1000		t0	-0.02			
4	a Recruit	=1/O2		a_Wt	0.0002			
5	b Recruit	=O1/O2		b_Wt	3			
6	R0	=O3		q	0.0005			
7	Rec_Nage	1	=B7*exp(-$B$2)	0.4493	0.3012	~	0.0408	0.0829
8	Rec_Biom	1E-06	=C7*C10/1000	0.153	0.1977	~	0.0601	0.1259
9	Length	1.8		73.6	119.4	148.6	~	194.6	196.5
10	Weight	0.0012		79.793	340.57	656.49	~	1473.5	1518.5
11	Selectivity	0		1	1	1	~	1	1
12	Year\Age	0		1	2	3	~	8	9
13	1967	=B6	=B13*exp(-$B$2)	720	482	~	65	133

The spawning biomass is defined as

$$B_y^s = \sum_{a=c}^{a_{max}} w_a N_{a,y}$$ (11.6)

where w_a is the average weight of an animal of age a, c is the age of sexual maturity, and a_{max} is the maximum age class. The weight can be determined empirically by market measuring or from a growth curve derived independently of any model fitting (Example Box 11.1)

$$w_a = a\left[L_\infty \left(1 - e^{-K(a-a_0)}\right)\right]^b \qquad (11.7)$$

where the constants a and b alter the von Bertalanffy growth curve into a curve relating to mass (see Chapter 8 on Growth). The fully selected fishing mortality rate can be determined if there is an estimate of the catchability coefficient and measures of the relative effort imposed each year

$$\hat{F}_y = \hat{q}E_y e^\varepsilon \qquad (11.8)$$

These estimated fishing mortality rates could be used when fitting the model.

One way of defining the recruitment terms was described in Chapter 9 on recruitment (Section 9.9.3 and Example Boxes 9.7 and 9.8; Example Box 11.2).

11.1.3 Fitting Age-Structured Models

The series of equations represented in Example Boxes 11.1 to 11.3 provide for a relatively simple simulation of an age-structure population starting from equilibrium. The end product is a matrix of numbers-at-age for the population after it has been exposed to natural and fishing mortality. Many of the ideas used to illustrate the simulation of an age-structured population are used when the objective is to fit an age-structured model to observations made on a fishery. As stated before, the minimum data required comprise the relative catch-at-age for a number of years of the fishery, plus some estimate of effort imposed upon the fishery through time. The aim, when fitting the model, will be to attempt to back-calculate a matrix of numbers-at-age that would have given rise to the observed catches, given the imposed fishing effort and estimates of the catchability coefficient and the selectivity coefficients. We will introduce the two main strategies used for conducting such analyses.

Virtual population analysis was the analytical strategy developed first, and this relies on the idea that if one has records of the catch-at-age of a set of cohorts until the cohorts all die off, then one should be able to literally back-calculate what the numbers-at-age must have been. In this way the numbers in the population can be projected backwards until estimates are obtained of the original recruitments. This requires an estimate of the natural mortality and estimates of the fishing mortality by age and year.

Example Box 11.2 The derivation of the recruitment parameters from the growth and mortality rates. This section of the worksheet should be added to that given in Example Box 11.1. The explanations behind these equations and their relationships are given in Section 9.9.3 and Example Boxes 9.7 and 9.8. The alpha and beta in column O are used to calculate the Beverton and Holt stock recruitment parameters in B4:B6.

	M	N	O
1	Alpha	=N4*(1-B1)/(4*B1*N3)	=O4*(1-B1)/(4*B1*O3)
2	Beta	=(5*B1-1)/(4*B1*N3)	=(5*B1-1)/(4*B1*O3)
3	R0	=N4/N5	=O4/O5
4	B0	=N5	=B3
5	A0	=sum(E8:K8)*exp(-B2)	=N6
6	A0/Rec	=N5/B7	

Example Box 11.3 Extension of Example Boxes 11.1 and 11.2. The years extend down to 1979 in row 25. Effort are arbitrary values between 50 and 1200. Row 13 is explained in Example Box 11.1. To generate the spawning biomass from each year, put =sumproduct(C13:K13,$C$10:$K$10)/1000 into L13 and copy down to row 25. Put =$E$6*O14 into N14 and copy down to row 25. To reflect Eq. 11.5, generate the expected recruitment from the spawning biomass, put =$B$4*$L13/($B$5+$L13)*loginv(rand(),0,M14) into B14, then, put =B13*exp(-($B$2+B$11*$N14)) into C14 and copy across to J14. Finally, put =J13*exp(-($B$2+J$11*$N14))+K13*exp(-($B$2+K$11*$N14)) into K14 to generate the plus group. Select B14:K14 and copy down to row 25 to complete the matrix of numbers at age. The sig_r introduces random variation into the stock recruitment relationship, to remove its effects make the values in column M very small. If you set all the effort values to zero only natural mortality will occur and the population will stay in equilibrium except for the effects of recruitment variability on the lower half of the matrix. If you set sig_r to 0.000001 and have a constant effort, then the whole population will attain an equilibrium. By plotting each year's numbers as a series of histograms vertically above each other (as in Fig. 11.1), relatively strong year classes should be visible through time (press F9).

	A	B	C	D	~	J	K	L	M	N	O
12	Yr\Age	0	1	2	~	8	9	Bs	sig_r	F	Effort
13	1967	=B6	1074	720	~	65	133	1822.6			
14	1968	2118	1074	720	~	40	81	1235.6	0.5	0.5	1000
15	1969	1914	1420	720	~	25	51	1014.7	0.5	0.45	900
16	1970	2380	1283	952	~	13	25	847.0	0.5	0.7	1400
17	1971	2078	1595	860	~	11	22	1080.8	0.5	0.15	300
18	1972	2516	1393	1069	~	10	19	1315.8	0.5	0.125	250

The data requirements for a VPA are fairly stringent in that there can be no years of missing information if the calculations are to continue uninterrupted. VPA models are sometimes referred to as cohort analysis, although Megrey (1989) indicates some differences. The number of parameters estimated equals the number of data points available so neither approach is fitted to fisheries data using an objective function such as by minimizing a sum of squared residuals or a negative log-likelihood. Instead, solutions to the model equations are determined through iterative, analytical methods. There are many alternative VPA fisheries models (Megrey, 1989) but we will only consider one of the simplest.

   The second analytical strategy appears to be the reverse of the first and goes under a number of names. Methot (1989) termed this approach the Synthetic analysis, but others have referred to this analytical strategy as Integrated Analysis (Punt *et al.*, 2001). Given equations describing the population dynamics, such that if we know $S_{a,y}$, the survivorship of age a to age $a+1$ from year y to $y+1$, and that $N_{a+1,y+1} = S_{a,y}N_{a,y}$, then, given knowledge of the initial population in year 1 and of the recruitments in each subsequent year, the numbers-at-age for the full population can be calculated. This approach requires the model to be fitted to data using some form of minimization routine. The parameter estimates include the initial population age-structure, the recruitment levels in each year, the selectivities by age-class, and often, other parameters as well. It is common to estimate tens of parameters and models exist that estimate hundreds of parameters. Of course, the number of parameters that can be estimated efficiently in any model will be at least partly determined by the number of independent data points available. The point is, however, that although Integrated Analyses share some equations with Cohort Analysis the methods differ fundamentally from each other (Fig. 11.4). A knowledge of both methodologies is helpful in fisheries stock assessment.

11.2 COHORT ANALYSIS

11.2.1 Introduction

Given that we are dealing with cohorts, the change in the numbers in each cohort can be derived each year from the numbers at age a at the start of year y, $N_{a,y}$, and the survivorship during that year, as in Eq. 11.4 (where M and F are the instantaneous rates of natural and fishing mortality, and s_a is the age-specific selectivity of the fishing gear used)

$$N_{a+1,y+1} = N_{a,y}e^{-(M+s_aF_y)} = N_{a,y}e^{-M}e^{-s_aF_y} \qquad (11.9)$$

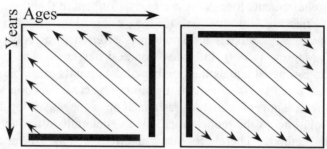

Figure 11.4 Alternative analytical strategies adopted in efforts to model commercial fisheries using age-structured information. The general objective is to estimate a vector of numbers-at-age or fishing mortalities-at-age for each year of the fishery. The left-hand panel relates to Cohort Analysis/VPA while the right-hand panel reflects Integrated Analysis or Statistical Catch-at-Age. In Cohort Analysis, the calculations proceed given knowledge of all ages in the last year and the last age-class in all years, and projecting backwards through time and ages. In Statistical Catch-at-Age, knowledge is assumed of all ages in the first year and the first age-class in all years (recruitment), projecting the cohorts forward through time and ages.

Unfortunately, the number of fish present in early years is unknown so forward projections cannot be made. However, there will be an age at which the number of fish remaining in the cohort is effectively zero. We can rearrange Eq. 11.9 to start at a known number of fish in the oldest age-class and project the population backwards until we reach the age of recruitment (which need not be age 0). Thus, if we have an estimate of natural mortality and an age-class in which the number of survivors is trivially small, then the fishing mortality for each age in each year can be estimated by back-calculating from the catches, C_y, and the natural mortality; this implies that there are as many parameter estimates as there are data points. Given

$$N_{a,y} - C_{a,y} = N_{a,y}e^{-s_aF_y} \tag{11.10}$$

rearrange Eq. 11.9 to give

$$N_{a,y} = \frac{N_{a+1,y+1}}{e^{-M}} + C_{a,y} \tag{11.11}$$

While Eq. 11.11 provides a mechanism for calculating the relative numbers at age in each year it is limited to complete cohorts (those that reach the maximum age-class considered (Fig. 11.5).

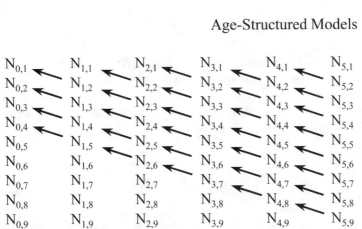

Figure 11.5 A table of numbers-at-age in each year showing the backward progression of cohorts from their last age-class. The columns represent age-classes (first subscript), while the rows represent nine years of data (second subscript). The arrows represent the direction of calculation using Eq. 11.11. Only completed cohorts, which have attained an age at which negligible animals remain, can be back-projected validly. This means that we are most uncertain about the most recent cohorts, the ones of most current interest.

One of the ways in which the many variants of VPA/Cohort Analysis differ is in how they address this problem of incomplete cohorts (Megrey, 1989). Given a_{MAX} age-classes, in the last year of data there will be $a_{MAX} -1$ incomplete year classes. Some way of estimating the fishing mortality rate in these age-classes in the final year of data is required to complete the table of estimates of numbers-at-age in each year. These fishing mortality rates are referred to descriptively as terminal-F estimates.

The data available include the catch-at-age and effort data. There is more than one way to fit a Cohort Analysis model (referred to by Megrey, 1989, as Sequential Population Analysis, SPA) to this data. We can either generate a fit directly to the fishing mortality rates for each age in each year, or fit to the numbers-at-age in each year. It is an odd fact that some of the seminal papers in fisheries modelling are published in obscure places (e.g., Gulland, 1965, cited in Megrey, 1989). Fortunately, alternative listings of these developments are available (Gulland, 1983; Megrey, 1989; Hilborn and Walters, 1992).

11.2.2. The Equations

The basic equation relating the numbers in a cohort in one year to those in the previous year is

$$N_{y+1} = N_y e^{-(M+F_y)} \tag{11.12}$$

where N_y is the population size in year y, and M and F are the instantaneous rates of natural and fishing mortality, respectively. This implies that $e^{-(M+F)}$ is the survivorship, the proportion of a population that survives from year to year. The survivorship itself can be determined simply as the ratio of numbers from one year to those in the year previous (starting from the last age-class whose numbers are known from the catch data)

$$\frac{N_{y+1}}{N_y} = e^{-(M+F_y)} = e^{-M}e^{-F_y} \tag{11.13}$$

The complement of survivorship would be the total loss from year to year and this can be represented a number of ways

$$N_y - N_{y+1} = N_y - N_y e^{-(M+F_y)} = N_y \left(1 - e^{-(M+F_y)}\right) \tag{11.14}$$

The catch each year would be the proportion of this total loss due to fishing

$$C_y = \frac{F_y}{M+F_y} N_y \left(1 - e^{-Z_y}\right) = \frac{F_y}{M+F_y}\left[N_y - N_{y+1}\right] \tag{11.15}$$

which is the total loss by the proportion of mortality due to fishing. It would be possible to remove explicit mention of F_y from Eq. 11.15 if we solved Eq. 11.13 for F_y

$$\frac{1}{e^{-F_y}} = \frac{N_y}{N_{y+1}} e^{-M} \tag{11.16}$$

remembering that $1/e^{-F}_y = e^F_y$, we can log-transform Eq. 11.16 to give F_y

$$F_y = \mathrm{Ln}\left(\frac{N_y}{N_{y+1}}\right) - M \tag{11.17}$$

This puts us into a position of being able to estimate the catch-at-age from knowledge of the catches, the final N_{y+1}, and a given value for M, and these estimates can be used to fit the model to the data

$$C_y = \frac{\mathrm{Ln}\left(N_y / N_{y+1}\right) - M}{\mathrm{Ln}\left(N_y / N_{y+1}\right) - M + M}\left[N_y - N_{y+1}\right] \tag{11.18}$$

Eq. 11.18 simplifies to

$$C_y = \left(1 - \frac{M}{\text{Ln}\left(N_y / N_{y+1}\right)}\right)\left[N_y - N_{y+1}\right] \qquad (11.19)$$

or

$$0 = \left(1 - \frac{M}{\text{Ln}\left(N_y / N_{y+1}\right)}\right)\left[N_y - N_{y+1}\right] - C_y \qquad (11.20)$$

There is no direct analytical solution to Eq. 11.20 but there are two ways in which the equation may be solved to produce a matrix of numbers-at-age that would balance Eq. 11.20.

11.2.3 Pope's and MacCall's Approximate Solutions

An approximate solution to Eq. 11.20 can be used to give the required matrix of numbers-at-age. Pope (1972) produced the following approximation (note that the exponents are positive)

$$N_y = N_{y+1}e^M + C_y e^{M/2} \qquad (11.21)$$

which derives from Eq. 11.12

$$\frac{N_{y+1}}{e^{-F}e^{-M}} = N_y = N_{y+1}e^M e^F \qquad (11.22)$$

Pope's (1972) advance was to introduce the discrete approximation of using the addition of $C_y e^{M/2}$ to be equivalent in effect to the multiplication by e^F. Being discrete, changes the multiplication in Eq. 11.22, into an addition but assumes that all fishing occurs instantaneously in the middle of the year (hence the $e^{M/2}$ to account for natural mortality acting before the fishery operated). Pope (1972) showed that his approximation was useable with values of M up to 0.3 and F of 1.2 over the time periods used in the model. Thus, if M or F are greater than these limits, cohort analysis can still be used if the catch-at-age data can be divided into shorter intervals than one year. MacCall (1986) provided an alternative approximate solution that was an improvement over Pope's (1972) equation

$$N_y = N_{y+1}e^M + C_y\left(\frac{M}{1 - e^{-M}}\right) \qquad (11.23)$$

MacCall's equation behaves rather better at higher values of M and is also less sensitive to the assumption that all the catch is taken halfway through the fishing year (Example Box 11.4).

It is no longer necessary to use the approximations by Pope (1972) and MacCall (1986), but they can still be used as the starting point for the second approach that can be used to solve Eq. 11.20.

11.2.4 Newton's Method

Classical analytical methods can be used to find values of N_y and N_{y+1} for each cohort that will solve Eq. 11.20 so that it approximates to zero. Newton's method provides a simple and powerful method for finding the roots of a function of the form $f(N_y) = 0$ (Jeffrey, 1969). This is an iterative method in which the solution is modified by the ratio of the function and the first differential of the function

$$N_y^{Updated} = N_y^{Orig} - \frac{f\left(N_y^{Orig}\right)}{f'\left(N_y^{Orig}\right)} \qquad (11.24)$$

Starting from the first iteration, N_y^{Orig} would be the individual elements of the numbers-at-age matrix derived from one of the approximations listed earlier. The modifier is made up of $f(N_y^{Orig})$, which would be Eq. 11.20 for each particular age and year being considered. This means that $f'(N_y^{Orig})$ is the differential of Eq. 11.20 with respect to N_y

$$f'\left(N_y^{Orig}\right) = 1 - \frac{M}{\text{Ln}\left(N_y / N_{y+1}\right)} + \frac{\left(1 - N_{y+1} / N_y\right)M}{\left(\text{Ln}\left(N_y / N_{y+1}\right)\right)^2} \qquad (11.25)$$

There are no N_{y+1} when the N_y are in the last age-class are being considered (e.g., in the North Sea plaice example there is no 11+ age-class) so Eq. 11.25 would fail if applied without modification. Thus, when N_{y+1} is zero one can use

$$f\left(N_y^{Orig}\right) = \left(1 - \frac{M}{\text{Ln}\left(N_y\right)}\right)N_y - C_y = 0 \qquad (11.26)$$

and

$$f'\left(N_y^{Orig}\right) = 1 - \frac{M}{\text{Ln}\left(N_y\right)} + \frac{M}{\left(\text{Ln}\left(N_y\right)\right)^2} \qquad (11.27)$$

Example Box 11.4 Two alternative approximate calculations of the matrix of numbers-at-age (Eq. 11.21 and Eq. 11.23) by Pope (1972) and MacCall (1986). This worksheet will become much larger than the number of columns and rows that can be fitted onto a single page so care is needed in its construction. Extend the column labels in row 5 across to column J with 4+, 5+,...., 10+. Fill in the catch-at-age data from Table 11.1. Copy the column and row labels across into L5:U5 and L6:L14 respectively. Name cell J17 as M, the natural mortality rate. In U6 put =V7*exp(M)+J6*exp(M/2) and copy it down to U14 to obtain Pope's approximation. Copy U6:U14 back across to column M to create the matrix of values. Select L5:U14 and copy the contents, select A19 and paste as values (Edit/Paste Special/Values; <Alt>ESV). Fishing mortalities and other statistics may be calculated from this matrix. This new matrix, denoting an approximation to the numbers-at-age, in B20:J28, will contribute to the extension of this worksheet in the Example Box 11.5, as will I15:J16. In A18 put the text "Numbers-at-Age." To compare the results with those obtained from MacCall's (1986) approximation, create another table with row labels in W5:W14 and column labels in X5:AF5. Column V must be left clear or else the calculations in column U will receive interference (trace the precedents of U6). In AF6 put =AG7*exp(M)+J6*(M/(1-exp(-M))) and copy down to AF14. Copy AF6:AF14 across to column W to generate the matrix. Compare this with the results from Pope's (1972) approximation. Notice that the values from MacCall's method tend to be slightly smaller than those from Pope's method. The absolute values are of less interest than the relative proportions by age, and while these still differ the differences are less marked. Consider the values in row 14 from column M to T. They also have no preceding data (no N_{y+1}) and yet are not the end of a cohort. Is this important?

	A	B	C	~	J	~	L	M	~	R	S	T	U
4	Catch-at-Age						Approximation to Numbers-at-Age						
5	Year\Age	2+	3+	~	10+	~	Y/A	2+	~	7+	8+	9+	10+
6	29/30	328	2120	~	48	~	29/30	15283.6	~	1409.2	414.6	153.1	54.4
7	30/31	223	2246	~	18	~	30/31	38952.0	~	348.0	419.7	113.7	20.4
8	31/32	95	2898	~	33	~	31/32	13196.1	~	276.6	177.5	167.1	37.4
9	32/33	77	606	~	51	~	32/33	11922.0	~	420.3	113.0	50.0	57.8
10	33/34	50	489	~	18	~	33/34	7664.2	~	284.9	205.5	51.0	20.4
11	34/35	44	475	~	14	~	34/35	11736.4	~	503.6	128.3	89.5	15.9
12	35/36	131	1373	~	28	~	35/36	4524.2	~	2060.9	251.0	55.8	31.7
13	36/37	38	691	~	15	~	36/37	1924.4	~	670.0	825.8	68.4	17.0
14	37/38	138	1293	~	26	~	37/38	156.4	~	441.9	184.7	258.4	29.5
15		Use Terminal F											
16			Limit										
17			M	0.25									

Obviously, if N_y is zero, there can be no older animals in that cohort so one should return a zero.

When Eq. 11.24 is completed to include both Eqs 11.20 and 11.25, along with the option of Eqs. 11.26 and 11.27, the result looks dauntingly long and complex. Fortunately, the elements are simple and with care it is easily implemented, even in Excel (see Example Box 11.5).

In operation, the initial guess at the numbers-at-age matrix (from one of the approximations) is updated using the function to be solved and its differential, as in Eq. 11.24. Then the original values are replaced by the updated, which leads to a new set of updated values. This iterative process is repeated until no perceptible difference between the original and the updated values is observed. Usually, a stopping rule needs to be defined that stops the iterations once some threshold similarity has been reached (see Example Box 11.5).

Example Box 11.5 Gulland's cohort analysis focused on numbers-at-age. This is an extension to the worksheet generated in Example Box 11.4. It assumes that the catch-at-age data are in B6:J14 and that Pope's approximation to the numbers-at-age, as values, are in B20:J28. The matrix described below describes the modifier in Eq. 11.24 i.e., it is the $f(N_y)/f'(N_y)$. The Excel equation looks terrific and great care is needed to put in the correct number of brackets. In U20 put =IF(J20>0,IF(K21>0, (((((1-(M/Ln(J20/K21)))*(J20-K21))-J6)/(1-M/Ln(J20/K21)) + (((1-K21/J20)*M) / (Ln(J20/K21)^2)))), (((1-(M/Ln(J20))) * (J20)-J6) / (1-(M/Ln(J20)) + (M/(Ln(J20)^2))))),0). If you have the data entered correctly and the Pope's approximation is correct, the value in U20 should be as below. If this is the case, then copy U20 down to U28 and copy U20:U28 across to column M. Remember that cell J17 is named M. The equation inside the If statement has two options, first is Eq. 11.20 divided by Eq. 11.25, second it is Eq. 11.26 divided by Eq. 11.27. The first option is for when both N_y and N_{y+1} are available while the second option is for when no N_{y+1} are available. Thus, option two will be used in column U and in row 28. Copy the column labels down into L31:U31 and the row labels from L20:L28 into L32:L40. In U32 put =J20-U20 and copy down to U40 and copy U23:U40 across to column M. This completes Eq. 11.24 and provides an updated version of the numbers-at-age. Notice that if the modifier matrix is negative in any cell this will increase the corresponding cell of the numbers-at-age matrix. Copy M32:U40 and paste as values over the top of the original numbers-at-age matrix in B20:J28. This updates the modifier which updates the derived numbers-at-age. If this copying as values onto the original numbers-at-age matrix is repeated enough times the cells in the modifier matrix will tend to zero. In J16, put =sum(M20:U28) to enable a watch on the sum of the modifier matrix. Repeat the copying of the updated M32:U40 onto the original B20:J28 until the J16 limit is very small. At this point the numbers-at-age matrix should have stabilized to a solution. Does it appear to be closer to Pope or MacCall's approximation?

Example Box 11.5 [cont.] Once the spreadsheet is constructed and working it can be improved by automating the copying and pasting required in the Newton's method iterations. Buttons can be added that set up the original numbers-at-age matrix as a set of values from either Pope's or MacCall's approximations. A final button can be added that automates the copying of the updated numbers-at-age matrix as values onto the original numbers-at-age. A map of the worksheet developed in Example Box 11.5 should appear to be very similar to Fig. 11.6.

	L	M	N	O	P	Q	R	S	T	U
19	Y\Age	2+	3+	4+	5+	6+	7+	8+	9+	10+
20	29/30	1.16	22.12	43.02	25.91	6.68	18.87	6.24	4.57	3.14
21	30/31	0.69	19.53	32.07	33.47	5.54	1.35	3.30	1.30	0.75
22	31/32	0.30	16.35	54.47	28.28	12.55	2.05	2.58	1.75	1.91
23	32/33	0.24	2.77	40.53	20.89	3.54	1.89	0.65	0.37	3.39
24	33/34	0.16	2.15	8.22	75.47	7.48	1.64	1.30	0.64	0.75
25	34/35	0.13	2.46	17.58	20.62	49.09	2.12	0.81	0.89	0.46
26	35/36	0.49	10.04	25.31	28.06	33.97	16.00	3.81	0.72	1.51
27	36/37	0.13	6.23	60.53	30.06	21.99	10.01	10.25	0.45	0.53
28	37/38	11.08	125.6	178.2	178.9	38.50	34.99	13.37	19.45	1.36
~										
31	Y\Age	2+	3+	4+	5+	6+	7+	8+	9+	10+
32	29/30	15282	8677	7461	2148	859	1390	408	=I20-T20	=J20-U20
33	30/31	38951	11594	4872	3355	692	347	416	=I21-T21	=J21-U21
34	31/32	13196	30123	7008	2081	1197	275	175	I22-T22	=J22-U22
35	32/33	11922	10191	20874	2817	624	418	112	50	54

11.2.5 Terminal F Estimates

In the last age-class the assumption is that the cohort will all die out. It is this termination of the cohort in the fishery that permits the backwards calculation of the numbers-at-age from the catch data and the natural mortality estimate. However, in the last year of data not all age-classes will represent completed cohorts. If there are A year classes then there will always be an A-1 x A-1 triangle of incomplete cohorts arrayed along the bottom left of the numbers-at-age matrix. If we are to calculate their relative abundance in a valid manner we must find a way to estimate the numbers-at-age in the unknown lower-left triangle of incomplete cohorts. Unfortunately, these incomplete cohorts are of most interest because they will affect future stock numbers, i.e., we know least about the year classes we need to know the most about.

Direct survey estimates of numbers or fishing mortality-at-age, perhaps using tagging, are possible but rarely made. The most common approach is to produce independent estimates of the fishing mortality experienced by the cohorts being fished. Given a value of F_y and the catch-at-age in the final year the numbers-at-age for the incomplete cohorts may be estimated using a rearrangement of Eq. 11.15

$$N_y = \frac{C_y}{1 - e^{-Z_y}} \frac{F_y + M}{F_y} \qquad (11.28)$$

the independent estimates of F_y are known as terminal F estimates, and used with Eq. 11.28 can give rise to improved estimates of the numbers-at-age in the bottom row of the numbers-at-age matrix.

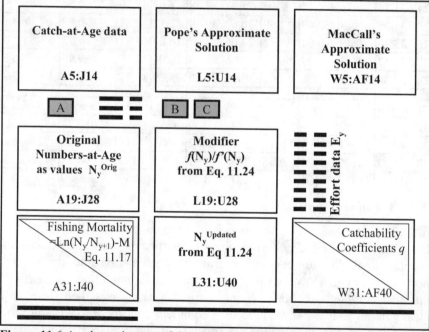

Figure 11.6 A schematic map of the worksheet developed in Example Boxes 11.4 and 11.5. Each block contains a matrix and its column and row labels. The contents are identified along with the cell positions on the sheet. The three grey objects are buttons attached to macros. The three sets of lines are text labels and numbers, including natural mortality, M. The rectangles with un-bolded text will be added later. The triangles imply that these are only upper-triangular matrices. The lines below MacCall's solution represent the effort data. The lines below the lower boxes are summary data such as average age-specific fishing mortality (see later). The effort data is used to calculate the terminal-F values.

The terminal F estimates may be obtained through some survey method or from the standard equation $F_y = q_a E_y$. The catchability coefficients for each age class a are commonly obtained by calculating the fishing mortality rate for complete cohorts, and then, using effort data in the years in which they were fished, the catchability by age can be determined from $q_a = F_y/E_y$. Given q_a and effort in the final year the terminal numbers-at-age may be determined from Eq. 11.28 (Example Box 11.6).

Example Box 11.6 Calculation of the instantaneous fishing mortality rates for complete cohorts. This extends Example Boxes 11.4 and 11.5. Using Eq. 11.17, and checking for zeros put =if(T32>0,if(U33>0,Ln(T32/U33)-M,"---"),"----") into I32. Copy this down to row 39, then copy I32:I39 across to column B. Delete the sub-diagonal elements. In B41 put =average(B32:B39) and copy across to column I to provide the estimates of average fishing mortality rate on each age-class. One might be tempted to use the average fishing mortality in Eq. 11.28 but this would ignore variations in fishing effort in each year. Instead, in B42 put =X41*$X$28 to estimate F_a as $q_y E_y$ and copy across to column I. Note that these only go across to where the incomplete cohorts apply. The effort data in the final year (X28) and average catchability for each age-class are described below. Estimates of the catchability coefficient by age-class and year. These rely on the effort data being placed in X20:X28 (from Table 11.1; it is best to label the column as effort in X19 and put year labels in W20:W28). The column labels, below, extend across to AF31 but that column does not produce numbers and is suppressed here for brevity.

	A	B	C	D	E	F	G	H	I	J
31	Year\Age	2+	3+	4+	5+	6+	7+	8+	9+	10+
32	29/30	0.025	0.325	0.547	0.885	0.664	0.964	1.050	1.766	---
33	30/31		0.249	0.596	0.781	0.676	0.428	0.681	0.881	---
34	31/32			0.662	0.958	0.807	0.647	1.020	0.835	---
35	32/33				0.643	0.541	0.470	0.551	0.657	---
36	33/34					0.590	0.550	0.590	0.922	---
37	34/35						0.452	0.587	0.803	---
38	35/36							1.055	0.945	---
39	36/37								0.607	---
40	37/38	---	---	---	---	---	---	---	---	---
41	Average F	0.025	0.287	0.602	0.817	0.656	0.585	0.791	0.927	
42	Terminal F	0.019	0.217	0.482	0.669	0.541	0.485	0.674	0.789	

Example Box 11.6 [cont.] In AE32 put =if(isnumber(I32),I32/$X20,"----") and copy down to row 40. Copy AE32:AE40 across to column X (and into AF for completeness). In X41 put =average(X32:X40) and copy across to column AE to generate the average catchability coefficient for each age-class. These get used, along with the last year's effort in X28 to generate the age-specific fishing mortalities for the incomplete cohorts (across in B42:I42). In X42 put =(B14/(1-exp(-(M+B42))))*((B42+M)/B42), which is Eq. 11.28, and copy across to column AE to generate the terminal N_y values. These are then ready for pasting into the left-hand side of the bottom row of the original numbers-at-age matrix. The algorithm is now to paste as values the approximation into the original matrix, paste as values the updated numbers-at-age over the original matrix, and then copy X42:AE42 and paste as values into B28:H28. This will produce a new updated numbers-at-age matrix along with a new estimate of the terminal F_y and N_y. This procedure is iterated until the limit of precision selected has been reached. When using the terminal F estimates the limit is found by summing the modifier matrix except the bottom row (*i.e.*, =sum(M20:U27) instead of =sum(M20:U28) in J16. To ensure no errors it is best to write a few short macros to do the copying and pasting of values. As in Fig. 11.6, create two buttons with attached macros, one of which copies and pastes Pope's approximation and the other MacCall's approximation into B20:J28. The third button should conduct the copy/pasting relating to the Newton's method iterations. If J15 is given an integer value greater than zero it would be possible to turn the use of the terminal F values on and off using a macro similar to that printed below. In this way it would be possible to easily compare the results obtained from the different starting points. How different are the answers if the terminal F values are used? Do the years with complete cohorts change or is it just the numbers-at-age for the incomplete cohorts?

	W	X	Y	Z	AA	AB	AC	AD	AE
30	Catchability Coefficient								
31	Y\Age	2+	3+	4+	5+	6+	7+	8+	9+
32	29/30	0.0043	0.0560	0.0942	0.1523	0.1142	0.1660	0.1807	0.3039
33	30/31	----	0.0427	0.1021	0.1337	0.1157	0.0733	0.1166	0.1508
34	31/32	----	----	0.1333	0.1928	0.1625	0.1301	0.2052	0.1680
35	32/33	----	----	----	0.1309	0.1103	0.0957	0.1122	0.1337
36	33/34	----	----	----	----	0.1137	0.1060	0.1136	0.1776
37	34/35	----	----	----	----	----	0.0914	0.1189	0.1625
38	35/36	----	----	----	----	----	----	0.2280	0.2041
39	36/37	----	----	----	----	----	----	----	0.1367
40	37/38	----	----	----	----	----	----	----	----
41	Average	0.0043	0.0494	0.1098	0.1524	0.1233	0.1104	0.1536	0.1797
42	Term N	8363	7468	5277	4135	1139	1136	370	465
43	Min N	370.4							

Example Box 11.6 [cont.] VBA macro to perform the Newton's method iterations with or without terminal F_y estimates leading to terminal N_y estimates for the incomplete cohorts. By placing 1 in J15 the terminal F estimates are made. If J15 is zero then the original approximations are used.

```
Sub do_vpa()
Dim termf As Integer
  termf = Range("J15").Value
  If termf > 0 Then
    Range("J16").Select  ' Alter the limit summation appropriately
    ActiveCell.FormulaR1C1 = "=SUM(R[4]C[3]:R[11]C[11])"
  Else
    Range("J16").Select
    ActiveCell.FormulaR1C1 = "=SUM(R[4]C[3]:R[12]C[11])"
  End If
  Range("M32:U40").Select   ' copy and paste the updated
  Selection.Copy            ' numbers-at-age
  Range("B20").Select
  Selection.PasteSpecial Paste:=xlValues
  Range("B19").Select
  If termf > 0 Then               ' If the terminal F option is selected then
    Range("X42:AE42").Select  ' Copy and paste terminal N values if the
    Selection.Copy
    Range("B28").Select
    Selection.PasteSpecial Paste:=xlValues
    Range("B12").Select
  End If
  Application.CutCopyMode = False
End Sub
```

11.2.6 Potential Problems with Cohort Analysis

The name "Virtual Population Analysis" or "Cohort Analysis" refers to a class of models each of which relies upon knowledge of final fishing mortality rates and the back-calculation of numbers-at-age in the fished population. The method is clearly sensitive to the estimates of the terminal F values that permit the analysis to be extended to the incomplete cohorts. If these estimates are flawed the analysis will be biased. Other sources of potential bias include the presence of ageing error in the determination of the catch-at-age. This is especially a problem if recruitment is highly variable. A 10% ageing error will not affect a small year class too badly but a large year-class could inflate the apparent numbers of smaller year-classes around it. Thus, ageing errors will tend to obscure recruitment variability (Richards *et al.*, 1992). A further problem relates to the idea of following particular cohorts. If there is significant immigration to a region then the numbers in a cohort may increase through time and thus lead to an over-estimate of the cohort size. This effect would be greatest if older fish are the ones that migrate the most. Finally, errors in the natural mortality estimate can have relatively complex impacts on the estimation of fishing mortality (Mertz and Myers, 1997).

11.2.7 Concluding Remarks on Cohort Analysis

Only a very introductory treatment of Cohort Analysis is presented here. Virtual Population Analysis is the preferred method of stock assessment in the European Community and other parts of the Atlantic. Not surprisingly there is an enormous literature, both gray and formal, dealing with the various techniques, ways to improve their performance, and how to "tune" the VPA to information from the fishery. The aim of this chapter was only to introduce the reader to the methodology so that a more detailed investigation of the primary literature would be more understandable. Megrey (1989) and Quinn and Deriso (1999) provide excellent reviews of the recent developments in the techniques used.

Without using an optimization technique (such as minimizing the sum of squared residuals) model selection and determination of uncertainty levels along with model projections for risk assessments present their own challenges. The alternative analytical strategy of Statistical Catch-at-Age or Catch-at-Age with Ancillary information, or Integrated Analysis, permits these options in a very straightforward manner.

11.3 STATISTICAL CATCH-AT-AGE

11.3.1 Introduction

Statistical Catch-at-Age will be referred to here as Integrated Analysis (Punt *et al.*, 2001). Unlike VPA, Integrated Analysis estimates fewer parameters than the available number of data points, although one can still be estimating tens of parameters. An objective function (least squares or maximum likelihood) is used to optimize the fit of the model to the available data. It requires catch-at-age data along with some information to tie the model to the stock size (either catch rates, or effort, or independent population estimates). The numbers-at-age at the start of the first year in the population being modelled are model parameters along with recruitment levels in each year of the fishery. With further parameters describing age-specific selectivity, it is possible to project each cohort forward to generate a matrix of numbers-at-age. From this it is possible to generate a matrix of predicted catch-at-age which can be compared with the observed data and the fit optimized. Catch-at-age data alone is usually insufficient to tie the model to reality so a further connection, either through effort or catch rates, is necessary.

11.3.2 The Equations

The equations behind Integrated Analysis are remarkably simple, the complex part is organizing the information and calculations. The fully selected fishing mortality rate in year y, F_y, is one of the foundations of the analysis and values for each year are treated as model parameters in the fitting process. The fishing mortality rate for each age, a, in each year y, $F_{a,y}$, is

$$F_{a,y} = s_a \hat{F}_y \tag{11.29}$$

where $\hat{F}_y$-hat is the fitted fishing mortality rate in year y and s_a is the selectivity of age a (Fig. 11.7). The fishing mortalities are combined with the natural mortality M, to generate the age- and year-specific survivorships, which are used to complete the matrix of numbers-at-age

$$N_{a+1,y+1} = N_{a,y}e^{-\left(M+s_a\hat{F}_y\right)} = N_{a,y}e^{-M}e^{-s_a\hat{F}_y} \tag{11.30}$$

where, as before $N_{a,y}$ is the numbers of age a in year y (Fig. 11.8).

Selectivity can be estimated either directly for each age or the parameters of an equation describing the shape of the selectivity curve can be estimated. A logistic equation is often used to describe selectivity. Given a as age, a_{50} as the age at which selectivity is 50%, a_{95} as the age at which selectivity is 95%, and s_a is the selectivity at age a (Fig. 11.7), then age-specific selectivities are

$$s_a = \frac{1}{1+e^{-\text{Ln}(19)\frac{(a-a_{50})}{(a_{95}-a_{50})}}} \tag{11.31}$$

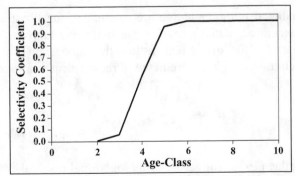

Figure 11.7 Selectivity curve generated by Eq. 11.31. It is not smooth because the different ages have been treated as integers. By varying the parameters a_{50} and a_{95} the steepness of the curve and its location along the age axis can be altered.

Fitting the model involves estimating the $N_{a,y}$ for all ages in year 1, and for the first age-class in all subsequent years (Fig.11.8).

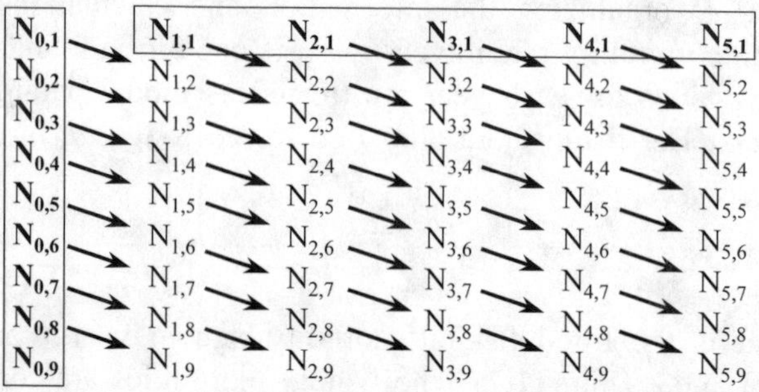

$N_{0,1}$	$N_{1,1}$	$N_{2,1}$	$N_{3,1}$	$N_{4,1}$	$N_{5,1}$
$N_{0,2}$	$N_{1,2}$	$N_{2,2}$	$N_{3,2}$	$N_{4,2}$	$N_{5,2}$
$N_{0,3}$	$N_{1,3}$	$N_{2,3}$	$N_{3,3}$	$N_{4,3}$	$N_{5,3}$
$N_{0,4}$	$N_{1,4}$	$N_{2,4}$	$N_{3,4}$	$N_{4,4}$	$N_{5,4}$
$N_{0,5}$	$N_{1,5}$	$N_{2,5}$	$N_{3,5}$	$N_{4,5}$	$N_{5,5}$
$N_{0,6}$	$N_{1,6}$	$N_{2,6}$	$N_{3,6}$	$N_{4,6}$	$N_{5,6}$
$N_{0,7}$	$N_{1,7}$	$N_{2,7}$	$N_{3,7}$	$N_{4,7}$	$N_{5,7}$
$N_{0,8}$	$N_{1,8}$	$N_{2,8}$	$N_{3,8}$	$N_{4,8}$	$N_{5,8}$
$N_{0,9}$	$N_{1,9}$	$N_{2,9}$	$N_{3,9}$	$N_{4,9}$	$N_{5,9}$

Figure 11.8 In Integrated Analysis the initial population age-structure and each year's recruitment (boxed off in the diagram) are estimated parameters. The survivorship of each age-class in each year is calculated ($e^{-(M+sF)}$) and these survivorships are used to complete the numbers-at-age matrix using Eq. 11.30.

Despite estimating a large number of parameters the calculations for an Integrated Analysis can still be conducted successfully on an Excel worksheet. We will construct an example of an Integrated Analysis using the data from Beverton and Holt (1957) that was used in the Cohort Analysis. Once again this will be a complex worksheet and care is needed in its construction. The map in Fig. 11.8 indicates the broad structure.

The algorithm begins by calculating the age-specific fishing mortality for each year from the selectivity equation parameters and the fishing mortality parameters (Eq. 11.29; Fig. 11.9). These are combined with the natural mortality rate to generate the age-specific survivorships, which are used, in turn, to complete the numbers-at-age matrix (Eq. 11.30).

11.3.3 Fitting to Catch-at-Age Data

Once the predicted numbers-at-age are calculated, the predicted catch-at-age can be generated, which provides the first opportunity to generate an objective function for use when fitting the model by comparing the observed catch-at-age with the predicted. Predicted catch-at-age is

$$\hat{C}_{a,y} = \frac{F_{a,y}}{M + F_{a,y}} N_{a,y} \left(1 - e^{-(M+F_{a,y})}\right) \tag{11.32}$$

Doubleday (1976) suggested using lognormal residual errors

$$SSR_C = \sum_a \sum_y \left(Ln\left(C_{a,y} / \hat{C}_{a,y} \right) \right)^2 = \sum_a \sum_y \left(Ln C_{a,y} - Ln\hat{C}_{a,y} \right)^2 \qquad (11.33)$$

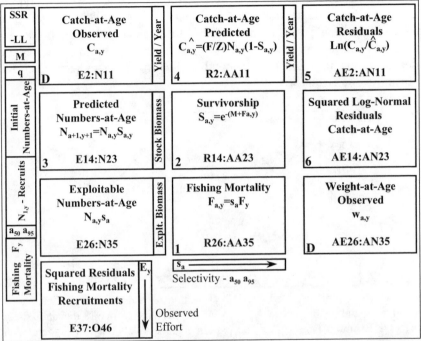

Figure 11.9 Schematic map of an Excel worksheet illustrating a possible layout for an Integrated Analysis. Each of the nine main boxes would have their upper edges labeled with ages and left-hand edges labeled with years. The small, upper, left-most box is where the minimizations occur. The model parameters are listed down the left-hand side of the worksheet below the natural mortality *M*. Each of the estimated parameters is natural-log transformed, which scales all parameters to similar sizes and makes all parameter changes proportional changes. In the worksheet they are back-transformed as appropriate. The Catch-at-Age residuals in the top-right are there for ease of plotting as a diagnostic relating to the quality of fit (Fig. 11.11). The heavy numerals in the lower-left of some of the boxes relates to the order of calculation in the algorithm. The capital D, in the same place, implies these are data matrices. See Example Boxes 11.7 and 11.8.

With the Beverton and Holt (1957) North Sea plaice data, an optimum fit gives rise to a relatively even spread of residuals (according to Eq. 11.33; Figs. 11.10 and 11.11; Example Box 11.7).

Doubleday (1976) was able to show that if one only fits the stock-assessment model using catch-at-age data, as with Eq. 11.33, then correlations between some of the parameters estimates could be so extreme that some of the parameter estimates can be effectively linear combinations of others. While this may be acceptable for estimating changes in the relative abundance, at least over short periods, catch-at-age data alone is insufficient to estimate absolute abundance. It is for this reason that Integrated Analysis is sometimes referred to as catch-at-age analysis with auxiliary data (Deriso *et. al.*, 1985; 1989). A common addition is to include observed effort into the model, combined with an extra parameter the catchability coefficient.

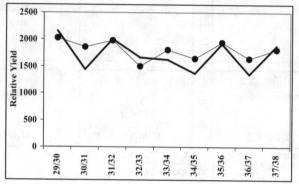

Figure 11.10 A plot of observed yield (fine line with dots) and predicted yield (thick line) against fishing year (April to March) deriving from a fit on catch-at-age alone (see Example Box 11.7). The pattern matches well but is exaggerated in 30/31, and is less pronounced in 33/34.

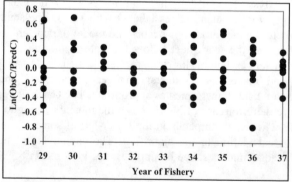

Figure 11.11 Catch-at-Age residuals against fishing year from a fit on catch-at-age alone (see Example Box 11.7). The overall fit is good with a regression through these residuals is essentially flat (Residual = 0.0156 − 0.00047 Year).

11.3.4 Fitting to Fully Selected Fishing Mortality

Including the observed fishing effort (E_y) adds a further nine data points, which offsets the fact that an estimate of the catchability coefficient (q) must be made. The two are combined to generate a semi-observed fishing mortality (F_y), which is compared with the fully-selected fishing mortality parameters (Example Box 11.8; Fig. 11.12)

$$F_y = \hat{q}E_y \qquad (11.34)$$

and thus

$$SSR_E = \sum_y \left[\operatorname{Ln}\left(\hat{F}_y\right) - \left(\operatorname{Ln}\left(\hat{q}\right) + \operatorname{Ln}\left(E_y\right)\right) \right]^2 \qquad (11.35)$$

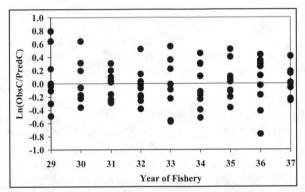

Figure 11.12 A plot of the residuals between the log of the observed catch-at-age and the log of the expected catch-at-age, from the North Sea plaice data (Beverton and Holt, 1957), fitted to catch-at-age and fishing mortality (Example Boxes 11.7 and 11.8). A regression line fitted to these data, as shown by the fine line) is essentially flat along the zero line (Residual=0.00389 - 0.00012Year). Note the differences between this residual plot and that shown in Fig. 11.11.

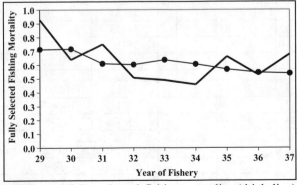

Figure 11.13   Estimated fully selected fishing mortality (thick line) vs. observed fishing mortality rates (as in Eq. 11.35).

Example Box 11.7 An Integrated Analysis of the North Sea plaice data (Beverton and Holt, 1957; see Table 11.1 and earlier example boxes). In preparation for later developments put =sum(AF15:AN23) in B2 and =sum(K38:K46) into B3. In B5 put =C2*B2+C3*B3 to get the total weighted sum of squared residuals. The log of the initial numbers-at-age estimates are given in B9:B16 (e.g., Ln_N_29_10 is the log of numbers-at-age 10 in 1929), the log-recruitments in each of the 9 years are given in B17:B25 (e.g., Ln_R_30_2 is the log-recruitment into age 2+ in year 1930). The logged parameters of the selectivity equation are given in B26:B27. Finally, the log-fishing mortality rates are given in rows 28 down to 36. Initiate all parameters (the logs of the first row and column of Pope's approximation will fill B9:B25, see Appendix 11.1), and then enter the catch-at-age data (from Table 11.1) in F3:N11 along with appropriate column and row labels (as in E3:E10 and E2:N2). Equivalent weight-at-age data should be entered into AF27:AN35 (with labels in row 26 and column AE); the weight data are in Appendix 11.1. In R37 put the label "Age", in S37:AA37 put the numbers 2 to 10, and in S38 put =1/(1+exp(-Ln(19)*(S37-exp($B$26))/(exp($B$27)-exp($B$26)))), which is Eq. 11.31. Copy S38 across to column AA to calculate selectivity (label as "Selectivity" in R38). In R25 put the label "Fishing Mortality by Age by Year" and label ages in S26:AA26 and years in R27:R35 as before. To calculate these fishing mortalities put =S$38*exp($B28) into S27, being sure to get the $ in the correct places. Copy S27 down to row 35, then copy S27:S35 across to column AA to generate the required matrix. Put =exp(-(S27+$B$6)) in S15 and copy down to row 23, then copy S15:S23 across to column AA to generate the survivorships by age and year. Again label for rows and columns appropriately.

	A	B	C	D	E	F	G	H
1	Source	Value	Wt		Catch At Age – Observed			
2	SSR$_C$	7.6363	1		Year\Age	2+	3+	4+
3	SSR$_E$	0.3789	1		29/30	328	2120	2783
4					30/31	223	2246	1938
5	SSR$_T$	8.0152			31/32	95	2898	3017
6	Natural M	0.2			32/33	77	606	4385
7	Parameter	Ln(value)			33/34	50	489	1121
8	Ln_q	-2.0981			34/35	44	475	1666
9	Ln_N_29_10	4.4570			35/36	131	1373	1595
10	Ln_N_29_9	5.0806			36/37	38	691	2862
~	~	~			~	~	~	~
17	Ln_R_29_2	9.3607			31/32	10807	25748	6614
18	Ln_R_30_2	10.3659			32/33	10684	8747	17685
19	Ln_R_31_2	9.2879			33/34	7478	8680	6359
~	~	~			~	~	~	~
25	Ln_R_37_2	9.5997			Exploitable Numbers-at-Age			
26	Ln_Sel50	1.2232			Year\Age	2+	3+	4+
27	Ln_Sel95	1.4784			29/30	177.6	1468.7	6398.4
28	Ln_F_29	-0.0739			30/31	485.2	2195.2	3545.5
29	Ln_F_30	-0.4475			31/32	165.1	6024.9	5671.0

Example Box 11.7 [cont.] From the worksheet map (Fig. 11.9), you should now have in place the parameters, both D (data) matrices and the fishing mortality (1) and survivorship (2) matrices. The predicted numbers-at-age matrix is constructed in two parts. In F15 put =exp(B17) and copy down to F23 to obtain the estimated recruitments. In G15 put =exp(B16), in H15 put = exp(B15), and similarly across to N15 with =exp(B9), to list the initial numbers-at-age. The matrix can then be completed by putting =M15*Z15 into N16 and copying down to N23, and then copying N16:N23 across to column G. Finally, to obtain the total biomass estimates the numbers in each age-class must be multiplied by their respective average weight and summed. Put =sumproduct(F15:N15,AF27:AN27)/1000 into O15 and copy down to O23. The division by 1000 is to keep the numbers manageable. The variation in recruitment from year to year is obvious, as are the strong year classes passing through the fishery. The stock biomass can be used as a diagnostic to indicate annual trends in relative abundance. By putting =sumproduct(F3:N3,$AF27:$AN27)/1000 into O3 and copying down to O11, an equivalent column can be added to the observed catch-at-age matrix; these values would be observed annual yield. From the matrix of numbers-at-age we calculate the predicted catch-at-age by putting =(S27/(S27+$B$6))*F15*S15 into S3 and copying down to S11 and across to column AA. This is Eq. 11.32 with S15 being the survivorship calculated previously. You should label the rows and columns appropriately. Copy O3:O11 and paste into AB3:AB11 to obtain the predicted annual yield. The final two matrices are made by putting =Ln(F3/S3) into AF3, copying down to AF11 and across to column AN. These are just the residuals and can be plotted against the year of the fishery to act as a diagnostic relating to how well the fitting process is proceeding (*cf.* Fig. 11.10). To obtain the squared residuals put =AF3^2 into AF15 and copy across to AN15 and down to row 23. This has the effect of completing the sum of squared residuals for the catch-at-age held in B2. With 8 in B9:B25, 1.25 in B26, 1.75 in B27, 0 in B28 and –0.4 in B29:B36 the SSR_C is about 174.08. Use the solver to minimize B2 by changing B9:B36 (28 parameters). If it does not converge to about 7.266 try a different starting point (*cf.* Figs. 11.10 and 11.11). Values here are for both SSR_C and SSR_E.

	E	F	G	H	I	J	K	L	M	N	O
14	Yr\Age	2+	3+	4+	5+	6+	7+	8+	9+	10+	Biomass
15	29/30	11623	6277	7462	2274	1087	1539	441	161	86	6721
16	30/31	31755	9382	4136	2755	741	352	498	143	52	9048
17	31/32	10805	25746	6614	1957	1197	320	152	215	62	8404
18	32/33	10685	8746	17684	2845	761	463	124	59	83	7483
19	33/34	7478	8680	6359	9370	1408	375	228	61	29	7034
20	34/35	11660	6076	6336	3419	4717	706	188	114	31	6967
21	35/36	9599	9480	4469	3501	1776	2442	366	97	59	6957
22	36/37	7889	7780	6648	2075	1488	751	1032	154	41	5777
23	37/38	14765	6407	5629	3460	1006	718	362	498	75	7093

Example Box 11.8 The inclusion of observed effort data (in O38:O46) into the work-sheet developed in Example Box 11.7. Copy J38:N38 down to row 46. Column N is Eq. 11.34, column J is the inner part of Eq. 11.35, deriving from the parameter estimate of F_y and the semi-observed in column M. Column K is simply the residual squared, which completes the sum of squared residuals in B3. Columns L and N are for plotting as a visual diagnostic of the quality of fit (Fig. 11.13), though one could equally well plot column J. If the worksheet has been solved for an optimum fit to catch-at-age and Ln_q = -2, then B3, the sum of squared residuals against Effort should be about 4.549. If both contributions to the total sum of squared residuals in B5 are given equal weight (*i.e.* 1 in C2:C3) then one can minimize B5 using the solver to modify cells B8:B36 (29 parameters). Given equal weight, the optimum balance between the two sources of squared residuals appears to be where $SSR_C = 7.936$ and $SSR_E = 0.378$. The parameters that give rise to this result are given in Appendix 11.1. It may be necessary to run the solver more than once, or from different starting points (the results from a VPA, or one of the approximations, would be a reasonable starting point – try the values obtained from Example Box 11.6. How robust are the answers when each SSR is given equal weighting? How sensitive are the results to the relative weighting of the two sets of squared residuals? The values below are for the optimum fit with equal weights.

	I	J	K	L	M	N	O
37	Yr	F_res	SSq	Pred_F	ObsLnF	ObsF	Effort
38	29	=B28-M38	=J38^2	=exp(B28)	=Ln(N38)	=exp($B$8)*O38	5.81
39	30	=B29-M39	=J39^2	=exp(B29)	=Ln(N39)	=exp($B$8)*O39	5.84
40	31	0.2080	0.0432	0.7506	-0.4949	0.6096	4.97
41	32	-0.1712	0.0293	0.5075	-0.5070	0.6023	4.91
42	33	-0.2610	0.0681	0.4904	-0.4516	0.6366	5.19
43	34	-0.2787	0.0777	0.4586	-0.5009	0.6060	4.94
44	35	0.1522	0.0232	0.6613	-0.5658	0.5679	4.63
45	36	-0.0304	0.0009	0.5283	-0.6077	0.5446	4.44
46	37	0.2305	0.0531	0.6781	-0.6190	0.5385	4.39

11.3.5 Adding a Stock-Recruitment Relationship

There is very little information relating to the last few cohorts, with the extreme being the single catch data point for the very latest recruits. Fitting the model to the catch-at-age and estimates of fully-selected fishing mortality could generate relatively uncertain estimates of the status of the affected years and age-classes. One suggested solution (Fournier and Archibald, 1982) is to impose a stock-recruitment relationship (see Chapter 9), to add extra constraints to the last few years of recruitment. This will also be necessary if risk assessment projections are to be made. Spawning stock size may be defined as the mature biomass, or the stock size times the relative fecundity at age, or some other available measure. In addition one can use a variety of stock-recruitment relationships (e.g., Beverton and Holt or Ricker). In general, as discussed in Chapter 9, the residual errors used would tend to be log-normal. Of course, if the age at recruitment is not 0+ then the implied time lag between spawning stock size and subsequent recruitment must be accounted for

$$\hat{N}_{r,y+r} = \frac{\alpha B_y^S}{\beta + B_y^S} e^{\varepsilon}$$
(11.36)

where r is the age at recruitment, B_y^S is the spawning stock size in year y, and $y+r$ is the year plus the time lag before the recruits join the fishery. The parameters α and β are from the Beverton and Holt stock recruitment relationship. Whether the predicted number of recruits in year $y+r$ are derived from a Beverton and Holt, a Ricker, or any other stock recruitment relationship does not affect the form of the squared residuals

$$SSR_R = \left(\mathrm{Ln} N_{r,y+r} - \mathrm{Ln} \hat{N}_{r,y+r} \right)^2$$
(11.37)

which compares the "observed" model parameters for recruitment in the rth year onwards (i.e the fitted recruitments in B17:B25) with those expected from the stock-recruitment relationship (Example Box 11.9; Fig. 11.14).

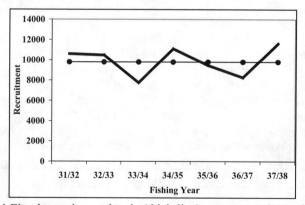

Figure 11.14 Fitted recruitment levels (thick line) vs. expected recruitment levels from a Beverton-Holt stock-recruitment relationship. The stock recruitment relationship predicts effectively constant recruitment.

Example Box 11.9 Implementation of a stock recruitment relationship into the worksheet developed in Example Boxes 11.7 and 11.8. Copy F40:G40 down to row 46. In A37 put "Alpha" and A38 "Beta" adding the logged initial value ready for fitting the model ($\alpha = 9$ and $\beta = -7$, are reasonable beginnings for the Beverton and Holt relationship, in B37:B38). If a Ricker relationship is preferred then put =exp($B$37)*O15*exp(-exp($B$38)*O15) into F40 and copy down to row 46. In this case starting values would need to be something like ($Ln(\alpha) = 1.6$ and $Ln(\beta) = -8.6$). Plot the recruitment parameters against those predicted from the stock recruitment relationship. Which appears best, the Beverton and Holt or the Ricker? F38:F39 are empty because of the lag of r years. There are now 31 parameters for the solver, B8:B38.

	E	F	G
37	**Year\Age**	**Beverton and Holt**	**Sum Squared Residual**
38	29/30		
39	30/31		
40	31/32	=exp($B$37)*O15/(exp($B$38)+O15)	=(B19-Ln(F40))^2
41	32/33	=exp($B$37)*O16/(exp($B$38)+O16)	=(B20-Ln(F41))^2
42	33/34	9625	0.059
43	34/35	9625	0.027
44	35/36	9625	0.002
45	36/37	9625	0.003
46	37/38	9625	0.038

Example Box 11.9 [cont.] The sum of squared residuals for all the additions to the Integrated Analysis. SSR_C relates to fitting the model to the catch-at-age data, SSR_E relates to fitting the model to the fully selected fishing mortality estimates, and SSR_R relates to the inclusion of the stock-recruitment relationship to the model fitting. The total sum of squared residuals SSR_T is the sum of each of the separate contributions after each is weighted according to the predefined weightings. As with all weighted least squares methods, ideally the weighting should relate to the variability in the estimates of the statistics involved; unfortunately, this is often unknown. Experiment with different relative weightings to see the impact upon the final model fit. If it is desired to turn off a particular component simply give it a weighting of zero. Does the impact of adding the stock-recruitment relationship differ depending upon whether a Beverton and Holt or a Ricker relationship is used?

	A	B	C
1	**Source**	**Value**	**Weighting**
2	SSR_C	=sum(AF15:AN23)	1
3	SSR_E	=sum(K38:K46)	1
4	SSR_R	=sum(G40:G46)	1
5	SSR_T	=C2*B2+C3*B3+C4*B4	
6	**Natural M**	0.2	

11.3.6 Other Auxiliary Data and Different Criteria of Fit

We have considered fitting the catch-at-age stock assessment model through using catch-at-age data, using relative effort (fishing mortality) data, and adding a stock recruitment relationship. Other possible sources that could be added include fishery independent surveys of stock size and fishery dependent catch/effort rates. If a series of fishery independent surveys is available then either they should derive from a standardized design or some measure of the relative efficiency of each survey (a relative catchability-coefficient) would be required. If commercial catch rates are to be used then, ideally, these should be standardized to remove noise unrelated to changes in stock size (Kimura, 1988).

With catch rates, log-normal residuals tend to be used. The expected catch rates derive from the simple relation

$$\hat{I}_y = qB_y^E = q\sum_a w_a s_a N_{a,y} \tag{11.38}$$

where q is the catchability coefficient, B^E_y is the exploitable biomass in year y, w_a is the average weight of fish of age a, s_a is the age-specific selectivity and $N_{a,y}$ is the numbers-at-age a in year y. The sum of squared residuals for this potential component of the total would be (Example Box 11.10)

$$SSR_I = \sum_y \left(\mathrm{Ln}\left(I_y\right) - \mathrm{Ln}\left(\hat{I}_y\right) \right)^2 \tag{11.39}$$

Example Box 11.10 Exploitable biomass and expected catch-rates. In E25 put the label "Exploitable Numbers-at-Age" and then label the columns as age-classes in F26:N26 and the rows as years in E27:E35. Put =F15*S$38 into F27 and copy down to row 35 and across to column N to generate the required matrix. Label O26 as "Biomass." Put =sumproduct(F27:N27,AF27:AN27)/1000 into O27 and copy down to row 35 to generate the exploitable biomass in each year of the fishery. Put =O27*exp($B$8)/1000 into P27, and copy down to row 35 to generate the expected catch rates (as per Eq. 11.38). Then a column of the squared residuals (as per Eq. 11.39) can be placed somewhere convenient on the worksheet and the weighted sum can be added to the total sum of squared residuals in B5.

A similar arrangement can be used for fishery independent surveys, although a separate q estimate would be required for the survey estimates of exploitable biomass.

While we have been consistently using the sum of squared residuals as the criterion of model fit we could equally well have used maximum likelihood methods and their extensions into Bayesian methods. Using the same weightings and the log-normal residuals we would expect to obtain essentially the same answers but, if any of the contributions fit very closely, the log-likelihoods for that component could go negative which will distort the impact of that component and a different result may occur. If the model is fitted to only one component, then the same parameters will be found as with the least squares method. The relative weightings can have a more marked effect with log-likelihoods than with least squares methods.

The multinomial distribution has been suggested as an alternative likelihood function for the fitting using the catch-at-age data (Deriso *et. al.*, 1985; 1989), especially where measurement errors are primarily due to ageing errors and sampling error in the catch sampling

$$\text{LL} = \sum_y \sum_a n_{a,y} \text{Ln} \left(\hat{p}_{a,y} \right) \qquad (11.40)$$

where LL is the multinomial log-likelihood, $n_{a,y}$ is the number of fish of age a, aged in year y, and $\hat{p}_{a,y}$ is the expected proportion of animals aged a in year y. Equation 11.40 does not actually involve the amount of catch so it has been suggested that it might be more stable to add a penalty term along the lines of $\Sigma(C'_y - C_y)^2$, weighted so that the expected catches (C'_y) are close to the observed (Quinn and Deriso, 1999). An alternative approach has been suggested by Schnute and Richards (1995) that does not require this extra term. If the use of multinomial likelihoods is relatively unstable it is also an option to solve the model initially using log-normal likelihoods, and then, when near the optimum fit, begin using multi-nomial likelihoods.

11.3.7 Relative Weight to Different Contributions

Irrespective of whether maximum likelihood methods, Bayesian methods, or sum of squared residuals are used as the criterion of optimum fit, when there are a number of categorically different sources contributing to the overall likelihood or sum of squares then each contribution will receive a relative weighting. No explicit weighting implies an equal weighting of one for each component of the total. With weighted least squares, and maximum penalized likelihood, it is usual to weight each component in relation to the degree of uncertainty associated with the data or statistic used. Thus, if some estimate of the variance or coefficient of variation for each data source is available, then the inverse of these would be used to ascribe relative weights to each component; the greater the variability the less the relative weight.

If each of the sources of information is consistent with the others, then fitting the model should not prove difficult. For example, in the Example Box 11.8, where a stock-recruitment relationship was added to the model, the impact of the Beverton and Holt relationship was minimal. At least it could be said not to be contradictory to the other data sources.

Problems can arise when the separate contributions to the overall criterion of fit are inconsistent or contradictory (Richards, 1991; Schnute and Hilborn, 1993). Under the schema described above if one had contradictory data sets, then the final outcome of the analyses would be largely influenced by the relative weights ascribed to the various components of the fit. The result will reflect the weighted average of the different conclusions deriving from the different components. As a minimum, under these circumstances, it is a good idea to conduct a series of sensitivity trials with different sets of weightings to determine the implications of whether one source of data is more reliable than the others (Fig. 11.15; Richards, 1991).

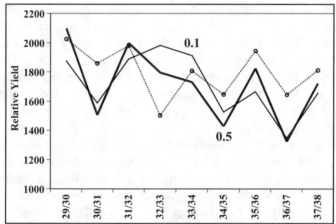

Figure 11.15 The impact of different relative weights (1.0, 0.5, and 0.1) being ascribed to the sum of squares contribution from the catch-at-age (SSR_C) when the model is fitted only to catch-at-age and effort data (*i.e.*, only the optimum stock-recruitment relationship included). The effort (fishing mortality) data is given a default weight of 1.0. The fine dotted line is the observed catch data, the fine line is where a weighting of 0.1 is used, and the thicker line where a weighting of 0.5 is used. A weighting of 1.0 produces a line almost coincident with the 0.5 line. The heavier weighting improves the fit in the first, fourth, seventh, and ninth year.

Schnute and Hilborn (1993) go somewhat further and suggest that, as with robust statistics (Huber, 1981), where individual data points that are really non-informative outliers may still have a non-zero likelihood, whole data sets may be non-informative in an analogous fashion. In effect, they recommend a more formal, structured investigation of the impacts on the quality of the overall fit of giving different emphases to the different contradictory data sets.

The issue of what weightings to use is one that will not go away and must be treated explicitly in any formal assessment. If a definite selection of relative weights is made instead of conducting a sensitivity analysis, then these weights require a formal justification. Further examples of the impacts of these weightings should be published. Further work along the lines suggested by Schnute and Hilborn (1993) may also prove helpful.

11.3.8 Characterization of Uncertainty

As with all other stock assessment analyses, one obtains a specific optimum model fit that will have a variety of management implications. From the Integrated Analysis one can obtain estimates of stock biomass, fully selected fishing mortality and average fishing mortality, along with other, possibly related, performance indicators. Although the analyses generate specific values for all of these parameters and outputs, a further step is required to obtain an indication of the level of uncertainty associated with each estimate. Depending upon how the minimization is conducted, some software will provide asymptotic standard errors around each parameter estimate. However, these standard errors rely on linear statistical theory and in all cases will be symmetrical (unless results are transformed). Confidence intervals derived from such standard errors are recognized as being only approximate.

A better way of obtaining approximate confidence intervals or of generating likelihood profiles around model estimates would be to use either Bayesian methods or bootstrapping. As described in Chapter 3 on parameter estimation, Bayesian methods retain the original data and describe uncertainty by determining how well different combinations of parameters fit the available data. Bootstrapping techniques recognize that the available data is only a sample of what was possible. By generating bootstrap samples from the data, re-fitting the model, and collating the resulting sets of parameter estimates it is possible to generate percentile confidence intervals. In this chapter we will only be considering bootstrap methods. See Punt and Hilborn (1997) for a description of Bayesian methods as applied to Integrated Analysis stock assessment models.

With catch-at-age data there are many serial correlations between age-classes and between years. When generating the required bootstrap samples it is best to resample the residuals from the optimum model fit and combine them with the expected catch-at-age data to form the bootstrap catch-at-age sample. Thus, the bootstrap samples (C^b) would be

$$C_{a,y}^b = \hat{C}_{a,y} \left(\frac{C_{a,y}}{\hat{C}_{a,y}} \right)^{boot} \tag{11.41}$$

where C-hat is the expected catch-at-age and the residual is a randomly selected residual from those available (Fig. 11.16).

Example Box 11.11 A macro for conducting bootstraps on the catch-at-age worksheet (see Example Boxes 11.7 to 11.10 and Fig. 11.16). Put =F3/S3 into AR15, copy down and across to AZ23. Select AR15:AZ23, copy and save as values onto itself. To create the bootstrap samples (see Fig. 11.16) Put =AR3*offset($AQ$14,trunc(rand()*9)+1,trunc(rand()*9)+1) into AR39 and copy down to row 47 and across to column AZ. Paste a copy of the original data in AR27:AZ35. Put =O15 into B39 and copy down to row 47, to make storing the model results slightly easier. Copy the optimum parameter values and outputs B8:B47, and Paste Special/Transpose into BE7:CR7. This will be used as the starting point for fitting each bootstrap sample to the model. The results will be pasted underneath these optimum values by the macro. It took just over 3.25 hours to conduct 1000 bootstraps on a 1.0Ghz Pentium III computer (Figs. 11.17, 11.18), when solving for all 31 parameters. Time how long it takes to conduct 10 bootstraps on your own machine before setting it off on a marathon.

```
Sub Do_Boot()
Dim i As Integer
Application.ScreenUpdating = False
 For i = 1 To 1000
   Range("AR39:AZ47").Select   ' replace original data with a bootstrap sample
   Selection.Copy
   Range("F3").Select
   Selection.PasteSpecial Paste:=xlValues, Transpose:=False
   Range("BE7:CI7").Select          ' Paste in Optimal parameters as best guess.
   Selection.Copy
   Range("B8").Select
   Selection.PasteSpecial Paste:=xlValues, Transpose:=True
   Application.CutCopyMode = False
   SolverOk SetCell:="$B$5", MaxMinVal:=2, ValueOf:="0",ByChange:="$B$8:$B$36"
   SolverSolve (True)               ' Run the solver twice to have more confidence
   SolverOk SetCell:="$B$5", MaxMinVal:=2, ValueOf:="0",ByChange:="$B$8:$B$36"
   SolverSolve (True)               ' in the solution.
   Range("B8:B47").Select
   Selection.Copy
   Range("BE8").Select
   ActiveCell.Offset(i, 0).Range("A1").Select   ' The use of i in the offset is important.
   Selection.PasteSpecial Paste:=xlValues, Transpose:=True
 Next i
 ActiveWorkbook.Save      ' just in case
 Application.ScreenUpdating = True
End Sub
```

The bootstrapping highlights the level of uncertainty in the various parts of the analysis and permits the investigator to make stronger statements about the management implications of the results. In this instance it also confirms the suspicions (Fig. 11.14) that the estimate of the fully selected fishing mortality in the fishing year 1929/1930 is biased (Fig. 11.16). The ability of the bootstrapping procedure to identify bias in the parameter estimates is a real advantage, although, as discussed in Chapter 3, once bias is detected it is difficult to know what to do about it.

In this case, it is only the first year that is exhibiting bias and this suggests a closer consideration of the first year's data. Perhaps the effort data used is more error prone than imagined. However the case may be, awareness of the problem is the first step to a better understanding of our perception of the fishery. Once again the stock assessment model synthesizes a wide range of information and provides a more convenient and defensible statement about the status of the stock. Standard diagnostic tests, such as plotting residuals and other visual indicators (Richards *et al.*, 1997) should always be made on the analysis results.

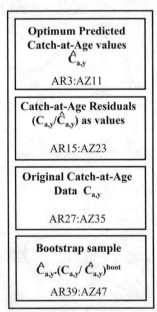

Figure 11.15 Schematic map of a worksheet arrangement for conducting bootstraps on the catch-at-age data and model. The cell ranges depict where the actual values go; it is assumed that the rows and columns will be labeled appropriately. The top three matrices are filled with values, only the bottom matrix contains equations. A macro is used to run the bootstrap (Example Box 11.11).

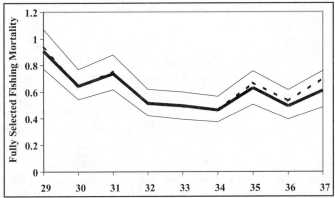

Figure 11.16 Bootstrap 95% percentile confidence intervals around the fully selected fishing mortality through the years of the fishery. The heavy line is the bootstrap average while the dotted line is the optimum fit. There is little evidence of bias except in the last three years where the optimum fit appears to be biased high (5.9% in 35, 6.9% in 36, and 12.8% in 37). That the most bias occurs in the final year is not surprising because that year has the least information available.

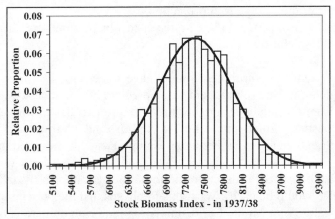

Figure 11.17 Bootstrap 95% percentile confidence intervals around the stock biomass estimate for the 1937/1938 fishing year. The optimum Index was 7,356t and the fitted normal curve has a mean of 7,352, so there is little bias.

11.3.9 Model Projections and Risk Assessment

As with surplus-production models, characterization of the uncertainty in an assessment model is only the first step. Ideally, it should be possible to decide on a management strategy (be it a certain catch level or fishing mortality level), impose that on the model fishery, and project the population forward in time to determine the consequences of different strategies. This would be the basis behind a formal risk assessment

(Francis, 1992). With the Integrated Analysis model this can be implemented relatively easily.

If a particular fully selected fishing mortality rate has been selected as the management strategy this might also entail imposing a particular selectivity curve. However, in the following discussion we will only consider imposing a particular fishing mortality rate. Assuming the model has been optimally fitted to catch-at-age, effort, and the stock-recruitment relationship has been included, the algorithm would be the following:

1) Calculate age-specific fishing mortality rates - from $s_a F_y$
2) Calculate age-specific survivorship rates - from $S_{a,y} = \exp(-(M + s_a F_y))$
3) Calculate the predicted numbers-at-age for age-classes above the age of recruitment - from $N_{a+1,y+1} = N_{a,y} S_{a,y}$, Eq. 11.30
4) Calculate predicted numbers-at-age for recruits from the stock-recruitment relationship and the biomass from r years previous using Eq. 11.36, including a stochastic residual term.
5) Calculate the predicted catch-at-age as per Eq. 11.32
6) Calculate the fishery performance measures
7) Repeat for the next year of projection.
8) Repeat a sufficient number of times to obtain summary information from the stochastic nature of the projections.

The predicted recruitment should have some intrinsic variation away from the deterministic recruitment value predicted from the spawning stock biomass from the requisite number of years prior to the recruitment year. This stochasticity could either be selected at random from the residuals available (analogous to a bootstrap sample) or, if the time series of residuals is short, should be selected at random from under a probability density function used to describe the recruitment residuals (usually a log-normal PDF would be used). What this implies is that the projections would be in the nature of a Monte Carlo simulation requiring numerous replicates to obtain the necessary summary information.

11.4 CONCLUDING REMARKS

In this chapter we have considered both of the main analytical strategies adopted for assessing age-structured fishery data. Which approach best suites a particular situation will depend upon circumstances. The fact that Integrated Analyses are less stringent in their data requirements means that this is a more useful method for many fisheries in countries that do not have long traditions of collected detailed age-structured information.

Both VPA (in all its forms) and Integrated Analyses are large fields of endeavour with many examples in the literature and many developments not covered in this chapter. Each of the analysis strategies would form the basis of a book by themselves. This introduction covers a number of the important issues but these models continue to be developed and articulated (Myers and Cadigan, 1995; Schnute and Richards, 1995).

In real life situations there are likely to be different stocks present within a single fishery, and possibly there will be more than one fishing fleet exploiting the various stocks (e.g., trawl and non-trawl) each with their own selectivity characteristics. These would require separate treatments, which implies parallel analyses for each fleet (the fleets are separated by their relative effort, their selectivity curves, and the particular stocked fished) and for each stock (recruitment being kept separate). These multiple analyses would imply the necessity for more extensive data sets relating to catch-at-age and related effort (Punt *et al.*, 2001).

The analyses in this chapter were conducted inside Excel. The fact that it only took just over two hours to conduct 1000 bootstraps in the Integrated Analysis means that the need for custom computer programs seems to have lessened. However, implementing a more sophisticated multi-fleet, multi-stock, catch-at-age model in Excel would be stretching the abilities of the Excel solver, although a stepped approach to solving sub-sets of the model would be possible. In the end, custom computer programs to conduct these stock assessments are still the optimum approach (Richards *et al.*, 1997). The ability to mimic such models in Excel is, however, a handy double check on the defensibility and reality of the results obtained from more sophisticated programs.

Once the basics have been absorbed, the best teacher relating to these methods is experience with different fisheries and problems. There are large numbers of options available when implementing one of these models. A good strategy is not to restrict oneself to a single model but to implement different versions of an assessment to investigate the implications of the different data sets and the different model structures.

APPENDIX 11.1 WEIGHT-AT-AGE DATA AND OPTIMUM FIT TO CATCH-AT-AGE MODEL

Table A11.1 Average weight in grams of North Sea plaice in each age-class measured from market samples taken in Lowestoft and Grimsby during 1929-1938. Data from Table 16.2 in Beverton and Holt (1957). In the original table the weights went up to age 20, but are truncated here at age 10 to match the ageing data. In addition, the weights were given for calendar years while the ageing was for fishing years (April 1st to March 31st); weights from 1929 were assumed to hold for the 1929/1930 fishing year, and equivalently for later years. Data used in Example Box 11.7.

Year\Age	2+	3+	4+	5+	6+	7+	8+	9+	10+
29/30	167	190	218	270	289	392	574	665	802
30/31	136	190	257	340	441	498	582	740	961
31/32	127	152	221	349	440	503	593	675	809
32/33	132	143	182	297	421	535	641	721	827
33/34	146	165	189	251	383	528	645	787	818
34/35	157	189	202	229	277	521	711	798	819
35/36	154	178	214	280	313	383	643	789	876
36/37	149	171	198	255	338	377	467	781	885
37/38	160	173	222	310	383	490	519	624	845

Table A11.2 Parameter values and sum of squared residuals for the optimum fit from Example Boxes 11.7 and 11.8. If the relative weights attributed to each squared residual contribution are altered, then the balance between the two sources of information will alter. The parameter values on the worksheet are all in a single column but have been staggered here for ease of presentation. SSQC is the sum of squared residuals from the catch-at-age data, SSQE is from the fully selected fishing mortality rate comparison, and SSQT is simply the sum of both sources weighted by their respective weights. Natural mortality is assumed to be constant.

Source	Value	Weight				
SSQC	7.6366	1		SSQT	8.0152115	
SSQE	0.3786	1		Natural M	0.2	
Parameter	Ln(value)					
Ln_q	-2.09828		N_30_2	10.3658	LnF_29	-0.0740
N_29_10	4.4570		N_31_2	9.2878	LnF_30	-0.4475
N_29_9	5.0807		N_32_2	9.2766	LnF_31	-0.2869
N_29_8	6.0900		N_33_2	8.9197	LnF_32	-0.6782
N_29_7	7.3389		N_34_2	9.3640	LnF_33	-0.7126
N_29_6	6.9909		N_35_2	9.1696	LnF_34	-0.7796
N_29_5	7.7294		N_36_2	8.9735	LnF_35	-0.4135
N_29_4	8.9176		N_37_2	9.6004	LnF_36	-0.6380
N_29_3	8.7447		Sel50	1.2232	LnF_37	-0.3885
N_29_2	9.3607		Sel95	1.4785		

Appendix A
The Use of Excel in Fisheries

1. INTRODUCTION

Excel has three main uses for fisheries scientists:

1) Entry and editing of data – formal tabulation of one's data.
2) The display and illustration of a data set's properties – data can be graphed or compared visually, which can be useful for detecting anomalous data. Tends to be preliminary to any detailed analysis.
3) Determination of a data set's properties – this can include simple analyses such as summary statistics and fitting linear regression lines, but can extend to estimating the parameters of non-linear equations and models, of varying complexity, to available data.

There always appears to be more than one way of doing things inside Excel. This may seem redundant, but by allowing greater flexibility, more people should be able to find ways of working with Excel. The simple trick is to find which way works best for you and stick with it.

In this appendix, aspects of the functionality of Excel, which the author has found useful will be introduced. This will not be an exhaustive list and functions or practices that others find indispensable may be omitted. This is, however, only an introduction. There are many excellent texts on the use of Excel and programming in the Visual Basic for Applications language (VBA) built into the program. We will introduce ideas relating to the basic use of the program, a set of the most useful functions, with their strengths and weaknesses, and a short introduction to VBA, sufficient for the needs of this book.

2. WORKBOOK SKILLS

2.1 Tools/Options, Auditing, and Customization

It is assumed the reader knows the Excel interface and has some knowledge of the functions of the bits on the Excel window (Fig. A1). Perhaps the first thing to do is to set up Excel with more of your preferences. Try pressing the Tools/Options menu item (Options is near the bottom of the list of menu items under the Tools menu). In the resulting dialog box you can set very many properties inside Excel. Under the "General" tab, you can set a default directory in which Excel should begin searching for files, plus the default font. Other than that, it is useful to go to the "Edit" tab and change at least two items. Excel often comes with the options "Edit directly in cell" and "Move Selection after enter" ticked as a default. The moving selection after pressing <Enter> is a matter of personal preference. I find it a nuisance, preferring to press the down arrow when entering data rather than relying on the software to move the cursor down a cell with each data record. But the "Edit directly in cell" option can be more troublesome. Apart from often being in a tiny font, when one enters a formula, one's typing obscures the contents of the cell(s) immediately to the right. This can be a nuisance if one wants to refer to that cell. Try turning these options off and seeing which set up you prefer. You can return to the Tools/Options dialog box at any time and alter the settings. It is worth experimenting with the various options just to see what is possible. Customization is not just playing about. Having a comfortable working environment can aid productivity.

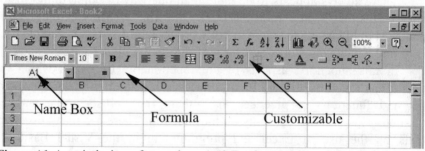

Figure A1 A typical view of a newly opened Excel spreadsheet. It is currently named Book2. The usual menu items are listed across the top. Note the Toolbars, which contain many shortcuts to actions such as saving the file, or the chart and function wizards. These toolbars can be customized to include the shortcuts of most use to individual users. Also highlighted is the "Name" box, which can be used to name cells for added convenience when creating equations, and the "Formula" area, where most of the editing of equations takes place.

It is valuable to familiarize oneself with the toolbars (Fig. A1) and what they can do for you. By placing your mouse cursor over a button (and not clicking) the button's name will appear, which generally describes it's function. You can also customize the toolbars as they appear on your computer so that its functionality reflects your individual needs.

Some of the most useful buttons available include the auditing buttons, these are useful for tracking down which cells are related to which. They can be added to the toolbars already available by clicking anywhere in the toolbar region using the right-hand mouse button, and selecting the Customize option at the bottom of the resulting menu (alternatively, View/Toolbars/Customize). This will give rise to a dialog box with three tabbed options. Select the middle Commands tab and then the Tools options under Categories. There are many useful tools listed but if you scroll down you will find the ones figured above. They are the Trace Precedents, Trace Dependents, and Remove All Arrows buttons. If a cell has an equation that refers to a number of other cells, pressing the Trace Precedents button will generate fine blue arrows to the cells involved. Similarly, pressing the Trace Dependents will identify any cells influenced by the active cell. To include a tool on a Toolbar, click and drag it from the Dialog box onto the selected Toolbar (if it just disappears you have probably missed placing it onto the Toolbar). To remove a button just click and drag it off the Toolbar into anywhere else. I also find the Zoom In and Zoom Out buttons to be very useful. These can be found under the View category. Further buttons to add would include the Paste Values button (under Edit) and the Button button (for generating active buttons for use with macros on the worksheet), which is found in the "Forms' category i.e., and plus many others that may suite your own style of working.

There are many other ways that a worker can customize his or her version of Excel and it is recommended that you search for the set up that suites your own uses and needs. If you are relatively new to Excel then do try experimenting with the different set up arrangements to see which you like the best. Personally, the author tries not to have more than two rows of toolbars visible so as to have a good balance between having numerous useful buttons and an extensive view of the active worksheet.

2.2 Data Entry

Data entry is straightforward. Get to know the number keypad and use it wherever you can for greater speed (laptop users will just have to suffer). If you enter data in columns, get used to typing a number and entering it by pressing the down arrow. There is no need to type a zero when entering a fractional number (e.g., 0.234 can be typed as .234).

If you need to copy a number down a column, enter the number, then place the cursor over the small dot, the copy-handle, at the lower right hand corner of the selected cell. The cursor should convert to a thick cross or addition sign. You can drag this down to copy the cell automatically into the selected area. If you require a series, such as 1, 1.5, 2, 2.5, ..., begin the series (e.g.,

1, 1.5) then select the two cells containing the two numbers, grab the copy handle, and drag in the direction (up/down or across) you require. You should see the count of the extent of the series to help know when to stop. If the increment you want is one, then just enter the starting number, grab the copy handle, and drag while holding the <Ctrl> key; this will extend a series from a single value in steps of one. If you enter text followed by a number, such as Year88, grab the copy handle and drag in the desired direction; the number following the text will increment by one in each cell dragged across. It should be emphasized that none of these shortcuts are necessary; they simply increase the speed with which data and worksheets can be worked with.

2.3 Movement Around Worksheets

There is a suite of keyboard commands for moving rapidly around a worksheet and these are generally some key combined with the <Ctrl> key (Table A1).

Table A1 Keyboard commands for moving rapidly around the worksheet and workbook. If more than one key is involved they must both be pressed together i.e., press the <Ctrl> key, hold it down, and then press the other key.

<Home>	Moves the active cell to the first column of the ActiveRow.
<Ctrl><Home>	Makes cell A1 the ActiveCell.
<Ctrl><PageUp>	Makes the WorkSheet to the left of the ActiveWorkSheet the ActiveWorkSheet.
<Ctrl><PageDown>	Makes the WorkSheet to the right of the ActiveWorkSheet the ActiveWorkSheet.
<Ctrl><End>	Makes the Cell at the bottom right of the ActiveArea the ActiveCell.
<Ctrl><←> <Ctrl><↑> <Ctrl><→> <Ctrl><↓>	If the ActiveCell is empty these key combinations move the ActiveCell in the direction of the arrow until it meets the first filled cell or a spreadsheet boundary. If the ActiveCell is full it moves until it meets an empty cell or a sheet boundary. Very useful for moving around large worksheets.
<F5>	Opens the "Goto" Dialog box into which you could type a cell reference and thereby move the ActiveCell.
DropList button on the NameBox	This provides a list of Named cells and ranges in the WorkBook. By selecting a Name from the list, the cell or range referred to becomes the ActiveCell or ActiveRange. Even if it is on a different Worksheet to the ActiveWorkSheet.

2.4 Range Selection

There are a number of options for selecting a range of cells:

1) If the range is small (completely visible on the screen) then one drags the mouse over the required range to make it the ActiveRange.
2) Alternatively, one could first press <F8>, which turns "extension" or "selection mode" on, and then proceed with one of the keyboard controls for moving the ActiveCell (Table A1). The use of <F8> is advantageous for one handed keyboard work, but if you are using two hands it is often simpler to hold down the <Shift> key when operating the keyboard movement commands. Thus, at the top of a column of 10000 filled cells, pressing <Shift><Ctrl><↓> selects the entire column of 10000 cells (assuming there are no missing data points).
3) Finally, if it desired to select non-contiguous ranges (e.g., when selecting data to be graphed) then one can proceed as in (2) above to select the first range. Then one has to hold down the <Ctrl> key when clicking the cursor at the start of the next selection, after which one can proceed as before as many times as required.

When creating formulae in cells, one very often refers to other cells (e.g., =A5*B$6). Instead of typing these addresses it is most simple to start typing the formula and when an address is required use the mouse pointer to select the required cell. This puts the correct address into the formula. Very often one wants some degree of absolute reference. Thus $A$5 refers to the cell A5 and even if this is copied somewhere else it will still refer to A5. Similarly, $A5 will always refer to column A when copied elsewhere, although the row number may change appropriate to the move. Instead of typing the dollar signs, once the cell reference or range has been input (either by typing or pointing), then assuming A5 has been entered, pressing the <F4> button will cycle through $A$5, A$5, $A5, A5.

2.5 Formatting and Naming Cells and Ranges

A range can be either a single cell or a block of cells. It is often convenient to give a cell a particular name (such as Gradient or M), which then makes referring to that cell in equations very simple. It is far more readable to type "=exp(-M)" then it would be to type "=exp(-$A$2)", assuming that cell A2 has been named M.

Naming a cell or range is most simply done by selecting the cell(s) required, typing the desired name into the Name box, and pressing <Return> or <Enter>. If one omits the last step, for example, by completing the action through clicking elsewhere with the mouse, the name will not be accepted. To see the name of a cell or range, make the cell(s) the ActiveCell or range and their name will be seen in the Name Box. If the name is selected in the droplist associated with the Name Box, the named cell(s) become the ActiveCell.

A range can be formatted by selecting the cells, clicking the right mouse button, and selecting the Format Cells option. Alternatively, the Format/Cells menu item can be selected and the required options selected. If it is desired to have multiple fonts in one cell then enter the text, select the particular letters and use the menu commands to alter the font. When the <Enter> key is pressed the text is entered into the cell and the multiple fonts are visible.

Under the Format/Conditional Formatting menu item it is possible to make the formatting of a cell dependent upon the value in the cell. This possibility can be surprisingly useful when manipulating data or model outputs.

2.6 Formulae

To use any data entered into an Excel Workbook some form of manipulation of the data will be necessary. This is where Excel really starts to show its power. To enter an equation start by selecting the cell into which the equation is to be placed. Start all equations with an "=" symbol. This is always the case unless there is a function or equation within an equation e.g., =if(isnumber(A2),1,0). If the equals sign is omitted, Excel will react as if you entered text and nothing interesting will happen. If on entering a formula then formula remains visible and not its value, check in Tools/Options under the View tab, that the Formulas option is not ticked. If you want to enter text but Excel keeps treating your entry as a formula or a number, then start the text with an ' symbol, which will not remain visible.

If you wish to concatenate text and numbers, which can be useful with functions such as =countif(), then, assuming cell A5 contained a value of 0.53, we could type =countif(A6:A1005,"<" & A5), which would be equivalent to =countif(A6:A1005,"<0.53"). The active agent in that is the "&" symbol. This can be very useful when manipulating cell references or creating text out of multiple non-text cells.

2.7 Functions

Two very useful Σ f_x buttons are the summation button which generates an =SUM() function, and the generalized Paste Function button. This latter can lead you to any one of the more than

200 worksheet functions that are available in Excel. It is strongly recommended that you explore what is available by looking through the contents of this Dialog Box. In the Example Boxes throughout the text of this book there are very many functions used whose entry can be simplified by using this shortcut button. From it you can obtain help in relation to any function as well as prompts for the parameters required by the various functions. We will consider some of the more useful to illustrate some of what is available. In general if there is a function you want it is likely to be already available. If it isn't, Excel has the flexibility to enable you to write your own custom functions (see, for example, Walkenback, J. (1999) *Microsoft Excel 2000 Power Programming with VBA*. IDG Books Worldwide Inc., Foster City, CA.). Thus, typical things such as =EXP(), =LN(), which is log to base e, =AVERAGE(), =STDEV(), =PI() are all available. There has been some criticism of Excel's use for statistical calculations. Some of these are correctly pointing out that some of the algorithms used are not necessarily the optimum available. However, in reality, this usually only influences the 8th or 9th significant digit in just a few functions, and, while this may not be acceptable for high-energy physics, it is not a problem for most ecological data (especially in fisheries).

So there are functions available for most of the obvious needs but some of the most valuable are not so obvious.

2.7.1 =SumProduct()

A very common thing to want to do is to take two columns of numbers, multiply the elements together, and sum to results. When we do a least squared-residual analysis in the Example Boxes we typically subtract one column's values from another, square the result, and then in a cell above the column we sum to the total. Assuming we have 1000 observed values in column C and the predicted values in column D. If we fill column E with =(C2-D2) (it's always a good idea to plot residuals) we could just put =SUMPRODUCT(E2:E1000,E2:E1000) into any cell in which we wished to obtain the sum of the squared residuals. In the examples given we generally calculated the squared residuals explicitly and summed those in an effort to be clear. You may wish to modify some of the Example Boxes to implement the simpler approach of the =sumproduct() function. In Chapter 11, we used columns of numbers-at-age and combined them with the average weight at age inside a =sumproduct() function to obtain total biomass.

2.7.2 =FREQUENCY()

Very commonly it is desired to convert a column of data into frequency counts relative to different categories. Thus, in the Example Box testing the Excel random number generator function, =RAND(), after generating a column of 1000 random numbers it is desired to count how many lie between 0.0 and 0.1, then between 0.1 and 0.2, and so on. One of the best ways of doing this is with the =frequency() function. Given 1000 random numbers in A5:A1004 and the numbers 0.0, 0.1, 0.2,....,1.0 in cells C3:C13, then to construct the frequency counts one: Selects D3:D13, types =frequency(A5:A1004,C3:C13), and then presses <Ctrl><Shift><Enter> all at once. This latter combination is required to enter an array function. This will generate the required numbers, which will approximate 100 in the cells opposite 0.1, 0.2,...., 1.0, and should be zero opposite 0. This may appear odd to biologists because they are usually trained to consider such frequency counts from the lowest category up. This means that the class 0.0, followed by 0.1, is often taken to mean everything greater than and equal to 0.0 but less than 0.1. However, for unknown reasons, in Excel, the frequency counts start from the top and work down. Thus, in the example, the category 1.0 refers to all values <= 1.0 but > 0.9. Similarly, the 0.9 category refers to all values <= 0.9 but >0.8, and so on down. It is important to realize that the 0.1 category refers to everything <= 0.1 and greater than 0.0. Thus, the 0.0 category refers to the number of zeros present and anything less than 0.0. This may all seem messy but it works well with integer data and, as long as one is aware of how this function works, it can be used successfully with real numbers as well.

2.7.3 =LINEST()

This is another array function, which can be used to fit a linear regression in real time on the worksheet. It can be used to provide just the parameters or the parameters along with an array of related statistics. Its format is =LINEST(Y-Values, X-Yalues, Boolean for Constant, Boolean for Statistics). If one enters TRUE for the first boolean (true/false) then the regression includes a constant; if one enters a FALSE value, then it is fitted through the origin. Given a true for the second Boolean will generate the related statistics (e.g., sum of squared residuals, number of observations, correlation coefficient). To use this function one selects a five row by two column range, types in the function, and then presses <Ctrl><Shift><Enter>, as usual with an array function.

2.7.4 =VLOOKUP()

The =VLOOKUP() function, along with the other reference functions [e.g., = offset()] enable the user to implement lookup tables within Excel. This is used in many of the computer intensive statistical routines. There is a slight quirk to the =vlookup() function that is worth knowing about.

If one is using the =vlookup() function to select from a series of real numbers then it is essential to use the full syntax. First, the =vlookup() function requires the Lookup_Value, which is the value to be found in the first column of the lookup-table, second, the Lookup-Table address, which is the range where the information columns are found, and third, the column_index_number, which is the column within the lookup_table from which to obtain the data, fourthly, the so-called range_lookup, which is an optional argument that defaults to true if omitted. This latter parameter has a significant quirk such that if this is set at true the action will be to find the closest match, if it is false it must find an exact match (or give an error message). If you require an exact match but put or leave true by default in this last, optional function argument, then even if the lookup_value is not present the closest match will be returned. Generally, in fisheries work, one would always wish to place "false" in the range_lookup argument.

2.7.5 Other Functions

There are too many useful functions to detail here but it may help to know that there are matrix algebra functions, text handling functions, statistical distribution functions, logical functions, lookup and reference functions, data and time functions, and others besides. Do not assume you understand how a function behaves. Just because it sounds just like something you do all the time does not imply that it does what you want in exactly the same way. Read through the help file relating to each function you use.

3. VISUAL BASIC FOR APPLICATIONS

3.1 Introduction

There are many places in which it is necessary to use a computer program to control or manipulate data beyond that which can be performed on the worksheets. Excel has Visual Basic built into it in a version that can interact with the objects and functions that make up the Excel program. Thus, it is possible to refer to cells on the worksheets, and interact with Visual Basic programs in real time.

Fortunately, it is possible to record large amounts of the code required by the language using the /Tools/Macro/Record new Macro menu item. Record as much of each macro as possible to save typing. In order to modify or articulate a program I often record small snippets of extra code and cut and paste them into my main sub-routine. Get used to the development environment and once again learn what set up and operational procedure best suites you.

A quick way into the VBA editor is to press <Alt><F11>. A quick way to record a macro is to place a macro button on the sheet, name the macro, and select the Record button on the dialog box.

I find it is best to type any VBA code in lower case (except names if you wish to use capitals) and without spaces because if the syntax is correct the code lines will be properly spaced out and recognized commands will be capitalized when the line is entered.

Always have Option Explicit at the top of your code, as this requires one to declare all variables, which helps prevent errors introduced through spelling a variable incorrectly somewhere in your code. When in the VBA editor window the Tools/Options menu items brings up a dialog box and when you first begin creating programs it is reasonable to tick all of the options under the Editor tab.

One good way to learn is to have both the Excel sheet in which one is recording and the VBA editor window visible on the sheet at the same time. By using the worksheet you will be able to see commands being generated inside the VBA window. Not everything recorded is required in the final program, any extraneous code can be deleted.

A note of warning is necessary. Macros are very powerful and can overwrite data on a worksheet without warning. You are in control but, as should always be the case, responsibility comes with control.

3.2 An Example Macro

An example of a subroutine is provided along with explanations for each line and each component. All the following subroutine does is copy one number from cell N7 down into a sequence of cells starting 5 cells ($4+i = 5$ when $i = 1$) below it. The macro works because the action of pasting anything on the active sheet causes the cell to recalculate (assuming calculation is set to Automatic). This resets all the calls to =**Rand()** which generates random numbers on the sheet. The sheet must be designed to automatically change the numbers used in a calculation whose final outcome is in N7 so the net effect is to calculate this statistic 1,000 times and store the results in column N. First we give the macro and then describe every line in the macro.

```
' Macro1 Macro    Operates the Bootstrap Routine
' Macro recorded by Malcolm Haddon
Sub Macro1()
   Randomize
   Calculate
   Application.ScreenUpdating = False
   For i = 1 To 1000
     Range("N7").Select
     Selection.Copy
     ActiveCell.Offset(4 + i, 0).Range("A1").Select
     Selection.PasteSpecial Paste:=xlValues, Operation:=xlNone, _
       SkipBlanks:=False, Transpose:=False
     Application.CutCopyMode = False
   Next i
' sort the bootstrap replicates
   Range("N12:N1011").Select
   Selection.Sort Key1:=Range("N12"), Order1:=xlAscending, Header:= _
     xlGuess, OrderCustom:=1, MatchCase:=False, Orientation:= _
     xlTopToBottom
   Application.ScreenUpdating = True
End Sub
```

Comments in VBA are signified by a single quotation mark '. When you create a VBA subroutine in a module by recording key-strokes (Tools/Macro/Record New Macro), the program is always prefaced by a five line comment. On the author's machine any text following on the same line turns green. Comments can start anywhere on a line:

' this is a comment
     for i = 1 to 1000 ' this is also a comment

would both work. Annotations are useful in understanding the program months after they were written; try to keep them brief but informative.

Randomize resets the computer's random number generator. The format : Randomize 12345 can also be used, where the number would be an initiator or seed to the pseudo-random number sequence. It cannot be recorded.

Calculate is the same as pressing F9 on a worksheet. It causes the worksheet to calculate all its equations. If you have many random numbers on a sheet it is a good idea to start the program in this way so as to randomize the data before any calculations are carried out. It can be recorded by pressing F9.

Application.ScreenUpdating = False is a very handy command which stops the screen updating until the next **Application.ScreenUpdating = True**. This is valuable because if the macro moves around the screen a good deal it can waste a huge amount of computer time rewriting the screen. It cannot be recorded.

For i = 1 to 1000 is a standard For loop in Basic which can enclose a block of code and thereby bring about a set of iterations (1000 in this case). The **i** is the counter variable, which can be used in equations inside the block of code. The block of code is defined by the for loop and by a **Next i** found below in the code. This command cannot be recorded.

Range("N7").Select - moves the cursor on the active spreadsheet to cell 'N7', so that "N7" becomes the ActiveCell.

Selection.Copy copies the value in the selection ready for pasting somewhere else.

ActiveCell.Offset(4 + i, 0).Range("A1").Select contrasts with Range("N7").Select which is the way of selecting a particular cell. ActiveCell.Offset.etc is the way to reference cells relative to the current ActiveCell. You will not need to type this, instead it is best to record it. The cell referenced is in position ActiveCell.Offset($4 + i, 0$), where the 4+i refers to how many rows down away from the current active cell to move, and the zero is how many columns across to move. You will need to edit the recorded text to include the counter variable **i** into the command.

Selection.PasteSpecial Paste:=xlValues, Operation:=xlNone, _
    SkipBlanks:=False, Transpose:=False is a two-line command, which again, is best recorded. It is simply telling Excel to Paste/Special/Values whatever was previously copied somewhere else in the program into the current active or selected cell. The default values for Operation, SkipBlanks, and Transpose can be deleted for clarity.

Application.CutCopyMode = False (again recorded) simply turns off the highlighting around a copied cell and prevents the accidental pasting of the clipboard into a cell by pressing return after a macro has finished running.

This is handy at the end of a macro but can be deleted from most places within the body of a macro.

Next i is just the end of the FOR loop which started a few lines before (see above). This command needs to be typed as it cannot be recorded.

'   sort the bootstrap replicates is just another comment inside the programme which describes the following lines of code

Range("N12:N1011").Select   Range("").Select can be used to select single or multiple contiguous cells. For a group of cells the syntax is to separate their addresses with a colon. In this case the program has generated 1000 bootstraps of some statistic and those 1000 numbers are to be sorted in order. First they must be selected. This command was recorded.

Selection.SortKey1:=Range("N12"),Order1:=xlAscending, Header:= _
    xlGuess, OrderCustom:=1, MatchCase:=False, Orientation:= _
    xlTopToBottom   is a three-line command that is recorded, and it simply sorts the selection into ascending order.

Application.ScreenUpdating = True turns screenupdating back on. If there was no initial turning off then this does nothing. Screenupdating automatically turns back on at the end of a macro.

End Sub complements the **Sub Macro1()** which the recorder places at the front of the macro. The two together define the body of the Macro and it is this name Macro1, which is used to identify the subroutine when it is called. You can change this name at any time (e.g., to Fred, or RandomRegress, or whatever you wish) but it is perhaps best to name the macro when it is first written or recorded.

3.3 Using the Solver inside a Macro

When using computer intensive methods it is common to want to apply the Solver inside a macro. This is simple to do but there are a couple of tricks that may cause problems unless care is taken. Using the solver can be recorded inside a macro like most other actions. In Excel 97 and 2000, one obtains something like:

```
SolverOk SetCell:="$B$12", MaxMinVal:=1, ValueOf:="0", ByChange:="$B$2:$B$5"
SolverSolve (true)
```

In order for this to proceed without calling up the "Solver has converged to a solution" dialog box one needs to add the text in italics (true). Of course, in that case you will need to ensure that an accepted solution really is acceptable without individual inspection.

Sometimes, after recording a macro that includes the solver, the macro fails to run when called and the program opens up the Visual Basic editor and complains that there is a "Compile Error, Cannot find Project or Library." You need to reset the program (the small blue square in the toolbar) and then go to the menu item Tools/References and tick the Solver box. If it says the Solver.xla is missing then you will need to browse for its location. This should fix the problem. One way of preventing this is to call the solver from the worksheet before running the macro.

4. CONCLUDING REMARKS

This very brief introduction to the use of Excel is only designed to encourage those people who are relative beginners and those others who feel that one should not use such a program for serious fisheries work. In the body of the book the Example Boxes range from very simple to highly complex. The ability to conduct a length frequency decomposition or a statistical catch-at-age analysis on a modern spreadsheet illustrates how powerful such software has now become. While, I am not recommending that this is a suitable vehicle for conducting large-scale serious assessments, Excel has huge advantages as a medium for teaching the details of different analyses. It is a very common, highly flexible program, it hides almost nothing, and a workbook can be audited by anyone. The addition of the built in programming language means that almost anything is possible, although the speed may not be great. As an aid to teaching I have found it invaluable. On the other hand, programmes written in some computer language, especially the more object-oriented languages such as C++ and Delphi, would make an appalling teaching medium. Hopefully, readers will find that the worked examples help them understand the algorithms and equations that make up the variety of quantitative methods illustrated in this book.

Excel is so easy to use that it is possible to fall into the trap of typing in a model before writing out the full set of equations. It is always best to know the full model structure before starting. In this way the most efficient layout can be selected. As implied in the Preface, the best way of learning a method is to use it, so I encourage you to implement your own problems using your own data.

Bibliography

Anderson, M.J. and P. Legendre (1999) An empirical comparison of permutation methods for tests of partial regression coefficients in a linear model. *Journal of Statistical Computing and Simulation*, **62**: 271-303.

Andrewartha, H.G. and L.C. Birch (1954) *The Distribution and Abundance of Animals*, The University of Chicago Press, Chicago. 782 p.

Annala, J.H., Sullivan, K.J., O'Brien, C.J. and N.W.McL. Smith (Comps.) (2001) Report from the Fishery Assessment Plenary, May 2001: stock assessments and yield estimates. 515 p. Ministry of Fisheries, New Zealand.

Australian Fisheries Management Authority (1999) Northern Prawn Fishery Statistics – Update. AFMA, Canberra, Australia.

Baranov, T.I. (1918) On the question of the biological basis of fisheries. Nauchn issledov. Ikhtiologicheskii Inst. Izv. 1: 81-128. (cited in Beverton & Holt, 1957).

Basu, D. (1980) Randomization analysis of experimental data: The Fisher Randomization Test. *Journal of the American Statistical Association*, **75**: 575-595.

Begon, M. and M. Mortimer (1986) *Population Ecology*. 2nd edn. Blackwell Scientific Publications, Oxford.

Bergh, M.O. and D.S. Butterworth (1987) Towards rational harvesting of the South African anchovy considering survey imprecision and recruitment variability. *South African Journal of Marine Science*, **5**: 937-951.

Bernard, D.R. (1981) Multivariate analysis as a means of comparing growth in fish. *Canadian Journal of Fisheries and Aquatic Sciences*, **38**:233-236.

Berryman, A.A. (1981) *Population Systems. A General Introduction.* Plenum Press. New York.

Beverton, R.J.H. (1962) Long-term dynamics of certain North Sea fish populations. pp 242-259 *In*: E.D. LeCren and M.W. Holdgate (*eds*), *The Exploitation of Natural Animal Populations*. Blackwell, Oxford.

Beverton, R.J.H. and S.J. Holt (1957) On the dynamics of exploited fish populations. *U.K. Ministry of Agriculture and Fisheries, Fisheries Investigations (Series 2)*, **19**: 1-533.

Bickel, P.J. and D.A. Freedman (1981) Some asymptotic theory for the bootstrap. *The Annals of Statistics*, **9**: 1196-1217.

Bickel, P.J. and D.A. Freedman (1984) Asymptotic normality and the bootstrap in stratified sampling. *The Annals of Statistics*, **12**: 470-482.

Birkes, D. and Y. Dodge (1993) *Alternative Methods of Regression*. John Wiley & Sons, New York.

Boerema, L.K. and J.A. Gulland (1973) Stock assessment of the Peruvian anchovy (*Engraulis ringens*) and management of the fishery. *Journal of the Fisheries Research Board of Canada*, **30**: 2226-2235.

Box, G.E.P. and G.C. Tiao (1973) *Bayesian Inference in Statistical Analysis.* Addison-Wesley, Reading, MA.

Brown, D. and P. Rothery (1993) *Models in Biology: Mathematics, Statistics, and Computing.* John Wiley & Sons, Chichester.

Brown, R.S., Caputi N., Barker E., and J. Kittaka (1995) A preliminary assessment of increases in fishing power on stock assessment and fishing effort expended in the western rock lobster (Panulirus cygnus) fishery. *Crustaceana*, **68(2)**:227-237.

Burnham, K.P. and D.R. Anderson (1989) *Model Selection and Inference. A Practical Information-Theoretic Approach.* Springer-Verlag, New York.

Caughley, G. (1977) *Analysis of Vertebrate Populations.* John Wiley & Sons, London.

Cerrato, R.M. (1990) Interpretable statistical tests for growth comparisons using parameters in the von Bertalanffy equation. *Canadian Journal of Fisheries and Aquatic Sciences*, **47**: 1416-1426.

Chanut, J.-P. and B. Pelletier (1991) STRATE: A microcomputer program for designing optimal stratified sampling in the marine environment by dynamic programming - I. Theory and method. *Computers and Geosciences*, **17**: 173-177.

Chapman, D.G. (1961) Statistical problems in dynamics of exploited fish populations. *Proceedings of the 4th Berkeley Symposium of Mathematics, Statistics and Probability*, **4**: 153-168. University of California Press, Berkeley, CA.

Chatterjee, S. and B. Price (1977) *Regression Analysis by Example.* John Wiley & Sons. New York.

Chen, Y. and D. Fournier (1999) Impacts of atypical data on Bayesian inference and robust Bayesian approach in fisheries. *Canadian Journal of Fisheries and Aquatic Sciences*, **56**: 1525-1533.

Chen, Y., Jackson, D.A., and H.H. Harvey (1992) A comparison of von Bertalanffy and polynomial functions in modelling fish growth data. *Canadian Journal of Fisheries and Aquatic Sciences*, **49**: 1228-1235.

Choat, J.H. and L.M. Axe (1996) Growth and longevity in acanthurid fishes; an analysis of otoliths increments. *Marine Ecology Progress Series*, **134**: 15-26.

Christiansen, F.B. and T.M. Fenchel (1977) *Theories of Populations in Biological Communities.* Springer-verlag, Berlin. 144 p.

Cochran, W.G. (1963) *Sampling Techniques.* 2nd edn. John Wiley & Sons, London.

Coutin, P., Walker, S. and A. Morison (Eds.) (1997) Black bream – 1996. Compiled by the Bay and Inlet Fisheries and Stock Assessment Group. *Fisheries Victoria Assessment Report*, No. **14**. (Fisheries Victoria: East Melbourne).

Curnow, R.N. (1984) Confidence intervals under experimental randomization. pp. 33-44 *In*: Hinkelmann, K. *(ed.) Experimental design, statistical models, and genetic statistics.* Marcel Dekker, Inc., New York.

Cushing, D.H. (1988) The study of stock and recruitment. pp. 105-128 *In*: Gulland, J.A. (*ed*) *Fish Population Dynamics*. John Wiley & Sons, New York.

Dann, T. and S. Pascoe (1994) *A Bioeconomic Model of the Northern Prawn Fishery*. ABARE Research Report **94.13**, Canberra.

Dennis, B. (1996) Discussion: should ecologists become Bayesians? *Ecological Applications*, **6**: 1095-1103.

Deriso, R.B. (1980) Harvesting strategies and parameter estimation for an age-structured model. *Canadian Journal of Fisheries and Aquatic Sciences*, **37**: 268-282.

Deriso, R.B., Quinn, T.J., and P.R. Neal (1985) Catch-age analysis with auxiliary information. *Canadian Journal of Fisheries and Aquatic Sciences*, **42**: 815-824.

Deriso, R.B., Neal, P.R. and T.J. Quinn (1989) Further aspects of catch-age analysis with auxiliary information. *Canadian Special Publication in Fisheries and Aquatic Sciences*, **108**: 127-135.

DiCiccio, T.J. and J.P. Romano (1988) A review of bootstrap confidence intervals. *Journal of the Royal Statistical Society*, **B 50**: 338-354.

Doubleday, W.G. (1976) A least-squares approach to analysing catch at age data. *Research Bulletin of the International Commission for Northwest Atlantic Fisheries*, **12**: 69-81.

Draper, N.R. and H. Smith (1981) *Applied Regression Analysis*. John Wiley & Sons, New York.

Easton, M.D.L. and R.K. Misra (1988) Mathematical representation of crustacean growth. *Journal du Conseil International pour l'Exploration de la Mer*, **45**: 61-72.

Edgington, E.S. (1987) *Randomization Tests*. 2nd edn. Marcel Dekker, New York.

Edgington, E.S. (1995) *Randomization Tests*. 3rd edn. Marcel Dekker, New York.

Edwards, A.W.F. (1972) *Likelihood: An Account of the Statistical Concept of Likelihood and its Application to Scientific Inference*. Cambridge University Press, London.

Edwards, E.F. and B.A. Megrey (*eds.*) (1989) *Mathematical Analysis of Fish Stock Dynamics*. American Fisheries Society Symposium, **6**.

Efron, B. (1979) Bootstrap methods: another look at the jackknife. *Annals of Statistics*, **7**: 1-26.

Efron, B. (1987) Better bootstrap confidence intervals. *Journal of the American Statistical Association*, **82**: 171-185.

Efron, B. (1992) Six questions raised by the bootstrap. pp. 99-126 *In*: Lepage, R. and L. Billard (*eds.*) *Exploring the Limits of Bootstrap*. John Wiley & Sons, New York.

Efron, B. and R. LePage (1992) Introduction to the Bootstrap. pp 3-10 *In*: Lepage, R. and L. Billard (*eds*) *Exploring the Limits of Bootstrap*. John Wiley & Sons, New York.

Efron, B. and R.J. Tibshirani (1993) *An Introduction to the Bootstrap*. Chapman & Hall, London.

Elliott, J.M. (1971) Some Methods for the Statistical Analysis of samples of Benthic Invertebrates. *Freshwater Biological Association, Scientific Publication*, **25**: 1-148.

Fabens, A.J. (1965) Properties and fitting of the von Bertalanffy growth curve. *Growth*, **29**: 265-289.

FAO, Fishery Information, data and Statistics Unit (1995) *World Fishery Production 1950-1993. Supplement to the FAO Yearbook of Fishery Statistics 1993.* **Vol. 76.** Catches and Landings. FAO. Rome.

Fisher, R.A. (1936) "The coefficient of racial likeness" and the future of craniometry. *Journal of the Royal Anthropological Institute of Great Britain and Ireland*, **66**: 57-63.

Fletcher, D. and R. Webster (1996) Skewness-adjusted confidence intervals in stratified biological surveys. *Journal of Agricultural, Biological, and Environmental Statistics*, **1**: 120-130.

Folks, J.L. (1984) Use of randomization in experimental research. pp 17-32 *In*: Hinkelmann, K. (*ed.*) *Experimental Design, Statistical Models, and Genetic Statistics*. Marcel Dekker, New York.

Fournier, D.A. and C.P. Archibald (1982) A general theory for analyzing catch at age data. *Canadian Journal of Fisheries and Aquatic Sciences*, **39**: 1195-1207.

Fournier, D.A. and P. Breen (1983) Estimation of abalone mortality rates with growth analysis. *Transactions of the American Fisheries Society*, **112**: 403-411.

Fox, W.W. (1970) An exponential surplus-yield model for optimizing exploited fish populations. *Transactions of the American Fish Society*, **99**: 80-88.

Fox, W.W. (1975) Fitting the generalized stock production model by least-squares and equilibrium approximation. *Fishery Bulletin*, **73**: 23-37.

Francis, R.I.C.C. (1988a) Maximum likelihood estimation of growth and growth variability from tagging data. *New Zealand Journal of Marine and Freshwater Research*, **22**: 42-51.

Francis, R.I.C.C. (1988b) Are growth parameters estimated from tagging and age-length data comparable? *Canadian Journal of Fisheries and Aquatic Science*, **45**: 936-942.

Francis, R.I.C.C. (1992) Use of risk analysis to assess fishery management strategies: A case study using orange roughy (*Hoplostethus atlanticus*) on the Chatham Rise, New Zealand. *Canadian Journal of Fisheries and Aquatic Science*, **49**: 922-930.

Francis, R.I.C.C. (1995) An alternative mark-recapture analogue of Schnute's growth model. *Fisheries Research*, **23**: 95-111.

Francis, R.I.C.C. (1997) Comment: How should fisheries scientists and managers react to uncertainty about stock-recruitment relationships? *Canadian Journal of Fisheries and Aquatic Sciences*, **54**:982-983.

Freedman, D.A. (1981) Bootstrapping regression models. *The Annals of Statistics*, **9**: 1218-1228.

Freund, R.J. (1980) The case of the missing cell. *The American Statistician*, **34**: 94-98.

Freund, R.J. and P.D. Minton (1979) *Regression Methods.* Marcel Dekker, New York.

Garcia, S.M., Sparre, P., and J. Csirke (1989) Estimating surplus production and maximum sustainable yield from biomass data when catch and effort time series are not available. *Fisheries Research*, **8**:13-23.

Garcia, S.M. and R. Grainger (1997) Fisheries management and sustainability: A new perspective of an old problem? pp. 631-654 *In*: Hancock, D.A., Smith, D.C., Grant, A., and J.P. Beumer (*eds.*) *Developing and Sustaining World Fisheries Resources. The State of Science and Management, 2ⁿᵈ World Fisheries Congress.* CSIRO Publishing, Collingwood, Australia.

Garstang, W. (1900) The impoverishment of the sea - a critical summary of the experimental and statistical evidence bearing upon the alleged depletion of the trawling grounds. *Journal of Marine Biological Association of the United Kingdom*, **6**: 1-69.

Gelman, A., Carlin, J.B., Stern, H.S., and D.B. Rubin (1995) *Bayesian Data Analysis.* Chapman & Hall, London.

Gilbert, D.J. (1997) Towards a new recruitment paradigm for fish stocks. *Canadian Journal of Fisheries and Aquatic Sciences*, **54**:969-977

Gillanders, B., Ferrell, D.J., and N.L. Andrew (1999) Ageing methods for yellowtail kingfish, *Seriola lalandi*, and results from age- and size-based growth models. *Fishery Bulletin*, **97**: 812-827.

Gleick, J. (1988) *Chaos: Making a New Science.* Penguin Books, London.

Gompertz, B. (1825) On the nature of the function expressive of the law of human mortality, and on a new mode of determining the value of life contingencies. *Philosophical Transactions of the Royal Society*, **115**: 515-585

Good, P. (1994) *Permutation Tests. A Practical Guide to Resampling Methods for Testing Hypotheses.* Springer-Verlag, Berlin.

Greaves, J. (1992) Population trends of the New Zealand Fur Seal *Arctocephalus forsteri*, at Taumaka, Open Bay Islands. B.Sc. Hons. Thesis. Victoria University of Wellington, New Zealand.

Gulland, J.A. (1956) On the fishing effort in English demersal fisheries. *Fishery Investigation Series 2*, **20**: 1-41.

Gulland, J.A. (1965) Estimation of mortality rates. Annex to Arctic Fisheries Working Group Report (meeting in Hamburg, January 1965). International Council for the Exploration of the Sea, Document 3 (mimeo), Copenhagen. (cited in Megrey, 1989)

Gulland, J.A. (1983) *Fish Stock Assessment: A Manual of Basic Methods.* John Wiley & Sons, New York.

Gulland, J.A. (*ed.*) (1988) *Fish Population Dynamics: The Implications for Management.* John Wiley & Sons, Chichester.

Gunderson, D.R. (1993) *Surveys of Fisheries Resources.* John Wiley & Sons, New York.

Haddon, M. (1980) Fisheries science now and in the future: a personal view. *New Zealand Journal of Marine and Freshwater Research*, **14**: 447-449.

Haddon, M. (1994) Size-fecundity relationships, mating behaviourm, and larval release in the New Zealand paddle crab, *Ovalipes catharus* (White, 1843) (Brachyura: Portunidae). *New Zealand Journal of Marine and Freshwater Research*, **28**: 329-334.

Haddon, M. (1998) The use of biomass-dynamic models with short-lived species: is B_0 = K an appropriate assumption? Pp. 63-78 *In*: *Risk Assessment. Otago Conference Series No 3* (*eds.*) *Fletcher, D.J. and B.F.J. Manly*. University of Otago Press, Dunedin.

Haddon, M. (1999) Discarding as unaccounted fishing mortality – when does bycatch mortality become significant? pp. 113-123 *In*: Buxton, C.D. and Eayrs, S.E. (*eds.*) *Establishing meaningful targets for bycatch reduction in Australian Fisheries*. Australian Society for Fish Biology Workshop Proceedings, Hobart, September 1998. Australian Society for Fish Biology, Sydney.

Haddon, M. and T.J. Willis (1995) Morphometric and meristic comparison of orange roughy (*Hoplostethus atlanticus*: Trachichthyidae) from the Puysegur Bank and Lord Howe Rise, New Zealand, and its implications for stock structure. *Marine Biology*, **123**: 19-27

Haddon, M., Willis, T.J., Wear, R.G. and V.C. Anderlini (1996) Biomass and distribution of five species of surf clam off an exposed west coast North Island beach, New Zealand. *Journal of Shellfish Research*, **15**: 331-339.

Hall, C.A.S. (1988) An assessment of several of the historically most influential models used in ecology and of the data provided in their support. *Ecological Modelling*, **43**:5-31.

Hammersley, J.M. and D.C. Handscomb (1964) *Monte Carlo Methods*. Chapman & Hall, London.

Hardy, A. (1959) *The Open Sea: Its Natural History Part II. Fish and Fisheries*. Collins, London.

Hastings, N.A.J. and J.B. Peacock (1975) *Statistical Distributions*. Butterworths and Co., London.

Higgins, K., Hastings, A., Sarvela, J.N., and L.W. Botsford (1997) Stochastic dynamics and deterministic skeletons - population behavior of dungeness crab. *Science*, **276**:1431-1435.

Hilborn, R. (1979) Comparison of fisheries control systems that utilize catch and effort data. *Journal of the Fisheries Research Board of Canada*, **36**:1477-1489.

Hilborn, R. (1992) Current and future trends in fisheries stock assessment and management. *South African Journal of Marine Science*, **12**: 975-988.

Hilborn, R. (1997) Comment: Recruitment paradigms for fish stocks. *Canadian Journal of Fisheries and Aquatic Sciences*, **54**:984-985.

Hilborn, R. and C.J. Walters (1992) *Quantitative Fisheries Stock Assessment: Choice, Dynamics, and Uncertainty*. Chapman & Hall, London.

Hinkley, D. (1980) Comment on Basu: Randomization analysis of experimental data. *Journal of the American Statistical Association*, **75**: 582-584.

Hinkley, D. (1983) Jacknife methods. pp. 280-287 *In*: *Encyclopedia of Statistical Sciences, Volume 4*. John Wiley & Sons, New York.

Hinkley, D.V. (1988) Bootstrap methods. *Journal of the Royal Statistical Society*, **B 50**: 321-337.

Hinkley, D.V. and S. Shi (1989) Importance sampling and the nested bootstrap. *Biometrika*, **76**: 435-446.

Huber, P.J. (1981) *Robust Statistics*. John Wiley & Sons, New York.

Huxley, T.H. (1881) The herring. *Nature*, **23**: 607-13.

Huxley, T.H. (1884) Inaugural Address. *Fisheries Exhibition Literature* **4**: 1-22.

Jeffrey, A. (1969) *Mathematics for Engineers and Scientists*. Thomas Nelson and Sons, London.

Jennings, E. and J.H. Ward, Jr. (1982) Hypothesis identification in the case of the missing cell. *The American Statistician*, **36**: 25-27.

Jensen, J.W. (1990) Comparing fish catches taken with gill nets of different combinations of mesh sizes. *Journal of Fish Biology*, **37**: 99-104.

Jinn, J.H., Sedransk, J., and P. Smith (1987) Optimal two-phase stratified sampling for estimation of the age composition of a fish population. *Biometrics*, **43**: 343-353.

Kempthorne,O. (1980) Comment on Basu: Randomization analysis of experimental data *Journal of the American Statistical Association*, **75**: 584-587.

Kennelly, S.J. (1992) Distributions, abundances and current status of exploited populations of spanner crabs *Ranina ranina* off the east coast of Australia. *Marine Ecology Progress Series*, **85**: 227-235.

Kent, J.T., Davison, A.C., Silverman, B.W., Young, G.A., Daniels, H.E., Tong, H., Garthwaite, P.H., Buckland, S.T., Beran, R., Hall, P., Koslow, S., Stewart, D.W., Tibshirani, R.J., Titterington, D.M., Verrall, R.J., Wynn, H.P., Wu, C.F.J., Hinkley, D., DiCiccio, T.J. and J.P. Romano (1988) Discussion of the papers by Hinkley and DiCiccio and Romano. *Journal of the Royal Statistical Society*, **B 50**: 355-370.

Kimura, D.K. (1980) Likelihood methods for the von Bertalanffy growth curve. *Fishery Bulletin*, **77**: 765-776

Kimura, D.K. (1981) Standardized measures of relative abundance based on modelling log (c.p.u.e.), and their application to pacific ocean perch (*Sebastes alutus*). *Journal du Conseil International pour l'Exploration de la Mer*, **39**: 211-218.

Kimura, D.K. (1988) Analyzing relative abundance indices with log-linear models. *North American Journal of Fisheries Management*, **8**: 175-180.

Kimura, D.K. and J.W. Balsinger (1985) *North American Journal of Fisheries Management*, **5**: 47-56.

Klaer, N.L. (1994) Methods for standardisation of catch/effort and data requirements. 86-90 *In*: Hancock, D.A (*ed*) *Population Dynamics for Fisheries Management*. Australian Society for Fish Biology Workshop Proceedings, Perth 24-25 August 1993. Australian Society for Fish Biology, Perth.

Knight, W. (1968) Asymptotic growth: An example of nonsense disguised as mathematics. *Journal of Fisheries Research Board of Canada*, **25**: 1303-1307.

Krebs, C.J. (1985) *Ecology: The Experimental Analysis of Distribution and Abundance*. 3rd edn. Harper & Row, New York.

Lai, H.L. and D.R. Gunderson (1987) Effects of ageing errors on estimates of growth, mortality and yield per recruit for Walleye Pollock (*Theragra chalcogramma*). *Fisheries Research*, **5**: 287-302.

Lakatos, I. (1970) Falsification and the methodology of scientific research programmes. pp 91-196 *In*: Lakatos, I. and A. Musgrave. *Criticism and the Growth of Knowledge*. Cambridge University Press, London.

Lane, D.A. (1980) Comment on Basu: Randomization analysis of experimental data. *Journal of the American Statistical Association*, **75**: 587-589.

Larkin, P.A. (1977) An epitaph for the concept of maximum sustainable yield. *Transactions of the American Fisheries Society*, **106**: 1-11.

Lauwerier, H. (1991) *Fractals. Endlessles Repeated Geometrical Figures*. Penguim Books, London.

Legendre, P. and M.J. Anderson (1999) Distance based redundancy analysis: Testing multispecies responses in multifactorial ecological experiments. *Ecological Monographs*, **69**: 1-24.

Lento, G.M., Haddon, M., Chambers, G.K., and C.S. Baker (1997) Genetic variation of Southern Hemisphere fur seals (*Arctocephalus* spp.): Investigation of population structure and species identity. *Journal of Heredity*, **88**: 202-208.

Lepage, R. and L. Billard (*eds.)* (1992) *Exploring the Limits of the Bootstrap*. John Wiley & Sons, New York.

Lindley, D.V. (1980) Comment on Basu: Randomization analysis of experimental data. *Journal of the American Statistical Association*, **75**: 589-590.

Lindley, D.V. and M.R. Novick (1981) The role of exchangeability in inference. *Annals of Statistics*, **9**: 45-58.

Ludwig, D. and C.J. Walters (1981) Measurement errors and uncertainty in parameter estimation for stock and recruitment. *Canadian Journal of Fisheries and Aquatic Sciences*, **38**: 711-720.

Ludwig, D. and C.J. Walters (1985) Are age-structured models appropriate for catch-effort data? *Canadian Journal of Fisheries and Aquatic Sciences*, **42**: 1066-1072

Ludwig, D. and C.J. Walters (1989) A robust method for parameter estimation from catch and effort data. *Canadian Journal of Fisheries and Aquatic Sciences*, **46**: 137-144

Ludwig, D., Walters, C.J., and J. Cooke (1988) Comparison of two models and two estimation methods for catch and effort data. *Natural Resource Modelling*, **2**: 457-498.

Lunney, G.H. (1970) Using analysis of variance with a dichotomous dependent variable: an empirical study. *Journal of Educational Meaurement*, **7**: 263-269.

MacCall, A.D. (1986) Virtual Population Analysis (VPA) equations for nonhomogeneous populations, and a family of approximations including improvements on Pope's cohort analysis. *Canadian Journal of Fisheries and Aquatic Sciences*, **43**: 2406-2409.

MacDonald, P.D.M. and T.J. Pitcher (1979) Age groups from size frequency data: a versatile and efficient method. *Journal of the Fisheries Research Board of Canada*, **36**: 987-1001.

Mace, P. (1997) Developing and sustaining World fisheries resources: The state of the Science and the Management. *In*: Hancock, D.A., Smith, D.C., Grant, A., and J.P. Beumer, (*eds.*) *Developing and Sustaining World Fisheries Resources. The State of Science and Management, 2nd World Fisheries Congress*. CSIRO Publishing, Collingwood, Australia.

Manly, B.F.J. (1991) *Randomization and Monte Carlo Methods in Biology*. Chapman & Hall, London.

Manly, B.F.J. (1997) *Randomization, Bootstrap and Monte Carlo Methods in Biology*. 2nd edn. Chapman & Hall, London.

Maunder, M.N. and P.J. Starr (1995) Rock lobster standardized CPUE analysis. *New Zealand Fisheries Assessment Research Document*, **95/11** 28 p.

May, R.M. (1973) *Stability and Complexity in Model Ecosystems*. Princeton University Press, Princeton, NJ.

May, R.M. (*ed.*) (1984) *Exploitation of Marine Communities*. Report of the Dahlem Workshop on Exploitation of Marine Communities, Berlin 1984, April 1-6. Springer-Verlag, Berlin.

Maynard-Smith, J. (1974) *Models in Ecology*. Cambridge University Press, Cambridge.

McAllister, M.K. and J.N. Ianelli (1997) Bayesian stock assessment using catch-age data and the sampling – importance resampling algorithm. *Canadian Journal of Fisheries and Aquatic Sciences*, **54**: 284-300.

McAllister, M.K., Pikitch, E.K., Punt, A.E., and R. Hilborn (1994) A bayesian approach to stock assessment and harvest decisions using the sampling/importance resampling algorithm. *Canadian Journal of Fisheries and Aquatic Sciences*, **51**: 2673-2687.

McArdle, B. (1990) Detecting and displaying impacts in biological monitoring: spatial problems and partial solutions. pp. 249-255 *In*: Z. Harnos, (*ed.*) *Proceedings of Invited Papers of the 15th International Biometrics Conference*. International Biometric Society, Hungarian Region, Budapest, Hungary.

McConnaughey, R.A. and L.L. Conquest (1993) Trawl survey estimation using a comparative approach based on lognormal theory. *Fishery Bulletin*, **91**: 107-118.

Megrey, B.A. (1989) Review and comparison of age-structured stock assessment models from theoretical and applied points of view. *American Fisheries Society Symposium*, **6**: 8-48.

Meinhold, R.J. and N.D. Singpurwalla (1983) Understanding the Kalman filter. *The American Statistician*, **37**: 123-127.

Mertz, G. and R.A. Myers (1997) Influence of errors in natural mortality estimates in cohort analysis. *Canadian Journal of Fisheries and Aquatic Sciences*, **54**: 1608-1612.

Methot, R.D. (1989) Synthetic estimates of historical abundance and mortality for Northern Anchovy. *American Fisheries Society Symposium*, **6**: 66-82.

Methot, R.D. (1990) Synthesis model: an adaptive framewirk for analysis of diverse stock assessment data. *International North Pacific Fisheries Commission Bulletin*, **50**: 259-277.

Misra, R.K. (1980) Statistical comparisons of several growth curves of the von Bertalanffy type. *Canadian Journal of Fisheries and Aquatic Sciences*, **37**: 920-926.

Moltschaniwskyj, N.A. (1995) Changes in shape associated with growth in the loligonid squid *Photololigo* sp.: a morphometric approach. *Canadian Journal of Zoology*, **73**: 1335-1343.

Moulton, P.L., Walker, T.I., and S.R. Saddlier (1992) Age and growth studies of gummy shark, *Mustelus antarcticus* Günther, and school shark, *Galeorhinus galeus* (Linnaeus), from Southern Australian waters. *Australian Journal of Marine and Freshwater Research*, **43**, 1241-1267.

Myers, R.A. (1991) Recruitment variability and range of three fish species. *NAFO Scienific Council Studies*, **16**: 21-24.

Myers, R.A. (1997) Comment and reanalysis: paradigms for recruitment studies. *Canadian Journal of Fisheries and Aquatic Sciences*, **54**: 978-81.

Myers, R.A. and N.G. Cadigan (1995) Statistical analysis of catch-at-age data with correlated errors. *Canadian Journal of Fisheries and Aquatic Sciences*, **52**:1265-1273.

Nelder, J.A. and R. Mead (1965) A simplex method for function minimization. *Computing Journal*, **7**: 308-313.

Neter, J., Kutner, M.H., Nachtsheim, C.J, and W. Wasserman (1996) *Applied Linear Statistical Models*. Richard D. Irwin, Chicago.

Nicholson, A.J. (1958) Dynamics of insect populations. *Annual Review of Entomology*, **3**:107-136.

Noreen, E.W. (1989) *Computer Intensive Methods for Testing Hypotheses: An Introduction*. John Wiley & Sons. New York.

Ólafsson, E.B. and T. Høisæter (1988) A stratified, two-stage sampling design for estimation of the biomass of *Mytilus edulis* L. in Lindåspollene, a land-locked fjord in western Norway. *Sarsia*, **73**: 267-281.

Overholtz, W.J., Sissenwine, M.P., and S.H. Clark (1986) Recruitment variability and its implication for manageing and rebuilding the Georges Bank haddock stock. *Canadian Journal of Fisheries and Aquatic Sciences*, **43**: 748-753.

Pauly, D. (1984) Fish population dynamics in tropical waters: A manual for use with programmable calculators. *ICLARM Studies and Reviews*, **8**: 325 p. International Centre for Living Aquatic Resources Management, Manila, Philippines.

Pauly, D., Palomares, M.L., Froese, R., Sa-a, P., Vakily, M., Preikshot, D. and S. Wallace (2001) Fishing down Canadina aquatic food webs. *Canadian Journal of Fisheries and Aquatic Sciences* **58**: 51-62.

Pearl, R. and L.J. Reed (1922) A further note on the mathematical theory of population growth. *Proceedings of the National Academyof Sciences of the U.S.* **8**: 365-368.

Pella, J.J. and P.K. Tomlinson   (1969) A generalized stock-production model. *Bulletin of the Inter-American Tropical Tuna Commission*, **13**: 421-458.

Pelletier, B. and Chanut, J.-P. (1991) STRATE: A microcomputer program for designing optimal stratified sampling in the marine environment by dynamic programming - II. Program and example. *Computers and Geosciences*, **17**: 179-196.

Penn, J.W. and N. Caputi (1986) Spawning stock-recuitment relationships and environmental influences on the tiger prawn (*Penaeus esculentus*) fishery in Exmouth Gulf, Western Australia. *Australian Journal of Marine and Freshwater Research*, **37**: 491-505.

Petersen, C.G.J. (1903) What is overfishing? *Journal of the Marine Biological Association of the United Kingdom* **6**: 587-594.

Pianka, E.R. (1974) *Evolutionary Ecology*. Harper & Row, New York. 356 p.

Pitcher, T.J. (2001) Fisheries managed to rebuild ecosystems? Reconstructing the past to salvage the future. *Ecological Applications* **11**: 601-617.

Pitcher, T. J. and D. Pauly (1998). Rebuilding ecosystems, not sustainability, as the proper goal of fishery management. pp 311-330 *In*: Pitcher, P., Hart P. J. B., and D. Pauly. *Reinventing fisheries management*. Kluwer Academic Publishers. London,

Pitcher, C.R., Skewes, T.D., Dennis, D.M. and J.H. Prescott (1992) Estimation of the abundance of the tropical lobster *Panulirus ornatus* in Torres Strait, using visual transect-survey methods. *Marine Biology*, **113**: 57-64.

Pitcher, T.J. and P.J.B. Hart (1982) *Fisheries Ecology*. Croom Helm, London.

Pitcher, T.J. and P.D.M. MacDonald (1973) Two models of seasonal growth. *Journal of Applied Ecology*, **10**: 599-606.

Pitman, E.J.G. (1937) Significance tests which may be applied to samples from any populations. *Supplement to the Journal of the Royal Statistical Society*, **4**: 119-130.

Pitman, E.J.G. (1937) Significance tests which may be applied to samples from any populations. II. The correlation coefficient test. *Supplement to the Journal of the Royal Statistical Society*, **4**: 225-232.

Polacheck, T., Hilborn, R., and A.E. Punt (1993) Fitting surplus production models: Comparing methods and measuring uncertainty. *Canadian Journal of Fisheries and Aquatic Sciences*, **50**: 2597-2607.

Pope, J.G. (1972) An investigation into the accuracy of virtual population analysis using cohort analysis. *Research Bulletin of the International Commission for Northwest Atlantic Fisheries*, **9**: 65-74.

Popper, K. (1963) *Conjectures and Refutations*. Routledge, London. 412 pp.

Pownall, P.C. (*ed.*) (1994) *Australia's Northern Prawn Fishery: the first 25 years*. NPF25, Cleveland, Australia.

Prager, M.H. (1994) A suite of extensions to a nonequilibrium surplus-production model. *Fishery Bulletin*, **92**: 374-389.

Press, W.H., Flannery, B.P., Teukolsky, S.A., and W.T. Vetterling (1989) *Numerical Recipes in Pascal: The Art of Scientific Computing*. Cambridge University Press, London.

Punt, A.E. (1990) Is $B_1 = K$ an appropriate assumption when applying an observation error production-model estimator to catch-effort data? *South African Journal of Marine Science*, **9**: 249-259.

Punt, A.E. (1994) Assessments of the stocks of Cape hakes *Merluccius* spps. off South Africa. *South African Journal of Marine Science*, **14**: 159-186.

Punt, A.E. (1995) The performance of a production-model management procedure. *Fisheries Research*, **21**: 349-374.

Punt, A.E. and R. Hilborn (1997) Fisheries stock assessment and decision analysis: the Bayesian approach. *Reviews in Fish Biology and Fisheries*, **7**: 35-63.

Punt, A.E., Kennedy, R.B., and S.D. Frusher (1997) Estimating the size-transition matrix for Tasmanian rock lobster, *Jasus edwardsii*. *Marine and Freshwater Research*, **48**: 981-992.

Punt, A.E., Smith, D.C., Thomson, R.B., Haddon, M., He, X., and J. Lyle (2001) Stock assessment of the blue grenadier *Macruronus novaezelandiae* resource off south-eastern Australia. *Marine and Freshwater Research*, **52**: 701-717.

Quenouille, M.H. (1956) Notes on bias in estimation. *Biometrika*, **43**: 353-360.

Quinn, T.J. II. and R.B. Deriso (1999) *Quantitative Fish Dynamics*. Oxford University Press, Oxford.

Rao, J.N.K. and C.F.J. Wu (1988) Resampling inference with complex survey data. *Journal of the American Statistical Association*, **83**: 231-241.

Reed, W.J. (1983) Confidence estimation of ecological aggregation indices based on counts - a robust procedure. *Biometrics*, **39**: 987-998.

Reed, W.J. and C.M. Simons (1996) Analyzing catch-effort data by means of the Kalman filter. *Canadian Journal of Fisheries and Aquatic Sciences*, **53**: 2597-2607.

Richards, F.J. (1959) A flexible growth function for empirical use. *Journal of Experimental Botany*, **10**: 290-300.

Richards, L.J. (1991) Use of contradictory data sources in stock assessments. *Fisheries Research*, **11**: 225-238.

Richards, L.J., Schnute, J.T., Kronlund, A.R., and R.J. Beamish (1992) Statistical models for the analysis of ageing error. *Canadian Journal of Fisheries and Aquatic Sciences*, **49**: 1801-1815.

Richards, L.J., Schnute, J.T., and N. Olsen (1997) Visualizing catch-age analysis: a case study. *Canadian Journal of Fisheries and Aquatic Sciences*, **54**: 1646-1658.

Ricker, W.E. (1944) Further notes on fishing mortality and effort. *Copeia* **1944 (1)**: 23-44

Ricker, W.E. (1954) Stock and recruitment. *Journal of the Fisheries Research Board of Canada*, **11**: 559-623.

Ricker, W.E. (1958) *Handbook of Computations for Biological Statistics of Fish Populations*. Fisheries Research Board of Canada, Bulletin No. **119**.

Ricker, W.E. (1973) Linear regressions in fishery research. *Journal of the Fishery Research Board of Canada*, **30**: 409-434.

Ricker, W.E. (1975) *Computation and Interpretation of Biological Statistics of Fish Populations.* Fisheries Research Board of Canada, Bulletin No. **191**.

Ricker, W.E. (1997) Cycles of abundance among Fraser River soceye salmon (*Oncorhynchus nerka*). *Canadian Journal of Fisheries and Aquatic Sciences* **54**: 950-968.

Robins, C.R., Wang, Y.-G., and D. Die (1998) Estimation of the impact of new technology on fishing power in the Northern Prawn Fishery. *Canadian Journal of Fisheries and Aquatic Sciences*, **55**: 1645-1651.

Roff, D.A. (1980) A motion to retire the von Bertalanffy function. *Canadian Journal of Fisheries and Aquatic Sciences*, **37**: 127-129.

Romano, J.P. (1988) A bootstrap revival of some nonparametric distance tests. *Journal of the American Statistical Association*, **83**: 698-708.

Romano, J.P. (1989) Bootstrap and randomization tests of some nonparametric hypotheses. *The Annals of Statistics*, **17**: 141-159.

Rothschild, B.J. and M.J. Fogarty (1989) Spawning-stock biomass: A source of error in recruitment/stock relationships and management advice. *Journal du Conseil International pour l'Exploration de la Mer*, **45**: 131-135.

Russell, E.S. (1931) Some theoretical considerations on the "overfishing" problem. *Journal du Conseil International pour l'Exploration de la Mer*, **6**: 3-20.

Russell, E.S. (1942) *The Overfishing Problem*. Cambridge University Press. London.

Saila, S.B., Annala, J.H., McKoy, J.L., and J.D. Booth (1979) Application of yield models to the New Zealand rock lobster fishery. *New Zealand Journal of Marine and Freshwater Research*, **13**: 1-11.

Sainsbury, K.J. (1980) Effect of individual variability on the von Bertalanffy growth equation. *Canadian Journal of Fisheries and Aquatic Science*, **37**: 241-247.

Schaefer, M.B. (1954) Some aspects of the dynamics of populations important to the management of the commercial marine fisheries. *Bulletin, Inter-American Tropical Tuna Commission*, **1**: 25-56.

Schaefer, M.B. (1957) A study of the dynamics of the fishery for yellowfin tuna in the Eastern Tropical Pacific Ocean. *Bulletin, Inter-American Tropical Tuna Commission*, **2**: 247-285

Schenker, N. (1985) Qualms about bootstrap confidence intervals. *Journal of the American Statistical Association*, **80**: 360-361. (Bootstrap methods)

Schnute, J. (1977) Improved estimates from the Schaefer production model: Theoretical considerations. *Journal of the Fisheries Research Board of Canada*, **34**: 583-603.

Schnute, J. (1981) A versatile growth model with statistically stable parameters. *Canadian Journal of Fisheries and Aquatic Science*, **38**: 1128-1140

Schnute, J. (1985) A general theory for analysis of catch and effort data. *Journal du Conseil International pour l'Exploration de la Mer*, **42**: 414-429.

Schnute, J. and R. Hilborn (1993) Analysis of contradictory data sources in fish stock assessment. *Canadian Journal of Fisheries and Aquatic Science*, **50**: 1916-1923.

Schnute, J.T. and L.J. Richards (1990) A unified approach to the analysis of fish growth, maturity, and survivorship data. *Canadian Journal of Fisheries and Aquatic Science*, **47**: 24-40.

Schnute, J.T. and L.J. Richards (1995) The influence of error on population estimates from catch-age models. *Canadian Journal of Fisheries and Aquatic Science*, **52**: 2063-2077.

Schnute, J.T., Richards, L.J., and N. Olsen (1998) Statistics, software, and fish stock assessment. *University of Alaska Sea Grant Program Report No.* 98-01: 171-184. University of Alaska, Fairbanks.

Seber, G.A.F. (1973) *The Estimation of Animal Abundance and Related Parameters.* Oxford University Press, High Wycombe, England.

Sigler, M.F. and J.T. Fujioka (1988) Evaluation of variability in sablefish, *Anoplopoma fimbria*, abundance indices in the Gulf of Alaska using the bootstrap method. *Fishery Bulletin*, **86**: 445-452.

Sigler, M.F. and J.T. Fujioka (1993) A comparison of policies for harvesting sablefish *Anoplopoma fimbria*, in the gulf of Alaska. *University of Alaska Sea Grant College Program Report No.* **93-02**: 7-19.

Sissenwine, M.P., Fogarty, M.J., and W.J. Overholtz (1988) Some fisheries management implications of recruitment variability. pp 129-152 *In*: J.A. Gulland (*ed*) *Fish Population Dynamics.* 2nd. edn. John Wiley & Sons, Brisbane.

Sitter, R.R. (1992a) A resampling procedure for complex survey data. *Journal of the American Statistical Association*, **87**: 755-765.

Sitter, R.R. (1992b) Comparing three bootstrap methods for survey data. *Canadian Journal of Statistics*, **20**: 135-154.

Slobodkin, L.B. (1961) *Growth and Regulation of Animal Populations.* Holt, Reinhart and Winston, New York. 184 p.

Smith, T.D. (1988) Stock assessment methods: the first fifty years. *In*: *Fish Population Dynamics: The Implications for Management.* (*ed.*) J.A. Gulland. pp 1-33. John Wiley & Sons, Chichester.

Smith, T.D. (1994) *Scaling Fisheries: The Science of Measuring the Effects of Fishing, 1855-1955.* Cambridge University Press, New York.

Snedecor, G.W. and W.G. Cochran (1967) *Statistical Methods.* 6th. edn. Iowa State University Press, Ames.

Snedecor, G.W. and W.G. Cochran (1989) *Statistical Methods.* 9th. edn. Iowa State University Press, Ames.

Sokal, R.R. (1979) Testing statistical significance of geographic variation patterns. *Systematic Zoology*, **28**: 227-232.

Sokal, R.R and F.J. Rohlf (1995) *Biometry.* 3rd. edn. W.H. Freeman and Company, New York.

Solow, A.R. (1989) Bootstrapping sparsely sampled spatial point patterns. *Ecology*, **70**: 379-382.

Soutar, A and J.D. Isaacs (1969) History of fish populations inferred from fish scales in anaerobic sediments off California. *California Cooperative Oceanology and Fishery Investigation Reports*, **13**: 63-70

Steele, J.H. and E.W Henderson (1984) Modeling long-term fluctuations in fish stocks *Science*, **224**: 985-987.

Sullivan, P.J., Han-Lin Lai, and V.F. Gallucci (1990) A catch-at-length analysis that incorporates a stochastic model of growth. *Canadian Journal of Fisheries and Aquatic Sciences*, **47**: 184-198.

Sullivan, P.J. (1992) A Kalman filter approach to catch-at-length analysis. *Biometrics*, **48**: 237-257.

Summerfelt, R.C. and G.E. Hall (*eds.*) (1987) *Age and Growth of Fish*. Iowa State University Press, Ames.

ter Braak, C.J.F. (1992) Permutation versus bootstrap significance tests in multiple regression and ANOVA. pp. 79-85 *In:*   Jöckel, K.-H., Rothe, G., and W. Sendler (*eds.*) *Bootstrapping and Related Techniques*. Springer-Verlag, Berlin.

Tukey, J.W. (1958) Bias and confidence in not-quite large samples. Abstract only. *The Annals of Mathematical Statistics*, **29**: 614.

Venzon, D.J. and S.H. Moolgavkar (1988) A method for computing profile-likelihood-based confidene intervals. *Applied Statistics*, **37**: 87-94.

Vignaux, M. (1992) Catch per unit effort (CPUE) analysis of the hoki fishery. *New Zealand Fisheries Assessment Research Document*, **92/14** 31 p.

Vignaux, M. (1993) Catch per unit od effort (CPUE) analysis of the Hoki fishery 1987-1992. *New Zealand Fisheries Assessment Research Document*, **93/14**: 1-15.

von Bertalanffy, L. (1938) A quantitative theory of organic growth. *Human Biology*, **10**: 181-213.

Walkenback, J. (1999) *Microsoft Excel 2000 Power Programming with VBA*. IDG Books Worldwide, Foster City, CA.

Walters, C.J. and R. Bonfil (1999) Multispecies spatial assessment models for the British Columbia groundfish trawl fishery. *Canadian Journal of Fisheries and Aquatic Sciences*, **56**: 601-628.

Walters, C.J., Christensen, V. and D. Pauly, (1997) Structuring dynamic models of exploited ecosystems from trophic mass-balance assessments. *Reviews in Fish Biology and Fisheries* **7**: 139-172.

Walters, C.J. and D. Ludwig (1981) Effects of measurement errors on the assessment of stock-recruitment relationships. *Canadian Journal of Fisheries and Aquatic Sciences*, **38**: 704-710.

Walters, C.J. and D. Ludwig (1994) Calculation of Bayes posterior probability distributions for key population parameters. *Canadian Journal of Fisheries and Aquatic Sciences*, **51**: 713-722.

Wang, Y. and D.J. Die (1996) Stock-recruitment relationships for the tiger prawns (*Penaeus esculentus* and *Penaeus semisulcatus*) in the Australian Northern Prawn Fishery. *Marine and Freshwater Research*, **47**: 87-95.

Wang, Y.-G. and M.R. Thomas (1995) Accounting for individual variability in the von Bertalanffy growth model. *Canadian Journal of Fisheries and Aquatic Sciences*, **52**: 1368-1375.

Watson, J.D. and F.H.C. Crick (1953) Molecular structure of nucleic acids. *Nature*, **171**: 737-738.

Weibull, W. (1951) A statistical distribution function of wide applicability. *Journal of Applied Mechanics*, **18**: 293-297

Weinberg, J.R. and T.E. Helser (1996) Growth of the Atlantic surfclam, *Spisula solidissima*, from Georges Bank to the Delmarva Peninsula, USA *Marine Biology*, **126**: 663-674.

Welch, W.J. (1990) Construction of permutation tests. *Journal of the American Statistical Association*, **85**: 693-698.

York, A.E. and P. Kozloff (1987) On the estimation of numbers of Northern Fur Seal, *Callorhinus ursinus*, pups born on St. Paul Island, 1980-86. *Fishery Bulletin*, **85**: 367-375.

Younger, M.S. (1979) *Handbook for Linear Regression*. Duxberry Press, Belmont.

Zar, J.H. (1984) *Biostatistical Analysis*. 2nd. edn. Prentice-Hall, Englewood Cliffs, NJ.

Subject Index